The American Bird Conservancy Guide to

Bird Conservation

Daniel J. Lebbin, Michael J. Parr, and George H. Fenwick

The University of Chicago Press
Lynx 🐾

AMERICAN BIRD
CONSERVANCY

SUPPORTED BY THE LEON LEVY FOUNDATION

The University of Chicago Press, Chicago 60637
The University of Chicago Press, Ltd., London
Lynx Edicions, Barcelona
© 2010 by The American Bird Conservancy
All rights reserved. Published 2010
Printed in China

20 19 18 17 16 15 14 13 12 11 10   1 2 3 4 5

ISBN-13: 978-0-226-64727-2 (cloth)
ISBN-10: 0-226-64727-7 (cloth)

With thanks to the Leon Levy Foundation

Library of Congress Cataloging-in-Publication Data

Lebbin, Daniel J. (Daniel Jason), 1978–
　　The American Bird Conservancy guide to bird conservation / Daniel J. Lebbin, Michael J. Parr, and George H. Fenwick.
　　　p. cm.
　　Includes bibliographical references and index.
　　ISBN-13: 978-0-226-64727-2 (cloth : alk. paper)
　　ISBN-10: 0-226-64727-7 (cloth : alk. paper) 1. Birds—Conservation—United States. 2. Birds—Habitat—United States. 3. Birds—
Conservation—North America. 4. Birds—Habitat—North America. I. Parr, Mike, 1962– II. Fenwick, George H. III. American Bird
Conservancy. IV. Title.
　　QL676.5.L395 2010
　　333.95′816097—dc22

2010007646

Illustrations by Chris Vest
Layout by George Opryszko and Lynx Edicions
Editing by Gavin Shire

Additional illustrations by: Richard Allen p. 76 top, p. 78 top; Norman Arlott p. 45 bottom, p. 46 top, p. 46 bottom, p. 47 top; Hilary Burn p. 47 bottom, p. 48 top, p. 73 bottom, p. 78 bottom, p. 79 top, p. 79 bottom, p. 82 top, p. 83 top, p. 88 bottom, p. 89 top, p. 89 bottom, p. 90 bottom, p. 94 bottom, p. 117 top, p. 117 bottom, p. 118 top; Ren Hathway p. 16 (3, 5), p. 17 (7), p. 94 top, p. 96 bottom, p. 97 top; Mark Hulme p. 17 (6), p. 82 bottom; Angels Jutglar p. 25 top, p. 25 bottom, p. 26 top, p. 26 bottom, p. 27 top, p. 30 top, p. 43 bottom, p. 48 bottom, p. 49 top, p. 49 bottom, p. 50 top, p. 50 bottom, p. 71 top, p. 72 top; Francesc Jutglar p. 17 (9), p. 22 top, p. 22 bottom, p. 23 top, p. 23 bottom, p. 24 top, p. 24 bottom, p. 31 top, p. 31 bottom, p. 32 top, p. 41 bottom, p. 42 bottom, p. 43 top, p. 51 top, p. 51 bottom, p. 52 top, p. 52 bottom, p. 53 top, p. 53 bottom, p. 54 top, p. 54 bottom, p. 55 top, p. 55 bottom, p. 56 top, p. 56 bottom, p. 57 top, p. 57 bottom, p. 58 top; Daniel Lebbin p. 16 (1); Ian Lewington p. 44 top, p. 59 bottom, p. 60 top, p. 60 bottom, p. 61 top, p. 61 bottom, p. 62 top, p. 62 bottom, p. 66 bottom, p. 74 bottom, p. 75 bottom, p. 84 top, p. 90 top, p. 95 top, p. 95 bottom, p. 96 top; Dave Nurney p. 75 top; Douglas Pratt p. 76 bottom, p. 77 top, p. 77 bottom, p. 91 top, p. 91 bottom, p. 92 top, p. 92 bottom, p. 118 bottom, p. 119 top, p. 119 bottom, p. 120 top, p. 120 bottom, p. 121 top, p. 121 bottom, p. 122 top, p. 122 bottom, p. 123 top, p. 123 bottom, p. 124 top, p. 124 bottom, p. 125 top, p. 125 bottom, p. 126 top, p. 126 bottom, p. 127 top, p. 127 bottom; David Quinn p. 88 top, p. 97 bottom, p. 98 top, p. 98 bottom, p. 99 top, p. 99 bottom, p. 100 top, p. 100 bottom, p. 101 top, p. 101 bottom; p. 102 top, p. 102 bottom, p. 103 top, p. 103 bottom, p. 104 top, p. 104 bottom, p. 105 top, p. 105 bottom; Chris Rose p. 66 top, p. 67 top, p. 67 bottom, p. 68 top, p. 68 bottom, p. 69 top, p. 69 bottom, p. 70 top; Lluís Sanz p. 17 (10), p. 42 top, p. 71 bottom; Brian Small. p. 16 (2, 4), p. 84 bottom, p. 85 top, p. 85 bottom, p. 86 top, p. 86 bottom, p. 87 top, p. 87 bottom, p. 106 top, p. 114 bottom, p. 115 top, p. 115 bottom, p. 116 top, p. 116 bottom; Lluís Solé p. 17 (8), p. 30 bottom; Juan Varela p. 32 bottom, p. 33 top, p. 33 bottom, p. 34 top, p. 34 bottom, p. 35 top, p. 35 bottom, p. 36 top, p. 36 bottom, p. 37 top, p. 37 bottom, p. 38 top, p. 38 bottom, p. 39 top, p. 39 bottom, p. 40 top, p. 40 bottom, p. 41 top; Etel Vilaró p. 27 bottom, p. 28 top, p. 28 bottom, p. 29 top, p. 29 bottom, p. 58 bottom, p. 59 top; Jan Wilczur p. 70 bottom, p. 72 bottom; Ian Willis p. 44 bottom, p. 45 top, p. 63 top, p. 63 bottom, p. 64 top, p. 64 bottom, p. 65 top, p. 65 bottom, p. 73 top, p. 80 top, p. 80 bottom, p. 81 top, p. 81 bottom; Tim Worfolk p. 74 top, p. 93 top, p. 93 bottom.

Illustrations of species from p. 106 (bottom) to p. 114 (top) reprinted with permission from the book *National Geographic Field Guide to the Birds of North America*, 5th edition (except for Baird's Sparrow, Black-chinned Sparrow, Brewer's Sparrow, Five-striped Sparrow, Henslow's Sparrow, Rufous-winged Sparrow, Lark Bunting, and Seaside Sparrow which are by Chris Vest based on photographs by Bill Hubick, Dominic Sherony, and Robert Royse). Illustration p. 83 (top) by Jonathan Alderfer ©National Geographic. Other National Geographic Society illustrations Copyright ©2006 National Geographic Society. All rights reserved.

This page: Costa's Hummingbird and ocotillo by Chris Vest, based on photography by Andre Noel (AndreNoelPhotography.com) and Clinton Steeds.

Lynx Edicions holds the copyright for the distribution maps on pages 22 to 127 of this book.

Recommended citation: Lebbin, D., Parr, M., and Fenwick, G. 2010. *The American Bird Conservancy Guide to Bird Conservation*. Lynx Edicions, Barcelona, and The University of Chicago Press, Chicago and London.

♾ The paper used in this publication meets the minimum requirements of the American National Standard for Information Sciences—Permanence of Paper for Printed Library Materials, ANSI Z39.48-1992.

The paper used in this book was sourced from responsibly managed forests.

FOREWORD

Human beings, like birds, are wired for short-term thinking. We evolved in conditions of scarcity and danger, moving from crisis to crisis, and today, more than ever, we're bombarded with news of fresh crises every week. If you're holding this book, you probably know that one of these crises is the state of birds and their habitats in the Americas. A third of all American bird species are in trouble—in many cases, they're in imminent danger of extinction—and this book is the clearest, most authoritative account ever published of the threats these species face. Until now, remarkably, no single book has tackled the issue comprehensively.

But human beings, at their best, are also capable of long-term thinking, of deep appreciation and compassion for other forms of life, and of understanding that their own well-being is inseparable from the well-being of the natural world. The particular beauty of this book is that it's not just another shout of crisis. It also lays out a concrete and achievable plan of long-term action to safeguard our country's rich bird life. It embodies the mission of the American Bird Conservancy itself, which is not only to protect the most critically endangered species but to recognize and address the totality of the threats that all birds face. Merely moving from crisis to crisis creates a kind of despair about the world's complexity. What this book offers is an argument for hope. Americans love their birds and are justly proud of them, and our country has the resources and ingenuity to find practical solutions to the more general crisis. We urgently need a clear and complete picture of the work to be done, along with the inspiration to do it, and this book, with its arresting artwork and photography, provides both.

Even if you're only a little bit interested in birds, and have never thought much about bird conservation, you'll appreciate the visual power and intellectual scope of these pages. Particularly valuable is the attention paid to our seabirds and Hawaiian birds, whose troubles are severe and often overlooked. But whatever part of the country you're from, and whatever birds you feel closest to, you should be heartened to learn about the collaborative, science-based steps that we can take to assure the continuing diversity and abundance of one of our great national treasures. This book helps us focus on the conservation priorities that will do the most good. I hope that you'll find it useful, and that if you're not already a supporter of American Bird Conservancy you'll decide to become one after you read it.

Jonathan Franzen

PREFACE

When American Bird Conservancy was begun 15 years ago, the main idea was to capitalize on the multitudes of people who loved birds; to coalesce the almost 1,000 U.S. bird clubs, perhaps millions of birders, and dozens of professional conservation organizations around solving the key issues affecting birds. By focusing on these, we hoped we might advance the cause of bird conservation significantly further than we would as thousands of groups and millions of individuals pulling in different directions. While ABC has organized or helped to organize numerous, successful partnerships, coalitions, and networks (such as the Bird Conservation Alliance, Alliance for Zero Extinction, National Pesticide Reform Coalition, Cats Indoors!, North American Bird Conservation Initiative) over the intervening years, we also came to learn much about our partners, our larger constituency, the opposition to bird conservation, and how each of these decides how and whether to become engaged.

For example, many in the bird community take pride in our numbers (the estimates of birders in the U.S. vary widely, probably depending on how the term birder is defined, usually ranging from one to seventy or so million people; we most often hear claims in the range of 50 million) and extrapolate from those numbers to conclude that they represent true political power for birds or conservation. In truth, that power is really much more about potential than true power. If avid birders are most likely to support conservation, and if they could be quantified as being roughly equivalent to the population of those who purchased a Sibley bird guide, then the number of these birders is probably closer to one- or two-million.

We wonder what kind of people these birders are, and to what degree can they be a force for conservation. It is always risky to slip into stereotyping, and this is just as true for birders as for most other cohorts. Nevertheless, it is worth considering that birders, taken as a group, may be a bit less aggressive, less willing to fight for their interests than others. We are people who prefer a walk in the woods. I have read or been told time after time by birders that they want their hobby to be just that – a relaxing pastime, and not a cause for which to fight or pay. This means that only a fraction of all birders wish to play a role in protecting the resource they enjoy—and that means that each birder engaged in conservation is proportionately more important than our total numbers might indicate.

Here is another observation that gives me pause about the significance of birding's power in society at large: A year ago, I was departing my local (chain) book store, disappointed that it did not carry a particular book on its four shelf feet of bird books. On the way out, I noticed a book section on manga, a simple Japanese comic form: I stepped off 52 shelf feet of something I did not know existed! If a book-intensive hobby such as birding gets only 1/13th the space of a Japanese comic form in a national chain book store, that opens to question the true degree of public interest in birds, and, importantly, the economic clout of birding, at least in book terms! Thus, we should not simply assume that our numbers make the case for bird conservation: instead, we will succeed only if we boldly speak up.

What have we learned about the opposition? For one thing, every point of view has its own science and economics to support its contentions, whether it be pro- or anti- pesticide use, free-roaming cats, bird collisions with glass or towers, conflicts with fisheries, land conversion, wind energy, mining, timbering, climate change, or any other issue we consider in addressing bird conservation. We see very few issues resolved through coming to consensus on the science. More often, science divides us. So, though we continue to need excellent science, conservationists also need to learn how to better persuade, use media effectively, and connect birds to the human condition, spirituality and ethical values. Facts alone will not win these battles, just try debating a creationist, but human connection might. Secondly, no matter the issue, our opposition will unfailingly question our priorities. So, they ask, "Why are you blaming _____ when we know habitat loss is the real problem?" When I am asked these questions, I reply first that we continue to add to our base of knowledge and thus improve decisions affecting bird populations; second, if winning the debate translates to financial gain for the debater, then that position cannot be truly unbiased; and third, isn't it analogous that if heart disease is the greatest mortality factor in our species, then why do we spend so much money combating cancer and other diseases, poverty and hunger? Solving any single problem—for birds or humans—is insufficient for our cause. Thoughtful people understand that it is critical that we fight for birds and humans across the full, broad front of issues.

After taking up the gauntlet for more than 100 issues in recent years, ABC has synthesized issues for birds into three basic approaches: saving the rarest species (primarily preventing extinction through reserve creation); managing and improving habitat for declining, rare and range-restricted species; and eliminating threats to protect all birds (by policy change, resource generation, and creating solutions to major human-caused threats). Each of these must, in turn, be supported by effective science, building capacity, strong partnerships, mission support, and funding. This is the ABC conservation approach and it is reflected throughout this book.

One last observation is that people in general seem to be risk averse with regard to becoming involved in bird conservation. Having a terrific cause and a call to action doesn't necessarily spring one into action, especially if that call comes from a relatively new and less well-known organization. People wait and watch, and they join in their own way and in their own time. This has been the case with many ABC conservation endeavors over the years. How can this be overcome? One way, we thought, would be to gather all the information about bird conservation—the birds, their habitats, the threats, and the potential solutions—into a single place so that those who were interested could explore these various aspects and consider what roles they might we want play. This reflection and following action is what we hope this book will inspire.

As this guide makes abundantly clear, a wide range of conservation actions are needed to keep our bird populations healthy—actions that can be undertaken by many kinds of people and organizations. A decade or so ago, a call was made for "citizen science", loosely defined as volunteer contributions to scientific endeavor. Most of this has come down to birders sending in their sightings to organizations who compile and analyze large volumes of data for scientific purposes—a painless way to be involved. While laudatory, this is not enough. Here, we call for a new era of "citizen conservation" in which those who care about the future of birds become involved in even more meaningful ways. Recently, I visited the Osa Biological Station in Costa Rica with Craig Thompson, a birder and administrator with the Wisconsin Department of Natural Resources. Craig made the case to his colleagues that many of Wisconsin's neotropical migrants were wintering in the Osa and needed both research and protection, and succeeded in getting them involved in the establishment of the reserve there. He did not stop there: he then organized a visit by supporters of the Natural Resources Foundation of Wisconsin and another for Wisconsin birders. In doing this—and turning the dial for birds, Craig is more than a citizen conservationist; he is a conservation hero. There is much more to be done, and you—like Craig—can personally make a difference. We hope his work and this book inspire you to become involved.

George H. Fenwick, President, American Bird Conservancy

ACKNOWLEDGMENTS

Many people and organizations contributed expertise and resources to this book and we are grateful for your support, and for every suggestion, idea, and question. Thank you!

Firstly, the Leon Levy Foundation provided virtually all the funding for this project, and we are extremely grateful both to the foundation, and individually to Shelby White, John Bernstein, and Judy Dobrzynski for their wonderful support and advice throughout the project. Thanks also to the Park Foundation, Nancy Raines, and Lynn Sheehan who also provided support.

Numerous ABC staff members reviewed or helped with portions of the text, including Bob Altman, Hugo Arnal, Dan Casey, Robert Chipley, Jane Fitzgerald, Rita Fenwick, Michael Fry, Jim Giocomo, Mary Gustafson, Steve Holmer, Todd Jones-Farrand, Sara Lara, Ed Laurent, Ann Law, Erin Lebbin, Casey Lott, Moira McKernan, Jessica Norris, David Pashley, Darin Schroeder, Christine Sheppard, Gavin Shire, Benjamin Skolnik, Brian Smith, George Wallace. Tomeka Davis helped with research on energy. Susannah Casey, Jessica Norris, George Opryzsko, and Gemma Radko assisted with map production and layout. Jay McGowan assisted with photo acquisition.

We would also like to thank Josep del Hoyo of Lynx Edicions for believing in the project and coming on board as our publisher, and Amy Chernasky for her enthusiasm and support. Also, thanks to the many wonderful Lynx illustrators whose paintings have been used for the book. Thanks also to Christie Henry of The University of Chicago Press for joining as our U.S. publisher. Thanks also to Ken Berlin and Mary Rasenberger of Skadden Arps for their *pro bono* legal help with contracts.

Ornithologists, conservation practitioners, and other experts external to ABC also volunteered their time to review accounts in this book, including Ken Berlin, Louis Bevier, Timothy Brush, John Cornely, Ashley Dayer, Jennie Duberstein, Laura Erickson, Paul Flint, Tom Gardali, Joelle Gehring, Geoff Geupel, Jim Goetz, Jaime Jahncke, Ian Jones, Alicia King, Gary Langham, David Lee, David Leonard, Patrick Magee, Annie Marshal, Greg McClelland, David Mizrahi, Hannah Nevins, Margaret Petersen, Jane Rubey, Mark Salvo, Nate Senner, Joel Schmutz, and Fred Zwickel. Cornell Laboratory of Ornithology provided electronic access to Cornell Library resources and Michael Powers provided additional literature. Thanks also to Rocky Mountain Bird Observatory and Partners in Flight for sharing their data. Thanks also to Rocky Mountain Bird Observatory and Partners in Flight for sharing their data. Species from p. 106 (second illustration) to p. 114 (first illustration) reprinted with permission from the book National Geographic Field Guide to the Birds of North America, 5th edition. Copyright ©2006 National Geographic Society. All rights reserved.

Many photographs in this book were generously donated by the photographers. Their credits appear adjacent to photos. We would especially like to thank the four photographers who provided the bulk of the photographic material for this book, Bill Hubick, Greg Lavaty, Robert Royse, and Glen Tepke. Chris Vest produced the digital vistas that grace the start of each Birdscape account (along with some of the sparrow illustrations). This digital artwork drew from photographs generously provided for this purpose by many photographers. Contributors found via Creative Commons sources (often Flickr) are denoted with an asterisk (*). We would like to acknowledge these sources, including: Ciro Ginez Albano, Muhammad Almeida*, Vince Alongi*, Luis Argerich (luisargerich.zenfolio.com)*, Nick Athanas, Justin Baeder*, Jimmy Baikovicius (jikatu on flickr.com), Michael "Mike" L. Baird (flickr.bairdphotos.com), Derek Bakken*, John Seb Barber* Omar Barcena*, David Baron*, David L. Bezaire*, Ken Bosma*, Adam Buchbinder*, Juan Carlos Bustamante*, Eddie Callaway (Birdfreak.com), Eunice Chang*, Jamie Chavez*, Margaret Anne Clark*, Sean Clawson*, clipart.com, Patrick Coe*, Chauncey Davis*, Anthony and Veronique Debord-Lazaro*, Christoph Diewald*, dumbonyc*, ECOAN, Kathryn Simon*, Charlotte Feld*, Peter Firminger*, Eoin Gardiner*, Sam Garza*, Arthur Grosset (www.arthurgrosset.com), Michel Gutierrez*, Chris Hartman*, Edwin E. Harvey*, John Haslam*, Bill Hubick (billhubick.com), Gary Irwin*, Andrew Ives*, Jaseman*, Jack Jeffrey (jackjeffreyphoto.com), Kevin Jones*, Ernie Jonker*, Kenjonbro*, Chad King*, Mac Knight (pbase.com/macknight), Sharon Komarow*, Frank Kovalchek*, Peter Latourette (birdphotography.com), David LaPuma*, Daniel J. Lebbin, Maureen Leong-Kee*, Jim Linwood*, Lyndi & Jason*, Nick M. (Satanoid)*, Greg Ma*, Susan Malzahn*, Alan Manson*, Ricardo Martins*, Natalie Maynor*, Bruce McAdam*, Brian McClure*, Mitchell J. McConnell (excellentbirds.com), Dave McMullen*, Jessica Merz*, Lyn Meter*, Michlt*, Andre Noel (AndreNoelPhotography.com), Michael O'Brien*, Jerry R. Oldenettel*, Chuck Painter*, Mike Parr, Anthony Patterson*, pbonenfant*, Jorge Mejia Peralta*, Andrea Pokrzywinski*, Sugar Pond*, Rob Pongsajapan*, Prayingmother*, ProAves Colombia (www.proaves.org), Psionicman*, Andy Purviance*, Alastair Rae*, Matt Riggott*, Royal Olive (theroyalolive.com), Adria Richards*, Charles and Clinton Robertson*, Steve Ryan*, Eli Sagor*, Dario Sanches*, Bob and Martha Sargent (hummingbirdsplus.org), Barth Schorre, Kyle Sharaf, Dean Shareski*, Trisha Shears*, Dominic Sherony*, Jonathan Sherrill (windcircle.org), Dwight Sipler*, Caleb Slemmons*, Marnin Somerman*, Kim and Forest Starr, Clinton Steeds*, Erin Stevenson O'Connor*, Scott Streit*, Will Sweet*, Steve Tannock*, Ingrid Taylar*, Teo*, Trees for the Future (http://treesftf.org)*, Nick Turland*, Julian Tysoe*, U.S. Department of Agriculture (USDA), U.S. Fish & Wildlife Service (FWS), Yolanda van Petten*, ©VascoPlanet.com (www.vascoplanet.com), Gavan Watson*, Elaine R. Wilson (naturespiconline.com), David Weidenfeld, Tim Wilson*, Michael Woodruff*, Lip Kee Yap*, Peter Young*.

Daniel J. Lebbin, Michael J. Parr, and George H. Fenwick, February 2010

THE AUTHORS

Dr. Daniel J. Lebbin, Conservation Biologist with ABC, received a Ph.D. in Ecology & Evolutionary Biology from Cornell University. His dissertation research investigated habitat specialization among Amazonian birds in Peru, where he spent a year as a Fulbright Scholar. He has also done field research in Jamaica, Costa Rica, Ecuador, and Venezuela.

Michael J. Parr, Vice President with ABC, graduated from the University of East Anglia in 1986. He worked as Development Officer at BirdLife International before joining American Bird Conservancy in 1996. He co-authored A Guide to the Parrots of the World published by Yale University Press, and The 500 Most Important Bird Areas in the U.S. He is Chair of the Alliance for Zero Extinction.

Dr. George H. Fenwick has served as President and CEO of ABC since its founding in 1994. Prior to that, he worked in a variety of capacities during 15 years with The Nature Conservancy, including Director of Science, and Chair of the Last Great Places Campaign Steering Committee. He received a Ph.D. in Pathobiology from Johns Hopkins University.

TABLE OF CONTENTS

GLOSSARY

Apart from "Important Bird Area", only legally enforceable land designations are usually cited (e.g., NWR, NP). The word "endangered" appears with a capital E when we refer to formal IUCN and ESA categories. Subspecies are indicated by inverted commas around the subspecies e.g., "Northern" Spotted Owl.

ABC American Bird Conservancy
AZE Alliance for Zero Extinction
BBS Breeding Bird Survey
BCR Bird Conservation Region
IBA Important Bird Area
BLM Bureau of Land Management
CRP Conservation Reserve Program
EPA Environmental Protection Agency
ESA Endangered Species Act
EEZ Exclusive Economic Zone
FWS U.S. Fish and Wildlife Service
IUCN IUCN-World Conservation Union
NABCI North American Bird Conservation
 Initiative
NCA National Conservation Area
NF National Forest
NG National Grassland
NGO Non-governmental Organization
NM National Monument (note that the state New Mexico also appears as NM)
NP National Park
NRCS Natural Resources Conservation Service
NS National Seashore
NWR National Wildlife Refuge
PIF Partners in Flight
Snags Standing dead trees
SP State Park
TNC The Nature Conservancy
U.S. United States of America
USDA U.S. Department of Agriculture
USFS U.S. Forest Service
WatchList U.S. WatchList of birds of
 conservation concern
WMA Wildlife Management Area

INTRODUCTION

In order to ensure that future generations continue to enjoy the abundance and diversity of birds we know today, it is important that our management of bird populations is based upon solid science, and that management decisions are planned with the long-term view in mind. This book sets out to outline current priorities in bird conservation throughout the Americas, to identify the places where gaps currently exist and need to be filled, and to point to areas where future problems may arise so they can be preempted. It also aims to communicate these priorities to facilitate a better understanding of bird conservation among a broad audience, and to encourage support for them.

Whether we live in cities, in rural suburbia, amid farm or range-land, or on the margins of true wilderness, birds are a ubiquitous feature of daily life in America, such that we often take them for granted. Unfortunately though, in recent decades, bird declines have reached the point where even casual observers have been able to notice that some species just aren't around in the numbers they once were. That awareness has spawned conservation action that has led to notable successes, including the recovery of some of the nation's most emblematic species such as the Bald Eagle, Brown Pelican, Whooping Crane, and Peregrine Falcon. Despite this, other species still remain in decline, and some are still on the brink of extinction. For them, there is still much more to be done.

In this book, we focus on bird species, habitats, and the threats facing birds. Although we concentrate on the U.S. (with an international chapter focused on the Neotropics), much of the information in this book applies equally to Canada (and

somewhat to northern Mexico), as many of the birds and threats, and some of the habitats are present throughout North America. Beyond this, we hope that the book's approach can serve as a valuable resource for bird conservationists throughout the Americas—and beyond. We begin by providing overview accounts for the 212 bird species in the U.S. most in need of conservation attention (those on the WatchList, see p. 10). Each account summarizes the threats affecting the species and the primary actions necessary for its conservation. We then describe habitats (grouped into 12 "Birdscapes"), and the conservation issues associated with them, along with fifty representative Important Bird Areas within these habitats. Thirdly, we provide a summary of the threats to birds. This is followed by an international chapter, which uses largely the same format as the preceding chapters to summarize bird conservation issues in Latin America and the Caribbean. Finally, we provide a list of strategies and tools that conservationists can use to help stabilize and restore priority bird populations and reduce threats. This final chapter includes information on some of the government agencies and private conservation groups that are working to conserve birds, and ornithological societies and research institutions that study and monitor birds. We also offer suggestions for how the reader can personally contribute to bird conservation.

WHY CONSERVE BIRDS?

With the multitude of priorities and problems faced by 21st Century American society, conserving birds might at first blush seem low down on the list, but birds matter. Perhaps most important-

PHOTO: ROBERT ROYSE

In the "Lower 48", the Bald Eagle has rebounded from c. 800 birds in 1963 to more than 19,000 today thanks to conservation action.

ly, we value them intrinsically, incorporating them as symbols in our religions, cultures, and national identities. They touch our hearts with their beauty, and fill our minds with wonder (and a hint of jealousy) at their ability to fly, sometimes for huge distances. Their disappearance is a loss for us, and every generation that succeeds us.

In addition to being objects of fascination and inspiration, however, birds are also an integral part of the natural ecosystems upon which we, and ultimately the life of this planet depend. Birds pollinate flowers, disperse seeds, and deposit nutrients, acting as essential drivers of ecosystems. They are both predators and prey. Some act as nature's garbage collectors by scavenging carrion. Others feed on insects and other animals that can become pests if not controlled. Some birds, such as woodpeckers, are even ecosystem engineers that construct cavities depended on by other organisms.

Beyond the caged canaries that once served as early warning signals of environmental danger in coal mines, wild birds have long served as indicators of larger environmental problems that affect people and the health of the Earth, including chemical pollution, global climate change, and emerging diseases. Birds are also a unique natural resource with huge economic bearing. According to the U.S. Fish and Wildlife Service, there are up to 70 million birdwatchers in the U.S. who spend over $25 billion per year feeding, watching, and traveling to enjoy birds—that's more than double the amount generated by the domestic music industry.

Despite the value and popularity of birds, our actions are now causing many of their populations to decline. It is our responsibility to reverse these declines, to preserve habitats, and to prevent future species extinctions. Saving all the pieces of our ecosystems, including birds, is also simply the right thing to do, so that we can pass on to future generations the wealth of life that existed when we ourselves arrived on the planet.

CHANGES IN BIRD POPULATIONS

Bird populations naturally grow, decline, and evolve over time, but human actions often have dramatic impacts beyond these natural fluctuations. Over the last five centuries, we have driven hundreds of bird species across the planet to extinction through excessive hunting, the introduction of exotic animals and non-native pathogens, habitat destruction, and a suite of other activities.

Prior to mass human exploitation, the Ice Ages were the major factor governing bird populations in temperate regions. These resulted in the ice caps expanding to force birds from previously occupied areas, especially in the northern latitudes. As the ice retreated, large numbers of birds were able to rapidly exploit the opening up of new habitats absent of competitors. More recently, the line between natural and anthropogenic global changes has blurred as our impact on the planet has altered increasing numbers of ecosystems and natural processes. The need to manage the negative effects of these impacts has never been greater, and birds are one of the best indicators for both environmental damage and our ability to redress it.

The impact of people on birds is not always negative, though. Some birds benefit from the opportunities we create, greatly increasing their populations and distributions as a result. But some of these successful birds, some gulls for example, threaten others that are less adaptable to human impacts.

Whether a bird population increases or declines depends on the balance of its reproductive and mortality rates. Essentially, populations decline when mortality exceeds reproductive recruitment. Declines can be localized, or can occur across the entire range of a species. Generally, species with

PHOTO: GREG LAVATY

Decimated by the feather trade, the Great Egret has now recovered to more than 250,000 birds in North America.

broader ranges are less vulnerable than species endemic to small geographic areas, since localized extinctions can be replaced by re-colonization if the threat is regional and temporary (e.g., DDT and the Bald Eagle).

The habitat requirements of a species can also influence its vulnerability to population decline. For example, birds such as the American Robin that have benefited from human-induced landscape changes are at an advantage, while more specialized species, such as the Wood Thrush, are more vulnerable to habitat loss or fragmentation.

Reproductive strategies can influence a bird population's ability to recover from a decline. Some birds, such as migrant warblers, are relatively short-lived, compensating for their high annual mortality with rapid reproductive rates. This strategy allows birds to recover rapidly when conditions improve. Other birds, such as albatrosses, are long-lived and have commensurately slower reproductive rates, meaning that they require more time to recover from population downturns.

The most vulnerable species are those with small population sizes and slow reproductive rates that occur in specialized habitats within small geographic ranges. Yet, even species with fast reproductive strategies occupying multiple habitats across large ranges will eventually become extinct if mortality remains elevated above reproductive output for long enough.

Although extinction is a natural process, humans have caused a massive increase over background rates of species loss. In many cases, both the process of natural selection and the evolution of new species are being impacted by the wholesale destruction of habitats.

Tracking Bird Declines

Some declines in bird populations have been so dramatic and obvious that no special effort was required to verify them. For example, over-hunting and the clearance of most old-growth forest in the latter part of the 19th Century caused the extinction or near extinction of several of North America's most spectacular and numerous birds. These included the Eskimo Curlew and Passenger Pigeon among others (see Modern Bird Extinctions in the Americas p. 16).

Not all population declines among birds are so visible, however. This is especially true for widespread and generally abundant species, those occupying remote habitats, or those that are nocturnal or otherwise difficult to detect. In much of the American Tropics, scientists are still documenting the distribution of birds, and we lack sufficient data to fully assess the conservation status of many tropical species (though this situation is rapidly improving).

Long-term studies provide the strongest data that can be used to measure changes in bird populations. In the U.S., two such information sources are the Christmas Bird Count (begun in 1900) and the Breeding Bird Survey (begun in 1965). Individual states (e.g., New York and Pennsylvania) have also repeatedly conducted Breeding Bird Atlas projects that provide useful data on local and regional changes. From these studies we know that many common birds, once thought to be safe, are actually suffering population declines. For example, in the last 40 years, species such as the Northern Bobwhite, Northern Pintail, Greater Scaup, Loggerhead Shrike, Boreal Chickadee, Evening Grosbeak, and Eastern Meadowlark have all declined by at least 70%.

In summary, of the approximately 830 bird species that are found in the 50 U.S. states on a regular basis, around 25% are currently considered to be of conservation concern due to their rarity, small ranges, high threats, or declining populations. Of the 4,400 species found in the Americas as a whole, close to 800 are considered to be either Near-Threatened or in a higher threat category under IUCN criteria. This international ranking system, however, focuses primarily on the threat of extinction. If a comparable analysis were completed for the whole of the Americas to that used for the WatchList, many more species would undoubtedly be found to be of conservation concern.

We may not know all the causes for bird population declines as yet, but we have detailed information on many of them. Now we must consider, with the limited resources at our disposal, which are the highest priorities.

Ranking Threats to Bird Populations

The destruction, degradation, and fragmentation of habitats are among the greatest threats to birds. In North America, vast areas of wetlands were drained prior to the middle of the 20th Century, most of the eastern forests were logged before 1900, and virtually all the tall-grass prairies have now been converted to agriculture, causing declines in birds dependent on those habitats.

Where people did not completely convert natural vegetation, our activities still frequently degraded or altered habitats for birds. People killed off the vast herds of American bison and many colonies of prairie dogs, both of which created important grassland habitats that defined much of the western landscape. Fire suppression policies reduced the frequency of low-intensity fires upon which healthy conifer forests in the West and Southeast once depended. This has increased the underbrush density to the point that when fires do strike, they can become so fierce that they destroy trees altogether. Consequently, fire-dependent habitats and their bird species have suffered major declines. Some species, such as the Red-cockaded Woodpecker that depend on southeastern longleaf pine forests, have suffered the twofold threats of timber harvesting and fire suppression, which compound these declines.

Other threats, such as over-hunting, are thankfully now behind us, but new ones have replaced them. For example, incidental bycatch in marine fisheries drown thousands of albatrosses and other seabirds on longline hooks annually; and in and around Delaware Bay, the overharvesting of horseshoe crabs reduces the availability of the crab's eggs which are a vital food source for migrating shorebirds.

Migrating birds must also now run a gauntlet of obstacles, including lighted buildings, window glass, communication towers, and increasingly, wind turbines. Collisions with these and other structures likely kill upwards of a billion birds each year.

People have also introduced countless invasive species, and facilitated the overabundance of some native species that now threaten rarer birds. Pathogens and predators brought to the Hawaiian Islands, first by Polynesians and later by western colonists, contributed to more than 50 bird extinctions in the archipelago, the last of which

occurred as recently as 2004. Our house cats, introduced from Europe, kill hundreds of millions of birds each year, and especially threaten beach-nesting birds and species that forage on or near the ground. The fragmentation and clearance of forests in the eastern U.S. has helped Brown-headed Cowbird populations to expand. This brood parasite lays its eggs in the nests of other species, causing host birds to raise the cowbird young at the expense of their own, contributing to the decline of many forest species in the East. Controlling cowbird parasitism is essential to recovering Endangered Kirtland's Warbler populations, for example.

Pesticides and pollution still kill birds as well. The pesticide DDT caused reproductive failure in many birds, including the Osprey, Bald Eagle, and Peregrine Falcon. Following a ban on DDT and related organochlorine chemicals, and with intensive management for their recovery, these species have largely rebounded in North America, but other pesticides and chemicals continue to threaten bird species, particularly birds of prey and those using grasslands and agricultural lands. Pesticides are now frequently more insidious and their effects harder to detect than those of DDT, but nevertheless, they are still harmful.

Other toxins are also killing birds. Lead shot poisons California Condors that feed on abandoned carcasses and "gut piles" left by deer hunters, while albatross chicks on Midway Atoll die from ingesting lead-based paint that is peeling from abandoned buildings. The sub-lethal effects of other heavy metals may also be having long-term impacts on species such as loons and other waterbirds. Finally, greenhouse gas emissions are driving climate change that is already beginning to impact bird populations—its effects will likely increase in coming years.

With such a variety of threats contributing to declining bird populations, conservationists have their work cut out in order to protect birds, and we need everyone who is concerned to lend their support.

PHOTO: ROBERT ROYSE

Thanks to wetland management programs, the population of the Northern Shoveler has more than doubled since the 1980s.

A BRIEF HISTORY OF BIRD CONSERVATION IN THE AMERICAS

North America's first anthropogenic bird extinctions likely occurred following the arrival of humans at the end of the last Ice Age, roughly 11,000 years ago. Armed with new hunting tools, these people killed off many of North America's largest animals, such as the giant ground sloths and mammoths. This may have removed the food supply of bird species that depended on these animals, such as giant *Teratornis* vultures. The arrival of Polynesians in the Hawaiian Islands, roughly 1,600 years ago, precipitated the extinction of flightless waterbirds that were hunted for meat, and raptors that depended on these birds as prey.

The first modern wave of bird extinctions in North America occurred after the arrival of European settlers who were equipped with firearms and saws. During the 1800s and early 1900s, mainland North America lost the Labrador Duck, "Heath Hen" (an eastern subspecies of the Greater Prairie-Chicken), Great Auk, and Carolina Parakeet. The Eskimo Curlew and Ivory-billed Woodpecker appear to have followed, though recent reports of the woodpecker, and other cryptic evidence of its potential continued existence, keep faint hope for its survival alive. Many Hawaiian songbirds also vanished, likely due to diseases spread by the mosquitoes that were brought to the islands during the 1820s.

Bird conservation first found its voice in America during the 1890s, when women across the country revolted against the slaughter of birds for their feathers and skins that were used in the fashion trade. The efforts of these women marked the first example of a large-scale grassroots environmental campaign. Eventually, they were successful in banning the plume trade, and in the process, formed numerous state Audubon and other organizations that supported legislation to create National Wildlife Refuges, protect endangered species, and launch a range of other conservation programs. Around the same time, market hunters had almost wiped out the Passenger Pigeon and slaughtered countless shorebirds and waterfowl to meet the demand for food in New York and other large cities. Legislation in the form of the Lacey Act of 1900 and the Migratory Bird Treaty Act of 1918 stopped this unregulated and unrelenting killing.

In the 1970s, 80 countries launched the Convention on International Trade in Endangered Species of Wild Fauna and Flora (known as CITES) to regulate the trafficking of threatened plants and animals, including many bird species. In 1992, the U.S. passed the Wild Bird Conservation Act, virtually halting the importation of wild birds into the country. This closed the world's largest caged bird market, and greatly reduced pressure on wild populations of parrots and songbirds in other countries. Concerns that this action would drive the trade underground and increase smuggling proved largely incorrect. Instead, domestic bird breeders filled the gap and now supply most U.S. demand for pet birds.

As the National Wildlife Refuge and National Park systems grew and the U.S. fought two world wars, bird conservation faded from public attention. A new threat catapulted birds back into the public eye, however, when Rachel Carson published *Silent Spring* in 1962. Her book focused attention on the population crashes of some of America's most visible birds, including the Bald Eagle, due to DDT and other organochlorine pesticides that were originally developed as nerve toxins for military purposes. The resulting outcry eventually led to the 1972 ban on DDT in America, demonstrating the potency of public

PHOTO: ROBERT ROYSE

In the 1990s, bird conservation attention turned to neotropical migrants such as the Baltimore Oriole.

conservation efforts, even in the face of vigorous industry opposition.

Fueled by this blossoming environmental movement, most of the major legislation protecting the environment in the U.S. was subsequently passed, including the National Environmental Policy Act (1969), Clean Air Act (1970), and Clean Water Act (1977). The Endangered Species Act was passed in 1973 to protect distinct populations of birds and other wildlife from extinction. Since its passage, the population trends of the majority of bird species listed under the Act have stabilized or increased. For example, the California Condor was down to 22 individuals in 1987, but now numbers 187 in the wild, with a further 161 in captivity. A new government agency, the Environmental Protection Agency, was also created (in 1970) to address the public's concerns about pollution and other threats to the environment.

The 1980s brought the threat of habitat loss to the forefront of bird conservation in North America.

In 1985, record low waterfowl numbers were attributed to drought and to the loss of wetlands in the continental U.S. (with similar losses in Canada), particularly in the "prairie pothole" region (e.g., the Dakotas) and other key waterfowl breeding areas. The economic and cultural importance of waterfowl to hunters and birders, combined with concern over declining populations, led to the creation of the North American Waterfowl Management Plan in 1986 by the U.S. and Canadian governments, who were joined in 1994 by Mexico. By 2006, $4.5 billion had been invested under the plan to protect and restore 15.7 million acres of waterfowl habitat, helping to recover waterfowl populations.

While efforts to recover waterfowl achieved successes in North America, new evidence was mounting that many songbirds migrating between North America and the tropics were suffering serious population declines. The loss of stop-over and wintering habitat was a suspected cause.

As the bird conservation movement grew in North America, the science and infrastructure for bird conservation in Latin America and the Caribbean lagged behind. Bird conservation in the Neotropics faced many distinct challenges. Among them, the limited number of trained conservation professionals, the paucity of natural history information, the sheer number of bird species to worry about, and even the fact that new bird species were still being discovered. Simultaneously, habitat conversion, overharvesting, and other threats increasingly impacted resident and migrant neotropical bird populations, threatening to repeat the North American history of bird extinctions but on a yet grander scale. For example, more than 80% of Brazil's Atlantic forests were cleared by 1990, and this region now hosts a huge number of threatened and endemic bird species that depend completely on the remaining forest fragments.

Driven by concerns over declines in neotropical migratory birds, conservation attention shifted south in the 1990s—eventually expanding to focus on resident birds in both the Northern and Southern Hemispheres, as well as birds that migrate between the two. In 1990, Partners in Flight (PIF) was formed to address the conservation of non-game species with a strong focus on neotropical migrants. PIF is a cooperative effort involving government agencies, foundations (such as the National Fish and Wildlife Foundation), conservation groups, industry, academia, and private individuals, working together to help save the bird species most at risk, and to keep common birds common across the Americas. In 1993, the Smithsonian Migratory Bird Center and the Cornell Lab of Ornithology (both PIF partners) launched International Migratory Bird Day, an educational campaign celebrated each spring to raise awareness of bird conservation issues.

American Bird Conservancy (ABC) was founded in 1994 and took on the role of pulling together the diverse constituencies in bird conservation, with the aim of making them add up to more than the sum of their parts. ABC took on the leadership of PIF, and later helped to launch the North American Bird Conservation Initiative (NABCI), which included many additional groups from across the U.S., as well as Canada and Mexico. ABC has since developed into a major force for bird conservation with hemisphere-wide programs.

During the late 1990s and the following decade, the development of the Internet and other information technologies created opportunities to connect birdwatchers and researchers across the Americas to share conservation-related information more efficiently. The emergence of e-mail, listserves, online journals, eBird, the Avian Incident Monitoring System, Avian Knowledge Network, and other online databases of bird sounds, images, and information all promise an explosion of data available to analyze bird populations and distributions.

In 2005, scientific analysis showed that extinction rates were poised to accelerate beyond sensitive island locations towards the planet's most biodiverse mainland regions—especially to tropical continental mountain chains such as the Andes. In order to preempt this new extinction wave, 67 biodiversity conservation organizations joined together in the Alliance for Zero Extinction (AZE). AZE identifies sites across the globe, each of which represents the sole refuge for one or more Endangered species. The list of AZE sites provides a roadmap for conservation that can create a frontline of defense against species extinctions globally.

SETTING TODAY'S PRIORITIES

As you will see below, bird conservationists have spent a great deal of time and effort setting up systems to prioritize bird species for conservation, but less time setting priorities for bird habitats, and for addressing threats to birds. One of the aims of this book is to better elucidate priorities for these two vitally important subjects, along with determining how resources can best be mobilized to address the top priorities for birds, sites, habitats, and threats.

Prioritizing Species

Scientists recognize approximately 10,000 bird species worldwide, more than one third of which occur in the Americas—approximately 3,200 species in South America, and 2,000 in Central and North America and the Caribbean. There is significant overlap of species between these regions such that around 4,400 total species are found in the Americas as a whole. The exact number is still growing slowly as our understanding of taxonomy improves and new species are described.

Below the species level, biologists distinguish some distinct populations of birds as subspecies. Subspecies are generally considered to be of lower conservation priority, even though they may be highly distinctive and endangered (see The Conservation of Subspecies p. 20). Regardless of taxonomy or formal status, there are many bird populations vying for conservation attention. How do we prioritize them?

Birds with small populations that face dire threats are the most prone to extinction. Safeguarding these rare species is clearly a top priority. At the global scale, IUCN is widely regarded as the authority for ranking and listing threatened species, including birds. BirdLife International places bird species in several categories of extinction risk (Least Concern, Near Threatened, Vulnerable, Endangered, and Critically Endangered) on behalf of IUCN. Beyond this, AZE has identified all the Endangered and Critically Endangered birds, mammals, amphibians, and reptiles that are each restricted entirely, or almost entirely, to single remaining sites.

The next set of bird species we need to assess are those that may not be on the verge of extinction yet, but still have declining populations, small ranges, or small populations. In 2007, scientists from ABC, the Cornell Lab of Ornithology, National Audubon Society, and the Rocky Mountain Bird Observatory joined together with PIF to create a consensus priority list for bird conservation called the WatchList (see www.abcbirds.org). The WatchList ranks conservation priority for all bird species that occur in the U.S. and Canada according to four factors: population size, range size, threats, and population trend. Species on the WatchList are classified into color-coded threat categories. Species with small population sizes, small range sizes, serious threats, and rapidly declining population trends receive the highest scores, and are considered to be of the Highest National Concern (Red WatchList). Species with fairly large populations and ranges, but that score high for threats and have negative population trends are considered as Declining Species (Yellow WatchList). Species with stable or less seriously declining populations, but small range sizes and/or populations are classified as Rare Species (also Yellow WatchList). Red WatchList birds require immediate conservation attention. Yellow listed birds also deserve attention to prevent their status from deteriorating further.

Common birds that do not qualify for inclusion on the WatchList also require monitoring, as declines among common birds may be difficult to

The habitat needs of non-game songbirds are often the same as those of game species such as this Spruce Grouse.

detect before drastic population reductions have occurred. The Passenger Pigeon was once perhaps the most abundant bird in North America, and hardly anyone believed that it could ever completely disappear. Yet, by 1914 it had been hunted to extinction. Keeping common birds common should therefore be one of our conservation goals given the difficulty and expense of bringing species back from the brink of extinction.

Aside from rating the threat level for each species, it is important to also focus on those birds for which our states, country, and region have the greatest global responsibility. For example, there are some globally abundant species that have small colonies or outposts in our region. While these species may be rare or threatened here, it would be wrong to divert more significant resources to their conservation than are allocated to species that depend completely on our region for their survival.

The actions needed to conserve species are almost as varied as the birds themselves. Each species ac-

count presented in this book describes the specific threats, and suggests ways these can be countered.

Prioritizing Places

At broad scales, ecologists divide the landscape into ecoregions or biomes—geographic areas united by similar geology and vegetation (referred to as "Birdscapes" in this book). But this alone is insufficient to prioritize the conservation of the specific sites that birds depend on for foraging and nesting, and during migration. ABC published the book: *The 500 Most Important Bird Areas in the United States*, to refine our understanding of the most important places to prioritize for bird conservation action. In this current book, we highlight 50 such sites, each placed in the context of a larger Birdscape that provides an overview of conservation needs and priorities for the broader ecosystem—each Birdscape is further divided into more specific bird habitats whose condition and threat level are ranked. By focusing our efforts on the most endangered habitats that harbor aggregations of declining species, we can often maximize conservation results. For the rarest of the rare, AZE identifies single sites that, if lost, would likely result in entire species extinctions; AZE sites have already been identified for the whole planet, including more than 70 for bird species in the Americas.

Working with landowners and managers to create the right habitat conditions for priority birds, working to set aside certain critical areas for conservation, and countering specific threats to sites and habitats as they arise, are the keys to effectively conserving "place" for birds.

Eliminating Threats

After identifying which species and habitats are most in need of conservation attention, we must identify the most serious threats to birds, and determine how we can achieve the most significant positive changes. There is no widely accepted ranking system for threats to birds as yet, but we have focused on those that cause the greatest habitat impacts and bird mortality in this book, highlighting those that cause more than one-million bird deaths each year (see Threats introduction on p. 278).

For species threatened by habitat destruction, degradation, and fragmentation, we must protect or restore sufficient habitat to maintain populations (see Prioritizing Places previous page). In the case of threats that cause large-scale mortality, such as predation by cats, pesticide poisoning, or collisions with man-made structures, we must change human behavior, enact conservation regulations, and influence industry to do the right thing.

The Future of Bird Conservation

We have reached a time in history where the continuing management of bird populations and their habitats will be required from this point forth,

to safeguard the rarest species, and to ultimately maintain the abundance and diversity of all birds. For species with the most endangered populations, emergency, species-specific restoration efforts will continue to be required to ensure that populations can recover to the point where they can endure future potential threats.

Minimizing threats that harm habitats and multiple bird species will also be crucial to preventing future emergencies. We will also need to control the numbers of the handful of species that benefit from human activities at the expense of other, less adaptable species. Some species will simply require monitoring to ensure that they do not warrant special conservation attention, though action should also be taken for these species should future problems arise. This effort will require engaging relevant stakeholders, developing productive partnerships that seek solutions to environmental problems, as well as significant, new financial resources to support conservation.

Among the top priorities for action, we need to:

- Prevent the extinction of any additional bird species in the Americas by protecting AZE sites and developing a sustainable network of protected areas that conserve all bird species listed as Endangered or Critically Endangered according to IUCN criteria.
- Ensure that all the species in the U.S. that need Endangered Species Act protection are receiving it, and that funding for the conservation of these species is adequate (e.g., list candidate species, increase funding for Hawaiian birds).
- Reduce threats to seabirds, which are especially vulnerable to invasive species at their nesting grounds, and to fisheries bycatch in their foraging ranges (e.g., by pressing fishing nations that sell their catch in the U.S. to adopt the same bycatch mitigation measures required by U.S. crews).
- Set population targets for all WatchList species and work with public and private landowners, managers, and policy makers to ensure that the U.S. landscape is designed to accommodate the

Since 2000, conservation attention has shifted to endangered endemic species such as the Jocotoco Antpitta.

PHOTO: MARK HARPER

port tolerances on pesticides that harm birds in Latin America and the Caribbean, require guidelines for the siting and operation of wind farms, change lighting on communication towers).

- Ensure that energy development and our response to the climate change challenge benefits rather than further threatens birds (e.g., "smart wind", native prairie biofuels, avoided deforestation, mine land restoration).
- Develop federal funding sources for bird conservation (e.g., ensure that funds from off-shore drilling go to conservation as mandated, instead of to other budget items; increase federal funding for Joint Ventures and the Neotropical Migratory Bird Conservation Act).

Despite the complexity and enormity of the challenge, opportunities to deliver effective bird conservation have never been greater than they are today, as we now know more about conservation science and practice than ever before. But while interest in birds seems to be at an all-time high, engaging the bird community (outside of hunting) in conservation at a scale that can actually affect bird populations at national or global levels is still proving to be a challenge. Today, there are millions of active birdwatchers and hunters across the country, and numerous local, state, national, and international initiatives that aim to conserve birds and their habitats. Pulling all these disparate parts together into a functioning greater whole is still one of the great opportunities and challenges for the future of bird conservation.

needs of these species (e.g., expand landowner programs that incentivize bird conservation; develop new zoning programs, especially for marine areas; end the persecution of prairie dogs that create short-grass habitat, restore natural fire regimes, rotate grazing to protect riparian woodlands, end federal programs that encourage coastal development).

- Develop effective programs that address the conservation needs of priority neotropical migrants on their wintering grounds.
- Ensure that National Wildlife Refuges and other key sites are properly funded and managed for priority birds (e.g., by funding programs to remove or adequately control invasive species).
- Reduce anthropogenic bird mortality such as from: pesticides and other toxics; collisions with glass, communication towers, and wind turbines; and predation by free-roaming cats (e.g., ban lead ammunition for hunting, revoke im-

Our collective efforts to conserve birds have already achieved some great successes, albeit set against a general backdrop of bird declines. We have prevented the extinction of several species, including the California Condor and Whooping Crane, and revived populations of many other species once decimated by market hunters, the plume-trade, DDT, and other threats. Our past victories give us hope for those species that are currently in decline. With a renewed commitment to birds and their conservation, we can prevent extinctions, conserve threatened habitats, and eliminate the most significant threats. Let's get to work!

HISTORIC LANDMARKS IN AMERICAN BIRD CONSERVATION

1886 George Grinnell founds the first Audubon Society.

1896 Harriet Hemenway, her cousin Minna Hall, and their friends start a campaign to urge women to stop wearing bird plumes.

1900 The Lacey Act puts limits on market hunting by making it illegal to hunt birds in one state and sell them in another.

1903 President Theodore Roosevelt designates Pelican Island, Florida, as a federal refuge, thereby establishing the National Wildlife Refuge System.

1916 Canada and the U.S. join in support of the Migratory Bird Convention, making it unlawful to take or sell any migratory bird, nest, or egg, unless under permit. The U.S. government passes the Migratory Bird Treaty Act soon after, in 1918.

1934 Waterfowl licenses are introduced and become the funding mechanism for wetlands acquisition, supporting habitat conservation for waterfowl and other birds through the "Duck Stamp" program.

1934 Roger Tory Peterson publishes his landmark *Field Guide to the Birds*. The book was the first to highlight readily noticeable visual field marks, making bird identification accessible to the layman and contributing to the understanding of birds by a broader audience.

1940 A Department of the Interior reorganization consolidates the Bureau of Fisheries and the Bureau of Biological Survey into one agency to be known as the Fish and Wildlife Service, with the mission of working with others to conserve, protect, and enhance fish, wildlife, plants, and their habitats for the continuing benefit of the American people.

1962 Rachel Carson publishes *Silent Spring*.

President Theodore Roosevelt

Brown Pelican:
Pelican Island protected in 1903

First Duck Stamp

Rachel Carson

1970 The U.S. Environmental Protection Agency is created, with the mission of protecting human health and safeguarding the natural environment—air, water and land—upon which life depends.

1972 DDT is banned in the U.S. This pesticide caused reproductive failure in some birds, particularly the Brown Pelican, Osprey, Bald Eagle, and Peregrine Falcon, resulting in severe population declines.

Bald Eagle: recovered, then delisted in 2007

1973 Congress passes the Endangered Species Act, which not only prohibits the unauthorized take, possession, sale, and transport of listed species, but also provides for comprehensive habitat protection, including a provision for the federal government to acquire land for endangered species conservation.

1973 The Convention on International Trade in Endangered Species of Wild Fauna and Flora (CITES) is signed, ensuring that the international trade in wild animals and plants does not threaten their survival. More than 150 nations now participate.

1989 The North American Wetlands Conservation Act is passed to authorize grants for the conservation of wetland bird habitat.

Red-and-Green Macaw: bird trade halted in 1992

1990 Partners in Flight is formed.

1992 The Wild Bird Conservation Act is passed, essentially halting U.S. importation of wild birds.

1994 American Bird Conservancy is founded.

2000 The Neotropical Migratory Bird Conservation Act designates funds for habitat protection, education, research, and the monitoring of neotropical migratory birds in Latin America, the Caribbean, and the U.S.

2005 The North American Bird Conservation Initiative is established.

2005 The Alliance for Zero Extinction identifies the last remaining sites for the most endangered bird species on Earth.

The State of the Birds report

2009 *The State of the Birds* report published.

MODERN BIRD EXTINCTIONS IN THE AMERICAS

1741-1840s The Spectacled Cormorant is described to science, but disappears from the islands of the Bering Sea, becoming extinct within 100 years of its discovery.

1837 Last reports of the Oahu Oo from Oahu, Hawaii.

1844 The Great Auk (**1**) becomes extinct when the last two are killed.

1859 The last known specimen of the Kioea is collected on Hawaii.

1864 Last known record of the Cuban Macaw (**2**). Extinction likely followed soon after.

1870s The Labrador Duck becomes extinct in North America.

1879 The last confirmed sighting of the Jamaica Petrel on the island of Jamaica.

1884 Last report of the Hawaiian Rail from the Island of Hawaii.

1892 The Ula-ai-Hawane (**3**) is last collected on the island of Hawaii. Extinction likely followed soon after.

1894 Last sighting of the Kona Grosbeak on the Island of Hawaii.

1896 Last record of the Greater Koa Finch from the Island of Hawaii. The Lesser Koa Finch, known only from specimens collected on the same island in 1891, likely becomes extinct soon after.

1899 Last record of the Hawaiian Mamo on the Island of Hawaii.

1901 The Greater Amakihi is last reported on the Island of Hawaii.

1904 The Molokai Oo is last reported from Molokai, Hawaii.

1907 The Black Mamo is last reported from Molokai.

1912 The Guadalupe Storm-Petrel is last reported from Guadalupe Island, Mexico.

1914 The Passenger Pigeon becomes extinct when the last individual dies in captivity at the Cincinnati Zoo.

1918 The last known Carolina Parakeet (**4**) dies in captivity.

1932 The Heath Hen (a subspecies of the Greater Prairie-Chicken) becomes extinct on Martha's Vineyard after a long decline in the northeastern U.S.

1934 Last report of a Hawaii Oo (**5**) from the Island of Hawaii.

1940s The Grand Cayman Thrush disappears from the Caribbean island of the same name.

1944 The Laysan Rail is recorded for the last time on Midway Atoll.

1956 The last confirmed report of the Imperial Woodpecker (**6**) is made when one is shot in Durango, Mexico. This species possibly survived into the 1990s, however, and rumors of its survival even persist today, though most consider it extinct.

1960s The brown tree snake accidentally arrives on the U.S. island territory of Guam, decimating native forest-dwelling bird species. Some of these species, including the Guam Rail and Micronesian Kingfisher persist in captivity, however, with new populations established on the nearby (currently) snake-free island of Rota.

1961 The Semper's Warbler is last recorded on the Caribbean island of St. Lucia, to which it was endemic; it likley became extinct soon after.

1962 An Eskimo Curlew is photographed on Galveston Island, Texas, the last individual to be confirmed in the U.S. A specimen was collected in Barbados a year later, but the species has not been recorded since with certainty.

1963 The Molokai Creeper is last seen on Molokai.

1967 Last report of an Akialoa from the Hawaiian island of Kauai, long after populations on other islands have disappeared.

1970s The Colombian Grebe is last reported. Surveys fail to find any in the early 1980s.

1972 The Socorro Dove likely becomes extinct in the wild on its island home off Mexico. Fortunately, remaining captive birds provide hope for the species' survival.

1979 The Bachman's Warbler (**7**) is likely extinct after no birds are found during an extensive systematic search between 1975 and 1979.

1980s The Atitlan Grebe (**8**) becomes extinct in Guatemala.

1980s The last wild Alagoas Curassow (**9**) is killed by hunters in northeastern Brazil. The species survives in captivity, however.

1980s The Kamao likely becomes extinct in the decade following a 1981 survey that reveals that 34 remain on the Hawaiian island of Kauai.

1987 The Kauai Oo is last heard on the Hawaiian island of Kauai

1987 The last individual of the "Dusky" subspecies of the Seaside Sparrow, endemic to eastern Florida, dies in captivity.

2000 The last know Spix's Macaw (**10**) in the wild, a lone male, dies, though the species survives in captivity.

2003 The Hawaiian Crow disappears from the wild on the Island of Hawaii, though more than 50 remain in captivity.

2004 The Poo-uli likely becomes extinct as the last known individual dies in captivity after a rapid decline in the wild on the Hawaiian island of Maui.

C H A P T E R   1

WATCHLIST BIRDS

The species is still the most widely accepted unit for conservation prioritization among birds (though see The Conservation of Subspecies p. 20). Most of the 800 plus bird species that occur in the U.S. are protected by federal or state laws. The Endangered Species Act (ESA) protects the rarest of the rare; the Migratory Bird Treaty Act, Lacey Act, and state wildlife and hunting laws protect almost all other birds—only non-native species such as the House Sparrow and European Starling have no protection at all (though animal cruelty laws may afford even these some protection depending on location and circumstance).

While these protections include both halting the killing of birds and the conservation of habitat, they are not fully enforced, and unfortunately many species are still declining due to a variety of threats. All birds listed under the ESA receive conservation attention, although in some cases this is still insufficient (e.g., some Hawaiian species), but until recently there was no single agreed list that also included the next most urgent priority bird species in the U.S. that NGOs, and federal and state governments could focus on for conservation action. The WatchList closes this gap by including species that are listed as federally Endangered, but also by providing an early warning system through the listing of additional species that, without action, may be headed in that direction. Many of these species co-occur as assemblages in rare or threatened habitats that can be conserved as suites of birds with similar needs (e.g., cavity-nesting woodpeckers and owls in fire-

suppressed forests). Others face threats that must be addressed one by one, however (e.g., rare seabirds that nest on islands invaded by rats).

In the following section we present species accounts for the 212 WatchList species: the priority species for conservation in the U.S. At the top of each species account is a blue fact bar (see opposite page). The common and scientific names for each species can be found on the left hand side of this bar. In the center column is a global population estimate, and the percentage of the population found in the U.S. If the species occurs only in the U.S., it will be noted as a U.S. Endemic. Species that only breed within the U.S., but that migrate outside its borders after breeding are noted as U.S. Breeding endemics. We also note those species whose entire populations winter in the U.S.

In the right hand column of the fact bar is the species' current population trend and WatchList status, and next to the WatchList status is the WatchList combined score within a colored box. Species are assessed on the basis of population size, range size, threats, and population trend. Species that score high in all categories are of "Highest National Concern" (Red WatchList), species that score high for threats and population trend are classified as "Declining", and those that score high for population and range size are "Rare" (latter two categories both Yellow WatchList). See Butcher et al. (2007) for more information. If the species is listed under the ESA in the U.S., or is proposed or a candidate for listing (not including foreign listed species), its listing status (Endan-

KEY TO SPECIES ACCOUNTS

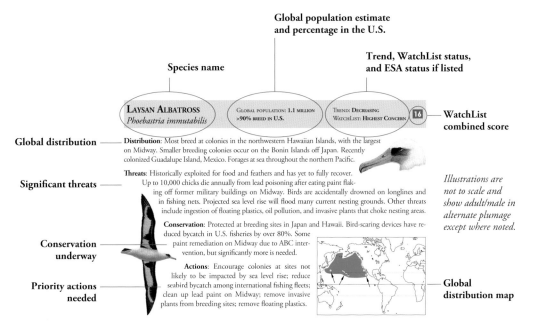

Global population estimate
and percentage in the U.S.

Species name

Trend, WatchList status,
and ESA status if listed

LAYSAN ALBATROSS
Phoebastria immutabilis

GLOBAL POPULATION: 1.1 MILLION
>90% BREED IN U.S.

TREND: DECREASING
WATCHLIST: HIGHEST CONCERN

16

WatchList
combined score

Global distribution

Distribution: Most breed at colonies in the northwestern Hawaiian Islands, with the largest on Midway. Smaller breeding colonies occur on the Bonin Islands off Japan. Recently colonized Guadalupe Island, Mexico. Forages at sea throughout the northern Pacific.

Significant threats

Threats: Historically exploited for food and feathers and has yet to fully recover. Up to 10,000 chicks die annually from lead poisoning after eating paint flaking off former military buildings on Midway. Birds are accidentally drowned on longlines and in fishing nets. Projected sea level rise will flood many current nesting grounds. Other threats include ingestion of floating plastics, oil pollution, and invasive plants that choke nesting areas.

Illustrations are not to scale and show adult/male in alternate plumage except where noted.

Conservation
underway

Conservation: Protected at breeding sites in Japan and Hawaii. Bird-scaring devices have reduced bycatch in U.S. fisheries by over 80%. Some paint remediation on Midway due to ABC intervention, but significantly more is needed.

Priority actions
needed

Actions: Encourage colonies at sites not likely to be impacted by sea level rise; reduce seabird bycatch among international fishing fleets; clean up lead paint on Midway; remove invasive plants from breeding sites; remove floating plastics.

Global
distribution map

gered, Threatened, Candidate, or proposed) will be noted beneath its WatchList status.

Beneath the fact bar, the text is divided into four sections. The first, **Distribution**, briefly describes the species' habitat and range. For migratory species, both breeding and non-breeding distributions are described. The second, **Threats** section succinctly lists major threats affecting the species. Some threats, such as climate change, cat predation, and glass collisions, may affect many species, and so we only document them for species that are (or could become) particularly adversely affected. The **Conservation** section highlights protected areas such as National Parks (NPs) or National Wildlife Refuges (NWRs) that protect the species, and provides information on major species-specific conservation projects that are underway. The last section, **Actions**, lists important actions that will benefit the species. Note that we typically do not include research or monitoring under "Actions", as these are needed for almost every species on the

WatchList. Some species are better known than others, however, and those with a trend that is unknown are priorities for such work.

Each species account features a distribution map and illustration of the species. On the map, areas inhabited year-round are colored green, areas inhabited only during the breeding season are yellow, and areas inhabited outside the breeding season are blue. Migratory routes are not depicted. Maps show the global distribution for species that are not restricted to the Americas.

Bird names follow the American Ornithologists' Union (AOU) Checklist for North America, Central America, and the Caribbean, and follow the South American Species Checklist Committee for birds in South America. We have made one exception to the AOU taxonomy by splitting Newell's Shearwater from Townsend's, which is commonly accepted by conservationists. See the Glossary on p. IX for a list of terms used.

THE CONSERVATION OF SUBSPECIES

Approximately 800 species of birds occur regularly in the continental U.S. and Canada, with an additional 30 or so native species breeding in the Hawaiian Islands. Many of these species have multiple distinct populations or races that are designated as subspecies, with each subspecies usually breeding in a different geographic region. Classifying subspecies is more problematic and controversial than species, and many subspecies classifications are likely invalid or out-dated, representing forms that are not truly distinct or diagnosable. At the other extreme, well defined subspecies or subspecies groups are sometimes reclassified as full species. Such "splits" are ongoing. For instance, the American Ornithologists' Union Checklist Committee split the Tufted Titmouse along subspecies lines, elevating the Black-crested Titmouse to species status in 2002; in 2006, the two major subspecies groups within the Blue Grouse were split into two full species, now called the Dusky Grouse and Sooty Grouse.

Due to the instability of subspecies classifications and their added complexity, subspecies are usually not included in broad analyses that prioritize bird populations for conservation, and the conservation status of many subspecies thus remains to be evaluated. Nevertheless, if any country has the capacity to conserve both species and subspecies, it is the U.S. In fact, the ESA already includes provisions for the protection of subspecies, and several birds, such as the "San Clemente" Loggerhead Shrike and "Cape Sable" Seaside Sparrow are already protected by the Act. Developing a broader-based conservation program for all subspecies, however, would benefit from an updated evaluation of bird taxonomy. It is hoped that work to tease apart this tricky issue in bird conservation can be prioritized in the near future. A few highly distinctive subspecies are highlighted here.

The ridgwayi *subspecies of the Northern Bobwhite ("Masked Bobwhite") was reintroduced to Arizona after extirpation, but is at high risk of disappearing again.*

*Eastern Bewick's Wrens (*bewickii *above, and* altus *subspecies) have both declined significantly and are listed by many eastern states as Endangered.*

*There are two rare island subspecies of the Loggerhead Shrike (*mearnsi *from San Clemente is pictured).*

U.S. ENDANGERED SPECIES, SUBSPECIES, AND TRENDS

This list includes bird species, subspecies, and populations in the U.S. that are currently protected under the ESA. Please refer to this page for the ESA status of species listed in the Habitats chapter. Note that conservationists have yet to systematically assess threat levels for all U.S. bird subspecies. Analysis shows that species that have been listed longer (including some that have been delisted) are faring better than those listed more recently, indicating that long-term conservation programs are working. Some species that were listed at the time the Act was passed were already likely extinct or so close to extinction that it was too late for conservation efforts to work, however (e.g., some Hawaiian species). Date of listing appears in parentheses after the species name. Trends are generalized estimates. For more details see: www.stateofthebirds.org (2009) report.

Birds	Trend since listing	Birds	Trend since listing
Hawaiian Goose (1973)	Increase	Ivory-billed Woodpecker (1973)	Likely extinct
Hawaiian Duck (1973)	Stable	Willow Flycatcher (Southwestern) (1995)	Stable
Laysan Duck (1973)	Increase	Loggerhead Shrike (San Clemente) (1977)	Increase
Steller's Eider (1997)	Stable	Bell's Vireo (Least) (1986)	Increase
Spectacled Eider (1993)	Increase	Black-capped Vireo (1987)	Decrease
Northern Bobwhite (Masked) (1973)	Decrease	Florida Scrub-Jay (1987)	Decrease
Greater Prairie-Chicken (Attwater's) (1973)	Decrease	Hawaiian Crow (1973)	Extinct in wild
Short-tailed Albatross (1973)	Increase	Elepaio (Oahu) (2000)	Decrease
Hawaiian Petrel (1973)	Decrease	Millerbird (1973)	Fluctuating
Newell's Shearwater (1975)	Decrease	California Gnatcatcher (Coastal) (1993)	Decrease
Wood Stork (1984)	Increase	Kamao (1973)	Likely extinct
California Condor (1973)	Increase	Olomao (1973)	Likely extinct
Snail Kite (Everglades) (1973)	Increase	Puaiohi (1973)	Increase
Bald Eagle (Sonoran Desert) (1973)	Increase	Kauai Oo (1973)	Likely extinct
Hawaiian Hawk (1973)	Stable	Bachman's Warbler (1973)	Likely extinct
Crested Caracara (Audubon's) (1987)	Stable	Golden-cheeked Warbler (1990)	Decrease
Aplomado Falcon (Northern) (1986)	Increase	Kirtland's Warbler (1973)	Increase
Clapper Rail (California) (1973)	Increase	California Towhee (Inyo) (1987)	Increase
Clapper Rail (Light-footed) (1973)	Increase	Sage Sparrow (San Clemente) (1977)	Stable
Clapper Rail (Yuma) (1973)	Stable	Grasshopper Sparrow (Florida) (1986)	Stable
Common Moorhen (Hawaiian) (1973)	Increase	Seaside Sparrow (Cape Sable) (1973)	Decrease
Hawaiian Coot (1973)	Stable	Laysan Finch (1973)	Fluctuating
Sandhill Crane (Mississippi) (1973)	Stable	Nihoa Finch (1973)	Fluctuating
Whooping Crane (1973)	Increase	Ou (1973)	Likely extinct
Snowy Plover (Western) (1993)	Increase	Palila (1973)	Decrease
Piping Plover (Atlantic Cst & Gt Plns) (1985)	Increase	Maui Parrotbill (1973)	Stable
Piping Plover (Great Lakes) (1985)	Fluctuating	Kauai Akialoa (1973)	Likely extinct
Black-necked Stilt (Hawaiian) (1973)	Increase	Nukupuu (1973)	Likely extinct
Eskimo Curlew (1973)	Likely extinct	Akiapolaau (1973)	Decrease
Least Tern (Interior) (1985)	Stable	Molokai Creeper (1973)	Likely extinct
Least Tern (California) (1973)	Increase	Hawaii Creeper (1975)	Decrease
Roseate Tern (Florida) (1987)	Decrease	Oahu Alauahio (1973)	Likely extinct
Roseate Tern (North-east) (1987)	Fluctuating	Akikiki (2010)	Stable
Marbled Murrelet (1992)	Decrease	Akekee (2010)	Stable
Spotted Owl (Northern) (1990)	Decrease	Akepa (Maui) (1973)	Likely extinct
Spotted Owl (Mexican) (1993)	Unknown	Akepa (Hawaii) (1973)	Stable
Mariana Swiftlet (1984)	Unknown	Akohekohe (1973)	Stable
Red-cockaded Woodpecker (1973)	Increase	Poouli (1975)	Likely extinct

EMPEROR GOOSE
Chen canagica

GLOBAL POPULATION: **92,000**
C.95% BREED IN U.S.

TREND: STABLE
WATCHLIST: RARE

15

Distribution: Breeds in coastal saltmarshes of the Yukon-Kuskokwim Delta, western Alaska, with small numbers in nearby Russia (where a large percentage of the population also stages to molt). Winters in rocky intertidal areas of western Alaska. Izembek NWR is a critical site for the species' staging and wintering.

Threats: Potential threats include oil pollution, subsistence hunting, and habitat loss due to sea level rise.

Conservation: Most breeding areas are protected within Yukon Delta NWR, and most wintering habitat is protected by the Alaska Maritime and Izembek NWRs.

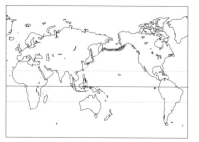

Actions: Reduce risk of oil pollution in wintering areas; continue reducing subsistence hunting.

HAWAIIAN GOOSE
Branta sandvicensis

GLOBAL POPULATION: **1,900**
U.S. ENDEMIC

TREND: INCREASING
WATCHLIST: HIGHEST CONCERN
ESA: ENDANGERED

17

Distribution: Once inhabited all major Hawaiian Islands and numbered c.25,000, but reduced to <30 by 1950s. Now restricted to grasslands and sparsely vegetated high volcanic slopes on Hawaii, Maui, Molokai, and Kauai. Kauai population faring better due to greater habitat availability and lack of mongooses.

Threats: Lack of suitable habitat resulting from agricultural development; predation by introduced mammals (mongooses, cats, rats). Other threats include collisions with vehicles, human disturbance, and behavioral and genetic issues.

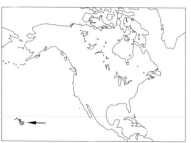

Conservation: Protected by the State of Hawaii, and occurs within many protected areas. Predator control and supplementary feeding are provided at several sites. 2,400 captive-bred birds were released from 1960-2006, though many of these died during periodic droughts.

Actions: Expand habitat management and restoration; continue predator control, especially outside Kauai; reduce road-kills. Additional large, predator-free reserves in the lowlands and cooperation with private landowners using Safe Harbor Agreements could make a major difference.

TRUMPETER SWAN
Cygnus buccinator

GLOBAL POPULATION: **35,000**
>75% BREED IN **U.S.**

TREND: **INCREASING**
WATCHLIST: **RARE**

15

Distribution: Breeds in freshwater marshes, ponds, and lakes of Alaska and western Canada, with smaller populations in northern U.S. and Ontario. Alaska population winters in estuaries in British Columbia, Canada, and Washington. Most of the Rocky Mountain population winters in Idaho, Montana, and Wyoming. Reintroduced birds in Wisconsin, Minnesota, and Michigan winter near nesting areas.

Threats: Hunting decimated populations in the lower 48 states and much of Canada during the 1800s. Current threats are loss and degradation of wetlands, lead poisoning from shot and fishing weights, and occasional collisions with power lines. Hunters may mistake these for the more widespread Tundra Swan in the Midwest.

Conservation: Most Alaskan breeding areas are secure. Red Rock Lakes NWR, Montana, protects the key site in the lower 48. Reintroduced to some of its former range.

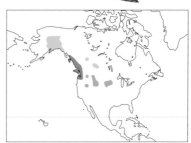

Actions: Increase quantity and quality of wintering habitat throughout range; protect additional breeding areas in the Boreal Forest and northern prairies; protect key breeding and wintering areas from development through easements; eliminate lead shot use in foraging fields and place visible markers on utility lines.

MOTTLED DUCK
Anas fulvigula

GLOBAL POPULATION: **170,000**
>75% BREED IN **U.S.**

TREND: **DECREASING**
WATCHLIST: **HIGHEST CONCERN**

16

Distribution: Inhabits freshwater and brackish wetlands in the southeastern U.S. and adjacent northeastern Mexico. Most of population is concentrated in Florida and in coastal wetlands of South Carolina, Louisiana, and Texas.

Threats: Major threat is wetland loss due to drainage for agriculture (total of more than nine million acres already lost from Florida, and one million from Louisiana and Texas), and conversion to open water through erosion and subsidence. Hybridization and competition with introduced domestic Mallards in Florida, hunting in Mexico, and potentially sea level rise, also threaten this species and its habitat.

Conservation: Many of its key sites are protected by NWRs in Florida, Louisiana, and Texas, and habitat restoration efforts are underway that should benefit this species.

Actions: Increase habitat through wetland restoration efforts, and discourage release of domestic Mallards.

HAWAIIAN DUCK
Anas wyvilliana

GLOBAL POPULATION: **2,400**
U.S. ENDEMIC

TREND: **STABLE**
WATCHLIST: **HIGHEST CONCERN** **19**
ESA: **ENDANGERED**

Distribution: Once inhabited a variety of wetland habitats on most Hawaiian Islands. Pure birds now restricted to Kauai, Niihau (together 90% of population), and parts of Hawaii; with Mallard hybrids predominating on Oahu and Maui.

Threats: Predation by introduced mammals, wetland habitat loss due to agriculture, urban development, and hunting reduced population to fewer than 500 birds in 1962. Hybridization with introduced Mallards is currently the most severe threat, and is now starting to affect birds on Kauai and Niihau.

Conservation: Captive-bred birds released on Oahu, Maui, and Hawaii from 1958-1989. Importation of Mallards to Hawaiian Islands was restricted in the 1980s. Research on identification of hybrids is occurring. The Hanalei and Huleia NWRs on Kauai are important protected areas.

Actions: Heighten public awareness of hybridization issues and educate site managers on hybrid identification; humanely remove feral Mallards and hybrids, especially on Kauai; restore wetland habitat and control invasive predators; reintroduce birds to appropriate sites on Maui and Molokai once the hybridization threat is removed.

LAYSAN DUCK
Anas laysanensis

GLOBAL POPULATION: **800**
U.S. ENDEMIC

TREND: **INCREASING**
WATCHLIST: **HIGHEST CONCERN** **17**
ESA: **ENDANGERED**

Distribution: Formerly distributed among wetlands of the Hawaiian Islands, but restricted to Laysan Island for most of the last 150 years. Recently reintroduced to Midway Atoll.

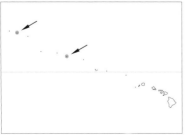

Threats: Introduced rats likely wiped out this species from most islands. Rabbits introduced to Laysan in the 1800s eliminated vegetation and nearly caused the species' extinction. Vulnerable to population crashes from drought and disease. Other threats include invasive ants on Laysan that may compete for insect prey, and invasive shrubs that degrade habitat. Botulism outbreaks, sea level rise, and the accidental introduction of mammalian predators are potential threats.

Conservation: Alien grass was eradicated from Laysan to allow native bunchgrass to recover, and the removal of invasive wetland shrubs will preserve foraging habitat there. Forty-two juvenile ducks were translocated to Midway NWR from 2004-2005, and have since bred successfully.

Actions: Continue habitat restoration on Laysan and Midway; disease surveillance; genetic management and botulism mitigation; establish additional populations on other islands with higher elevations to reduce extinction risk from sea level rise and other disasters.

STELLER'S EIDER
Polysticta stelleri

GLOBAL POPULATION: **110,000-125,000**
>70% WINTER IN **U.S.**, C. **1,000** BREED

TREND: **STABLE**
WATCHLIST: **HIGHEST CONCERN**
ESA: **THREATENED**

17

Distribution: Breeds in arctic tundra near freshwater ponds in western and eastern Siberia, along Alaska's North Coast, and sparsely in the Yukon-Kus-kokwim Delta. Western Siberian populations winter in extreme Northern Europe, but most birds winter in coastal areas of the Alaska Peninsula, the Aleutian Islands, and Russia's Kamchatka Peninsula.

Threats: Subsistence hunting and poisoning from lead shot may be problematic in Alaska and Russia. Other threats may include oil pollution, and habitat loss to sea level rise associated with climate change.

Conservation: Legally protected in the U.S. and Russia. Most important areas in the U.S. are protected within refuges, including Izembek (>17,000 staging), Kodiak Island, Alaska Peninsula, Yukon Delta, and Alaska Maritime NWRs. Primary breeding grounds along the Arctic Coastal Plain are within the National Petroleum Reserve, which is open for oil and gas extraction, however.

Actions: Ensure oil and gas development within National Petroleum Reserve avoids affecting key breeding areas; decrease hunting pressure in Alaska and Russia; assess the impacts of oil pollution and accidental mortality in fisheries.

SPECTACLED EIDER
Somateria fisheri

GLOBAL POPULATION: **330,000**
U.S. WINTER ENDEMIC, C. **7,500** BREED

TREND: **INCREASING**
WATCHLIST: **HIGHEST CONCERN**
ESA: **THREATENED**

18

Distribution: Breeds in arctic tundra lakes, ponds, and rivers in Alaska's Yukon-Kuskokwim Delta and Arctic Coastal Plain, plus three known areas in Siberia. Molts at staging areas in Norton Sound and Ledyard Bay, Alaska, and in two areas of coastal Siberia. Entire population winters south of Alaska's St. Lawrence Island in the Bering Sea.

Threats: In western Alaska, poisoning by ingestion of lead shot reduces survival rates, but direct effects of subsistence hunting appear minimal. Other threats include the expansion of human settlements near breeding areas that boost populations of gulls and foxes, increasing predation of eggs and young. Climate change appears to be affecting the relative abundance of different clam species the eiders feed on outside the nesting season, which could harm winter survival rates.

Conservation: Wintering, molting, and breeding areas outside the National Petroleum Reserve designated as Critical Habitat.

Actions: Ensure oil and gas exploration within the National Petroleum Reserve does not harm this species; reduce greenhouse gas emissions to combat climate change; eliminate use of lead shot near breeding areas.

MOUNTAIN QUAIL
Oreortyx pictus

GLOBAL POPULATION: **160,000**
>95% BREED IN U.S.

TREND: UNKNOWN
WatchList: RARE

15

Distribution: Inhabits dense shrub, chaparral, and forests of the Pacific Coast and coastal ranges, and Northern Rockies, ranging from Vancouver Island south to northern Baja California, Mexico, from sea level to c.10,000 feet. In many areas, birds migrate down slope to avoid heavy winter snow. Declines and range contraction during 20th Century most severe in eastern portion of range in Idaho and eastern Oregon.

Threats: Habitat loss and degradation from logging, over-grazing by cattle, and (in California) urban development.

Conservation: Hunting banned as a precaution in Idaho and Oregon, where declines have occurred. Nevada and Oregon are using translocations to restore birds to former range, but the success of these efforts is thus far unknown. Much of its range occurs within NFs and other federally managed lands, especially in Oregon and the Sierra Nevada.

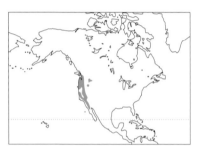

Actions: Determine optimum fire-habitat relationships to provide management guidance.

SCALED QUAIL
Callipepla squamata

GLOBAL POPULATION: **1.2 MILLION**
50% BREED IN U.S.

TREND: DECREASING
WatchList: DECLINING

15

Distribution: Inhabits grasslands with scattered shrubs in the Chihuahuan Desert and short-grass prairie regions, from southeastern Colorado and Arizona through New Mexico, western Kansas, Oklahoma, and Texas, south into Mexico

Threats: Habitat degradation due over-grazing by cattle.

Conservation: Providing brush piles for cover has enhanced habitat on some ranches. This species occurs within many protected or federally managed areas including Comanche National Grassland, Colorado; Cimarron National Grassland, Kansas; Mascalero Sands, New Mexico; Buenos Aires NWR, Arizona; and Guadalupe Mountains NP and the Davis Mountains, Texas.

Actions: Research effects of livestock grazing, water improvements, climactic factors, and timing of hunting seasons to provide management guidance.

MONTEZUMA QUAIL
Cyrtonyx montezumae

GLOBAL POPULATION: **1.5 MILLION**
10% BREED IN U.S.

TREND: **DECREASING**
WATCHLIST: **DECLINING**

14

Distribution: Resident in oak-savanna, open oak and pine forests, and grassland. Around 90% of the population occurs in Mexico, reaching its northern range limit in southeastern Arizona, southern New Mexico, and south-central Texas.

Threats: Habitat loss and degradation by cattle grazing is the main threat, as this removes important grasses and can alter fire regimes. Small disjunct populations in U.S. may also be threatened by inbreeding, urban development, and hunting. Habitat in Mexico is also threatened by logging.

Conservation: Within the U.S., this species occurs in several protected areas including Buenos Aires NWR and Coronado NF, Arizona; Animas Foundation's Gray Ranch and Lincoln NF, New Mexico; and Guadalupe Mountains NP and other protected areas in Texas.

Actions: Restore and increase acreage of suitable habitat using prescribed burns and reducing cattle grazing; increase research effort to monitor population trends of this difficult-to-observe species.

GREATER SAGE-GROUSE
Centrocercus urophasianus

GLOBAL POPULATION: **142,000**
>99% BREED IN U.S.

TREND: **DECREASING**
WATCHLIST: **DECLINING**
ESA: **CANDIDATE**

16

Distribution: Inhabits sagebrush interspersed with native grasses. Found from Washington to Colorado, and in Alberta and Saskatchewan, Canada.

Threats: Habitat loss due to grazing, conversion to cropland, altered fire regimes, off-road vehicles, invasive cheatgrass, and conifer encroachment. Sage-grouse also abandon habitat near wind farms; and oil and gas wells, tens of thousands of which have been drilled within the species' range. Collisions with barbed wire fences and West Nile virus have also caused mortality, and the latter may endanger isolated populations.

Conservation: FWS decided in February 2010 that listing of the species under the ESA was warranted but precluded by higher priorities. Habitat restoration and management projects have been conducted at Parker Mountain, Utah; the DOD Yakima Training Center, Washington; and TNC's Crooked Creek Ranch, and Curlew National Grassland, Idaho.

Actions: List under the ESA; control cheatgrass; limit and minimize impacts of wind, oil, and gas development; mark barbed wire fences to reduce fatal collisions; include effective sage-grouse habitat objectives in BLM land leases where important populations occur; protect private lands using easements.

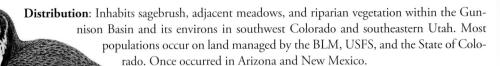

GUNNISON SAGE-GROUSE
Centrocercus minimus

GLOBAL POPULATION: C. **2,000**
U.S. ENDEMIC

TREND: **DECREASING**
WATCHLIST: **HIGHEST CONCERN**

20

Distribution: Inhabits sagebrush, adjacent meadows, and riparian vegetation within the Gunnison Basin and its environs in southwest Colorado and southeastern Utah. Most populations occur on land managed by the BLM, USFS, and the State of Colorado. Once occurred in Arizona and New Mexico.

Threats: Habitat loss, degradation, and fragmentation from housing development, roads, grazing, agriculture, and off-road vehicle use. Other threats include fire suppression, invasive cheatgrass, severe winters and drought, and inbreeding within isolated populations.

Conservation: Local, state, and federal agencies are cooperating to conserve this species; several local working groups have been established to draft conservation plans. Private landowners have granted conservation easements on thousands of acres within its range.

Actions: Continue habitat protection and restoration on public and private lands, especially on BLM lands where Gunnison Sage-Grouse objectives are not yet incorporated into lease agreements; list under the ESA.

SOOTY GROUSE
Dendrapagus fuliginosus

GLOBAL POPULATION: C.<1 MILLION
>50% BREED IN U.S.

TREND: **DECREASING**
WATCHLIST: **HIGHEST CONCERN**

16

Distribution: Inhabits coniferous forest, forest edge, and openings, in both early- and late-successional habitats, often moving to higher elevations in winter. Ranges from southeastern Alaska south through British Columbia, Canada, to Washington, Oregon, and California. Until 2006, this species was lumped with the Dusky Grouse (together known as the Blue Grouse).

Threats: Habitat loss and degradation appear to be the primary threats. Over-grazing by cattle also degrades meadows used by this species in the Sierra Nevada.

Conservation: Occurs within Misty Fiords NM and Tongass NF, Alaska; Olympic NP, Olympic and Wenatchee NFs, Washington; Siuslaw NF, Oregon; and several protected areas within the Sierra Nevada, California.

Actions: Improve forest management to mimic natural disturbance; maintain large functioning ecosystems so that clear-cuts grow into valuable grouse habitat, rather than unsuitable tree plantations.

GREATER PRAIRIE-CHICKEN
Tympanuchus cupido

GLOBAL POPULATION: **690,000** TREND: **DECREASING**
U.S. ENDEMIC (FORMERLY WATCHLIST: **HIGHEST CONCERN**
ALSO CANADA) ESA: **ENDANGERED ("ATTWATER'S" SUBSPECIES)**

17

Distribution: Once distributed widely in grasslands and oak savannas from Canada to Texas, most now in the northern Great Plains, with the largest populations in Colorado, Kansas, Nebraska, and South Dakota. A disjunct Atlantic Coast population: the "Heath Hen," went extinct on Martha's Vineyard, Massachusetts, in 1932.

Threats: Loss of grasslands due to agriculture, over-grazing by cattle, housing developments, mining and gas extraction, and fire suppression have reduced habitat by >99%; predation by fire ants is a threat to the "Attwater's" subspecies.

Conservation: Habitat management has raised Colorado's population from 600 to 10,000 since the 1970s. Translocation has been used to boost small populations. The Attwater's subspecies (c.50 wild birds) occurs within the Attwater Prairie Chicken NWR and TNC's Texas City Preserve, with c. 200 more in captivity. Since 1996, 83 birds have been released annually there to stabilize the wild population, but have suffered low survival rates.

Actions: Plan and coordinate habitat restoration across entire range; prevent further habitat loss where birds occur; strengthen the Conservation Reserve Program.

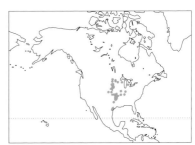

LESSER PRAIRIE-CHICKEN
Tympanuchus pallidicinctus

GLOBAL POPULATION: **32,000** TREND: **DECREASING**
U.S. ENDEMIC WATCHLIST: **HIGHEST CONCERN**
 ESA: **CANDIDATE**

20

Distribution: Resident within short- and mixed-grass prairie with sagebrush and shinnery oak in Kansas (50% of birds), New Mexico, Colorado, Texas, and Oklahoma.

Threats: Habitat loss and degradation have reduced its range by 92% since the 1800s. Threats include over-grazing, agriculture, droughts, and heavy winter snows. Birds avoid areas close to paved roads, and tall structures such as power lines and wind turbines. Collisions with barbed wire fences cause high mortality in some areas.

Conservation: Limited hunting is allowed only in Kansas and Texas but the impacts of even this need to be monitored. The Kansas Ornithological Society petitioned the state to list this species as threatened, but listing has yet to be approved. The High Plains Partnership has enrolled >80,000 acres of private land in Oklahoma and Texas, and TNC acquired several properties in three states to benefit this species. Attempts at establishing new populations through translocation have been unsuccessful.

Actions: List under the ESA; continue to restore, manage, and protect large areas of habitat on public and private lands; mark barbed wire fences to reduce collisions.

YELLOW-BILLED LOON
Gavia adamsii

GLOBAL POPULATION: **22,000-25,000**
>17% BREED IN U.S.

TREND: **UNKNOWN**
WATCHLIST: **RARE**
ESA: **CANDIDATE**

15

Distribution: Nests at edges of arctic lakes of North America and Russia. Alaskan and Canadian high arctic breeders winter predominantly in East Asian waters, with some along the Alaska Peninsula and Aleutians. Scattered individuals occur during the winter along the Pacific Coast from Kodiak Island to California. Alaskan breeders number c.4,000, mainly between the Colville and Meade Rivers.

Threats: Risk of oil pollution and contaminants and potentially gillnet bycatch; expansion of oil and gas drilling in the National Petroleum Reserve, and sea level-rise due to climate change.

Conservation: Small numbers breed within the Bering Land Bridge NM and Selawik NWR. The FWS recently determined that this species warranted listing under the ESA but was precluded from listing by higher priorities.

Actions: List under the ESA; identify important breeding, staging, and wintering areas; assess impacts of subsistence harvest, bycatch in subsistence gillnet fisheries, and contaminants; safeguard key breeding sites from impacts of oil development within the National Petroleum Reserve.

CLARK'S GREBE
Aechmophorus clarkii

GLOBAL POPULATION: **11,000-21,000**
C.75% BREED IN U.S.

TREND: **UNKNOWN**
WATCHLIST: **RARE**

15

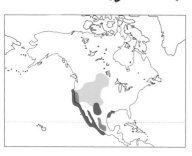

Distribution: Breeds mainly on large open freshwater lakes bordered by bulrush and other marshes throughout much of the western U.S. (barely reaching southern Canada), and throughout much of central Mexico. Migrants stage on large lakes, and winter along sea coasts, bays, estuaries, and sometimes on freshwater lakes and rivers. Range overlaps widely with the similar Western Grebe, from which this species was split in 1985.

Threats: Tens of thousands were killed for feathers by market hunters between 1890 and 1906. Current threats include the drainage of wetlands for agriculture, disturbance at breeding sites by recreational boaters (leading to nest loss or abandonment), accidental mortality in fishing gear (gillnets, lures, discarded fishing line), oil spills, and pesticide poisoning.

Conservation: The Upper Souris NWR in North Dakota is an important refuge for this species.

Actions: Manage water levels to maintain habitat; restrict boating activity near colonies, or protect colonies from boating wakes with buoy fences; implement range-wide monitoring to better track population status, and site-specific threats.

LAYSAN ALBATROSS
Phoebastria immutabilis

GLOBAL POPULATION: **1.1 MILLION**
>90% BREED IN **U.S.**

TREND: **DECREASING**
WATCHLIST: **HIGHEST CONCERN**

16

Distribution: Most breed at colonies in the northwestern Hawaiian Islands, with the largest on Midway. Smaller breeding colonies occur on the Bonin Islands off Japan. Recently colonized Guadalupe Island, Mexico. Forages at sea throughout the northern Pacific.

Threats: Historically exploited for food and feathers and has yet to fully recover. Up to 10,000 chicks die annually from lead poisoning after eating paint flaking off former military buildings on Midway. Birds are accidentally drowned on longlines and in fishing nets. Projected sea level rise will flood many current nesting grounds. Other threats include ingestion of floating plastics, oil pollution, and invasive plants that choke nesting areas.

Conservation: Protected at breeding sites in Japan and Hawaii. Bird-scaring devices have reduced bycatch in U.S. fisheries by over 80%. Some paint remediation on Midway due to ABC intervention, but significantly more is needed.

Actions: Encourage colonies at sites not likely to be impacted by sea level rise; reduce seabird bycatch among international fishing fleets; clean up lead paint on Midway; remove invasive plants from breeding sites; remove floating plastics.

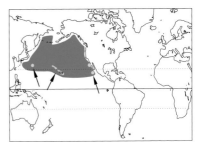

BLACK-FOOTED ALBATROSS
Phoebastria nigripes

GLOBAL POPULATION: **148,000**
>95% BREED IN **U.S.**

TREND: **DECREASING**
WATCHLIST: **HIGHEST CONCERN**

17

Distribution: Most of population (>95%) breeds in the Northwestern Hawaiian Islands, with the largest colony on Midway, a large colony on Laysan, and smaller colonies on islands off Japan. Forages at sea throughout the northern Pacific.

Threats: Rapid declines (-19% from 1995-2000) likely due to accidental mortality in longline fisheries, that at one time were killing 19,000 annually. Other threats include changes in fish populations due to overfishing, climate change, ingestion of floating plastic debris, oil pollution, and invasive golden crownbeard that chokes nesting areas. Projected sea level rise will flood many current nesting colonies.

Conservation: Protected at breeding sites in Japan and Hawaii. Bird-scaring devices have reduced seabird bycatch in Alaskan and Hawaiian fisheries by over 80%.

Actions: List under the ESA; encourage colonies at sites not likely to be impacted by sea level rise; reduce seabird bycatch among international fishing fleets; clean up lead paint on Midway; remove invasive plants from Hawaiian breeding sites; remove floating plastics.

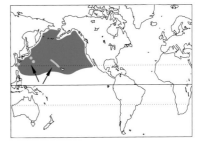

SHORT-TAILED ALBATROSS
Phoebastria albatrus

GLOBAL POPULATION: **2,600**
% IN U.S. UNKNOWN (BUT REGULAR OFFSHORE IN ALASKA)

TREND: **INCREASING**
WATCHLIST: **HIGHEST CONCERN**

20

Distribution: Breeding colonies located on only a few Japanese volcanic islands, with occasional individuals occurring on Midway (recent nesting attempt), Hawaii. Forages at sea throughout the northern Pacific.

Threats: Birds are accidentally hooked and drowned on longlines. Soil erosion threatens breeding colonies, as does the risk of volcanic eruptions at main breeding site on Torishima. Other threats include ingestion of floating plastics, and oil pollution. Historic exploitation for food and feathers that killed five million birds on Torishima from 1885-1903 is no longer a threat.

Conservation: Protected at breeding sites in Japan. Population increasing since reaching historic low of c.20 birds in 1953. Project to establish new colony at Mukojima will translocate 100 chicks by 2013 (25 chicks fledged there from 2008-09). Bird-scaring devices and other techniques reduced seabird bycatch in U.S. fisheries by over 80%.

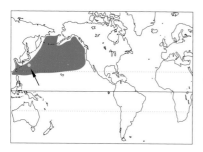

Actions: Increase population size by reducing seabird bycatch among international fishing fleets throughout the northern Pacific; establish new breeding populations on at least two predator-free islands and unoccupied portions of Torishima.

BERMUDA PETREL
Pterodroma cahow

GLOBAL POPULATION: **250**
% IN U.S. UNKNOWN (OCCASIONAL OFFSHORE)

TREND: **INCREASING**
WATCHLIST: **HIGHEST CONCERN**

20

Distribution: Breeds on small islands off Bermuda. Forages in warm waters of the Gulf Stream in the western Atlantic.

Threats: Exploitation of adults and eggs for food by early settlers, and introduced rats, pigs, and other predators reduced the population from over one-million to near extinction. Current threats include inadequate nesting habitat, hurricanes, rat predation, and light pollution at night. Longline fisheries, offshore oil extraction, and sea level rise are also potential threats.

Conservation: Nesting islands protected in Bermuda. Nesting success increased from 5% to 50% by use of concrete nest burrows with baffles to prevent use by tropicbirds.

The number of nesting pairs increased from 18 in 1951 to 71 in 2005. Chicks were translocated to Nonsuch Island beginning in 2004 to establish a new breeding colony. The first wild chicks fledged there in 2009.

Actions: Increase population on Nonsuch Island through continued artificial nest site construction, translocations, playing voice recordings to attract breeding adults, and rat control; continue management at other colonies.

BLACK-CAPPED PETREL
Pterodroma hasitata

GLOBAL POPULATION: **5,000**
% IN U.S. UNKNOWN (BUT REGULAR OFFSHORE)

TREND: **DECREASING**
WATCHLIST: **HIGHEST CONCERN**

20

Distribution: Once nested on islands across much of the Caribbean, but now only a few colonies are known on Hispaniola, and possibly Dominica and Cuba. Forages over warm pelagic waters from northern South America through the Caribbean to the southeastern U.S.

Threats: Deforestation and introduced predators at breeding sites are likely the most severe threats. Communication towers pose a collision risk near Sierra de Bahoruco. Also threatened by accidental drowning in fisheries, by toxic heavy metals and other pollutants, and potentially by offshore oil drilling. Historically hunted at colonies.

Conservation: Known colonies occur within NPs in Haiti, and at the Sierra de Bahoruco NP in the Dominican Republic (though still threatened there). ABC has assisted Sierra de Bahoruco in securing and expanding its boundaries, improving infrastructure, and training guards.

Actions: List under the ESA; search for remaining colonies outside Hispaniola; secure park boundaries and control invasive predators where nesting colonies occur; end all exploitation by people.

HAWAIIAN PETREL
Pterodroma sandwichensis

GLOBAL POPULATION: **c.19,000**
U.S. BREEDING ENDEMIC

TREND: **DECREASING**
WATCHLIST: **HIGHEST CONCERN**
ESA: **ENDANGERED**

19

Distribution: Nests in burrows on vegetated cliffs, volcanic slopes, and lava flows in the Hawaiian Islands. Forages widely across the Pacific, often following schools of large predatory fish. Formerly lumped as single species with Galapagos Petrel. Population status and trends are poorly known as nesting sites are often inaccessible. Populations on Maui and Kauai appear stable, but it is likely declining elsewhere. Probably extirpated from Oahu. Nesting site on Lanai recently surveyed and may contain several thousand individuals.

Threats: Exploited for food by early Polynesians, and then decimated by invasive predators on nesting islands. Collides with lights and power lines in urban areas at night. Impacted by decline of tuna and other large fish.

Conservation: Most nesting sites are within protected areas, including Volcanoes NP on Hawaii, and Haleakala NP on Maui. The National Park Service modified fences at Haleakala to reduce collisions.

Actions: Continue invasive species eradication programs on Maui and expand to other breeding locations; establish colonies in predator free areas or islets; shield street lights and realign power lines to reduce collisions.

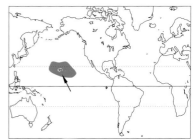

CORY'S SHEARWATER
Calonectris diomedea

GLOBAL POPULATION: **600,000-1.2 MILLION**
% IN U.S. UNKNOWN (BUT REGULAR OFFSHORE)

TREND: **UNKNOWN**
WATCHLIST: **RARE**

14

Distribution: Breeds on barren rocky islands in the Mediterranean and western Atlantic (e.g., Berlengas Island, Azores, Canary Islands). Forages widely across both North and South Atlantic, including in waters off the U.S. East Coast.

Threats: Historically, market hunters exploited thousands of chicks annually for food and fishing bait. Nests are also predated by introduced mammals. At sea, this species faces similar threats to other seabirds, including potential mortality in fishing gear, oil spills, and ingestion of floating plastics.

Conservation: Some breeding islands (e.g., key sites in Madeira) have been declared reserves where birds are protected from harvesting.

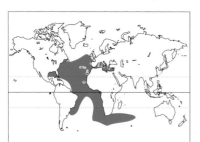

Actions: Protect breeding sites in eastern Atlantic and Mediterranean, including removal of invasive mammalian predators; in U.S., determine whether fisheries bycatch is a serious threat.

PINK-FOOTED SHEARWATER
Puffinus creatopus

GLOBAL POPULATION: **40,000-80,000**
% IN U.S. UNKNOWN (BUT REGULAR OFFSHORE)

TREND: **DECREASING**
WATCHLIST: **HIGHEST CONCERN**

18

Distribution: Nests in burrows in forest on two of the Juan Fernández Islands, and on Isla Mocha, off the coast of Chile. Otherwise ranges widely across the eastern Pacific north to the Gulf of Alaska, with major concentrations off Baja California, Mexico; California, and Oregon.

Threats: Cat predation; hunters remove chicks from nest burrows at unsustainable levels on Isla Mocha; rats predate nests, and rabbits, goats and cattle degrade habitat. Other potential threats include accidental mortality in fishing gear, oil contamination, and ingestion of floating plastics.

Conservation: All breeding sites are protected within NPs and reserves in Chile. Rabbits were eradicated from Santa Clara in 2003, increasing active nesting burrows from 43% to 60%. Efforts to implement cat control and sterilization are underway in the Juan Fernández Islands. ABC is working with local groups to reduce chick harvesting.

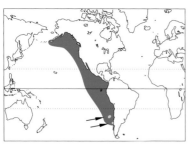

Actions: Remove cats; reduce chick harvest to sustainable levels at colonies in Chile; determine level of threat from bycatch in fisheries, and mortality from oil spills and floating plastics.

FLESH-FOOTED SHEARWATER
Puffinus carneipes

GLOBAL POPULATION: **650,000-1.5 MILLION**   TREND: **UNKNOWN**
% IN U.S. UNKNOWN (BUT REGULAR OFFSHORE)   WATCHLIST: **RARE**

14

Distribution: Nests in burrows on St. Paul Island in the southern Indian Ocean, in coastal southern Australia, on Lord Howe Island off eastern Australia, and around New Zealand's North Island. Ranges widely outside the breeding season, including into the western Pacific north to the Gulf of Alaska, with small numbers reaching south to California waters.

Threats: Historically exploited by people, now threatened by deforestation, tourism, urbanization, and introduced predators (rats, cats, lizards, foxes, and historically, pigs) at colonies. Accidental mortality in fishing gear, oil contamination, and ingestion of floating plastics are all threats. As many as 4,486 are killed annually in tuna and billfish fisheries off eastern Australia.

Conservation: Legally protected on Lord Howe Island; efforts to remove invasive mammals are underway.

Actions: All breeding colonies require formal protection and restricted access; eradicate invasive predators at breeding sites; reduce mortality due to fisheries bycatch, oil spills, and floating plastics.

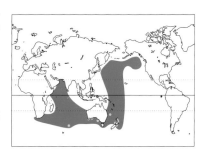

GREATER SHEARWATER
Puffinus gravis

GLOBAL POPULATION: **15 MILLION**
% IN U.S. UNKNOWN (BUT REGULAR OFFSHORE)

TREND: **UNKNOWN**
WATCHLIST: **RARE**

14

Distribution: Nests in burrows and crevices in grasslands and forests, predominantly on three of the Tristan da Cunha Islands (Nightingale, Inaccessible, Gough) in the southern Atlantic, with small numbers breeding in the Falkland Islands. Forages widely across the Atlantic, including pelagic waters off the U.S. East Coast.

Threats: Exploitation of thousands of adults and chicks annually on Nightingale Island for food may not be sustainable. Predation by cats could become a problem on East Falkland. Accidental mortality in fishing gear (especially pair trawlers in the northwest Atlantic), oil contamination, and ingestion of floating plastics are all potential threats.

Conservation: Marine protected areas near Machias Seal Island in New Brunswick, Canada, and in coastal Maine support important concentrations of this species.

Actions: End hunting of adults and chicks; control introduced predators near breeding sites; reduce mortality from fisheries bycatch, oil spills, and floating plastics.

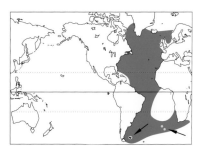

BULLER'S SHEARWATER
Puffinus bulleri

GLOBAL POPULATION: **2.5 MILLION**
% IN U.S. UNKNOWN (BUT REGULAR OFFSHORE)

TREND: **UNKNOWN**
WATCHLIST: **RARE**

15

Distribution: Nests in burrows on the islands of Aorangi and Tawhiti Rahi, and nearby islets, close to New Zealand's North Island. Forages widely across the Pacific, including in pelagic waters off western North America.

Threats: Like many other seabirds, accidental mortality in fishing gear (especially nets, but also potentially on longlines), oil contamination, and ingestion of floating plastics are all potential threats. The small number of breeding locations makes this species vulnerable to any accidental introduction of invasive predators in the future.

Conservation: Breeding islands are protected within the Poor Knights Islands Marine Reserve, where fishing was banned in 1996. The breeding population on Aorangi increased greatly from approximately 200 to 200,000 pairs from 1938 to 1981, following the eradication of introduced pigs, but current trends are unknown.

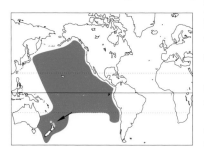

Actions: Monitor populations at breeding islands and at sea; reduce impact of fisheries bycatch, oil spills, and floating plastics.

SOOTY SHEARWATER
Puffinus griseus

GLOBAL POPULATION: **20 MILLION**
% IN U.S. UNKNOWN (BUT REGULAR OFFSHORE)

TREND: **DECREASING**
WATCHLIST: **DECLINING**

15

Distribution: Nests in burrows on islands in southern Chile, the Falklands, Australia, and New Zealand. Forages widely across cold pelagic ocean waters globally.

Threats: Predation by introduced mammals (rats, cats, pigs, stoats) and Weka (a rail native to mainland New Zealand) at colonies, bycatch in longlines and gillnets, and ingestion of floating plastics are all threats. Nesting burrows have declined 37% over 30 years at the largest colony in New Zealand, and numbers off California have declined 90% in 20 years. Driftnets formerly killed 350,000 birds annually, and 250,000 chicks were once commercially exploited each year for food, oil, and soap.

Conservation: The Rakiura Maori Community, University of Otago, Oikonos, and the New Zealand Department of Conservation have worked to eradicate rats from several breeding islands and are monitoring the species' population.

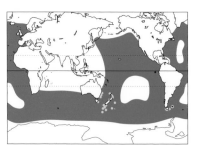

Actions: Reduce mortality due to fisheries bycatch, oil spills, and floating plastics; continue efforts to eradicate introduced mammals on breeding islands and monitor population response.

MANX SHEARWATER
Puffinus puffinus

GLOBAL POPULATION: **1.1 MILLION**
<1% BREED IN U.S.

TREND: **UNKNOWN**
WATCHLIST: **RARE**

15

Distribution: Nests in burrows and crevices, mainly on offshore, predator-free islands of the North Atlantic from Iceland south through Great Britain and Europe. A few hundred birds also nest in North America, mainly on Middle Lawn Island off Newfoundland (discovered in the 1970s), plus Penikese Island off Massachusetts, and Maine's Matinicus Rock. Winters along the continental shelf off eastern South America.

Threats: Introduced mammals such as rabbits, rats, and cats are the most severe threats (several British colonies extirpated by rats). Harvested extensively prior to the early 1900s in Europe for oil, food, fertilizer, and lobster bait. Mortality from fishing gear, ingestion of plastic garbage, collisions with lighted structures at night, and pesticide and heavy metal contamination have all been recorded.

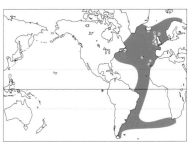

Conservation: Artificial burrows constructed in Newfoundland to increase nest site availability. Reducing grazing helped populations at some sites in Great Britain, but success of transplant efforts yet to be determined.

Actions: Reduce collisions with lighted structures by replacing steady-burning lights with flashing lights or by shielded lights pointed upward.

NEWELL'S SHEARWATER
Puffinus newelli

GLOBAL POPULATION: **38,000-89,000**
U.S. BREEDING ENDEMIC

TREND: **DECREASING**
WATCHLIST: **HIGHEST CONCERN**
ESA: **THREATENED**

18

Distribution: Once widespread throughout the Hawaiian Islands, but breeding now restricted to steep vegetated slopes on Hawaii, Molokai, and Kauai. Highly pelagic, ranging across warm, waters of the Pacific outside the breeding season.

Threats: Loss of c.75% of forests to urban development and agriculture, predation by introduced mammals, and lava flows have greatly reduced breeding habitat. Birds collide with overhead wires and cars at night. Accidental introduction of mongooses from other Hawaiian islands or brown tree snakes from Guam remain serious risks.

Conservation: In the late 1970s, eggs from highland colonies were placed in the nests of Wedge-tailed Shearwaters at Kilauea Point NWR, resulting in a small colony that persists today. "Save Our Shearwaters" has collected, banded, and released more than 30,000 Newell's Shearwaters and other seabirds found exhausted or that had collided with cars, power lines, and other structures.

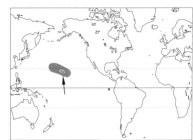

Actions: Reduce artificial lighting at night, bury or realign power lines, and identify flight paths to and from colonies to reduce collisions; expand control of invasive predators near breeding colonies.

BLACK-VENTED SHEARWATER
Puffinus opisthomelas

GLOBAL POPULATION: **153,000**
% IN U.S. UNKNOWN (BUT REGULAR OFFSHORE)

TREND: INCREASING
WATCHLIST: HIGHEST CONCERN

17

Distribution: 95% of the population nests in burrows or rocky crevices on Natividad Island, Mexico; the remainder nest along the Pacific coast of Baja California, Mexico. Forages over coastal waters from Monterey Bay south to central Mexico.

Threats: Introduced feral cats likely eradicated this species from several islands, and introduced herbivores degraded vegetation. Significant breeding habitat on Natividad has been destroyed by urban development. Collisions with buildings and power lines, and mortality in gillnet fisheries have been recorded.

Conservation: Natividad Island is legally protected. Removing 25 cats from Natividad reduced shearwater mortality by over 90%. Goat eradication and habitat restoration now underway on Guadalupe Island. Monterey Bay is protected as a National Marine Sanctuary.

Actions: Prevent future introductions of mammalian predators, halt habitat loss, reduce collisions with lighted structures, and restrict road access through colonies on Natividad; expand legal protection of breeding sites to islands along the coast of Baja California, Mexico; continue restoration of Guadalupe to enable reintroduction there.

AUDUBON'S SHEARWATER
Puffinus lherminieri

GLOBAL POPULATION: **500,000**
% IN U.S. UNKNOWN (BUT REGULAR OFFSHORE)

TREND: DECREASING
WATCHLIST: DECLINING

16

Distribution: Nests in burrows or rock crevices on islands, and forages widely over tropical oceans of the world. In the U.S., can be found in warm waters from the Gulf of Mexico, along the southeast coast, and following the Gulf Stream north to Cape Cod. They breed on islands in the Caribbean, and formerly on Bermuda.

Threats: Populations in the West Indies declining due to loss of breeding habitat; predation by introduced cats, rats, and mongooses at nesting sites, collisions with power lines, and with lighted buildings at night.

Conservation: Some important breeding sites in the Caribbean are protected, but generally little conservation attention is directed specifically at this species.

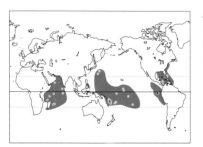

Actions: Legally protect and restore breeding sites; reduce exposure to floating plastics, oil spills, and other contaminants in the environment.

ASHY STORM-PETREL
Oceanodroma homochroa

GLOBAL POPULATION: **5,200-10,000**
>**95% BREED IN U.S.**

TREND: **DECREASING**
WATCHLIST: **HIGHEST CONCERN**

19

Distribution: Nests in deep rocky crevices on talus slopes, mostly on the Farallon and Channel Islands, California; and on the Coronados Islands off Baja California, Mexico. Forages in the California Current up to 100 miles from shore.

Threats: Predation by introduced mammals and over-abundant gulls at breeding sites are the main threats, though the population is likely also limited by suitable nesting habitat. Populations dramatically declined on South Farallon Island as gull populations increased. Introduced grasses may limit breeding on Farallons. Oil pollution also a risk.

Conservation: All U.S. breeding sites and many at-sea congregation areas protected within Farallon NWR; the Gulf of the Farallones, Cordell Bank, and Monterey Bay National Marine Sanctuaries; Channel Islands NP and National Marine Sanctuary, and California state marine protected areas. Rabbits were eradicated from the Farallons in 1974, and rats from Anacapa within the Channel Islands in 2002.

Actions: Limit gull predation in Farallons; eradicate invasive species from breeding islands.

BAND-RUMPED STORM-PETREL
Oceanodroma castro

GLOBAL POPULATION: **150,000-200,000**
<**1% BREED IN U.S.**

TREND: **DECREASING**
WATCHLIST: **HIGHEST CONCERN**
ESA: **CANDIDATE**

17

Distribution: Nests in rocky crevices and burrows on remote islands in the eastern Atlantic and Pacific (including the Hawaiian Islands). Forages widely across warm pelagic waters, including along the East and Gulf Coasts of the U.S.

Threats: Introduced rats at breeding sites; introduced grazing animals damage nests and degrade vegetation on nesting islands. Historically exploited by Polynesians for food; exploitation may continue on Madeira. High levels of mercury found in birds in the Azores. Collisions with buildings, power lines, and towers occur on Kauai and Hawaii. Oil spills and other contaminants are also risks. One large colony in Japan declined >95% between the 1960s and mid-1990s.

Conservation: The Hawaiian population is protected by the state and is a candidate for listing under the ESA. The Pacific population is sometimes considered a separate species from the Atlantic population.

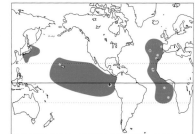

Actions: List under the ESA; identify breeding areas in Hawaii to direct predator and grazing animal control efforts; reduce collisions with lighted structures at night.

BLACK STORM-PETREL
Oceanodroma melania

GLOBAL POPULATION: **500,000**
<1% BREED IN U.S.

TREND: **DECREASING**
WATCHLIST: **HIGHEST CONCERN**

16

Distribution: Nests in rocky crevices or burrows on talus slopes of islands in southern California, along the Pacific coast of Mexico's Baja California, and on Channel Islands in the Gulf of California (largest colony in the San Benito Islands). Forages over warm waters, with most wintering off the Pacific coast of Central and northern South America, but some ranging north along the coast of California.

Threats: Historic population declines followed the introduction of mammalian predators to breeding islands; rats and cats remain greatest problems. Other threats include oil spills, and other contaminants.

Conservation: Breeding islands protected by Channel Islands NP and National Marine Sanctuary, and Gulf of California Biosphere Reserve, but the San Benito Islands, Mexico, lack legal protection. Introduced mammals such as rabbits and goats have been eradicated from the San Benito Islands, and 23 potential breeding islands.

Actions: San Benito Islands and others off Baja California require legal protection; continue efforts to remove introduced mammals from potential breeding islands and re-establish colonies on predator-free islands; limit risk from contaminants at sea.

TRISTRAM'S STORM-PETREL
Oceanodroma tristrami

GLOBAL POPULATION: **>22,000**
C.70% BREED IN U.S.

TREND: **UNKNOWN**
WATCHLIST: **HIGHEST CONCERN**

16

Distribution: Nests in burrows on flat, sandy atolls, and in rocky crevices on cliffs of volcanic islands in the Northwestern Hawaiian Islands, as well as in the Izu and Bonin Islands off southern Japan. Forages widely across the Pacific between Hawaii and Japan.

Threats: Rats decimated populations nesting on Midway Atoll, and rats and cats decimated Japanese colonies. Ingested plastic debris recorded in birds from Laysan and Nihoa. Collisions with lighted structures may be an issue on Midway, but most nesting islands are not developed. Colonies on low-lying islands could be threatened by sea level rise.

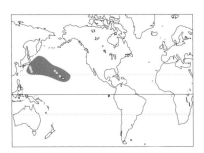

Conservation: Rats were eradicated from Midway by 1996 and a few storm-petrels have returned to breed. Nest boxes on Tern Island could potentially increase nest site availability there. Nesting areas in Hawaii are protected.

Actions: Eradicate rats from Japanese breeding islands; reduce plastic debris, oil, and other contaminants.

LEAST STORM-PETREL
Oceanodroma microsoma

GLOBAL POPULATION: **100,000-1** MILLION
% IN U.S. UNKNOWN (BUT REGULAR OFFSHORE)

TREND: **DECREASING**
WATCHLIST: **HIGHEST CONCERN**

16

Distribution: Nests in crevices or burrows on rocky islands off the Pacific coast of Baja California, Mexico, and within the Gulf of California, including the San Benito Islands. Forages over warm waters, with most wintering off the Pacific coast of Central and northern South America, but some ranging north along the California coast.

Threats: Historic population declines followed the introduction of rats and cats to breeding islands; habitat degradation by grazing animals, oil spills, pesticides, and other contaminants are all potential threats.

Conservation: Many breeding islands are protected within the Gulf of California Biosphere Reserve and other marine protected areas, but the largest colony on the San Benito Islands lacks legal protection. Introduced rabbits, goats, and donkeys have been removed from San Benito West, however.

Actions: San Benito Islands and others off Baja California require legal protection; remove introduced mammals from additional potential breeding islands, and re-establish colonies on predator-free islands; limit risk from contaminants at sea.

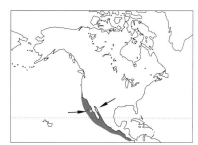

MASKED BOOBY
Sula dactylatra

GLOBAL POPULATION: **200,000**
c.**3.5%** BREED IN U.S.

TREND: **DECREASING**
WATCHLIST: **DECLINING**

15

Distribution: Nests on flat, sandy ground or rocky cliffs on islands throughout the world's tropical oceans, and forages far from land following schools of tuna and dolphin. In the U.S., roughly ten pairs nest within the Dry Tortugas NP, Florida, with around 7,000 in Hawaii. The *dactylatra* subspecies occurs in the Caribbean and South Atlantic; *personota* breeds in Australasia and the central Pacific.

Threats: Feral pigs can decimate breeding colonies (e.g., on the Pacific island of Clipperton). Accidental mortality in fishing gear recorded, but the extent of the bycatch problem is unknown. In 2008, Hurricane Ike depleted sand from Hospital Key, Florida, making it unsuitable for breeding, but the colony relocated to Middle Key. Sea level rise could affect low-lying breeding islands.

Conservation: In U.S., Atlantic breeding population is protected within Dry Tortugas NP, and most Hawaiian colonies are protected within the Papahānaumokuākea Marine NM.

Actions: Ensure that sufficient nesting habitat persists in the Dry Tortugas.

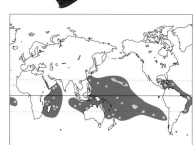

RED-FACED CORMORANT
Phalacrocorax urile

GLOBAL POPULATION: **200,000**
33% BREED IN U.S.

TREND: **UNKNOWN**
WATCHLIST: **RARE**

16

Distribution: Resident in northern Pacific and Bering Sea, breeding on steep cliffs from the Gulf of Alaska west through the Alaska Peninsula, Pribilof Islands, Aleutian Islands, and along the eastern coast of Russia's Kamchatka Peninsula to northern Japan. Forages close to the coast. In winter, some birds move away from breeding areas that ice-over.

Threats: Arctic foxes introduced to many Aleutian Islands through the 1940s reduced populations. Human disturbance is limited on Pribilof and Komandorski Islands, and most other breeding areas are far from people. Oil spills and accidental mortality in fisheries are potential threats in the eastern portion of its range.

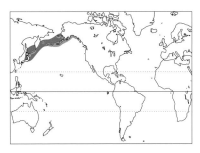

Conservation: Most U.S. breeding sites are protected within the Alaska Maritime NWR where fox eradication efforts continue.

Actions: Eradicate foxes from Aleutian Islands; limit risk of oil spills.

MAGNIFICENT FRIGATEBIRD
Fregata magnificens

GLOBAL POPULATION: **200,000**
<1% BREED IN U.S.

TREND: **UNKNOWN**
WATCHLIST: **HIGHEST CONCERN**

16

Distribution: Nests in low scrub or trees on small Pacific islands from Baja California, Mexico, to Ecuador, with colonies also scattered throughout the Caribbean, and from Venezuela to Brazil. Also breeds on Cape Verde Islands. One U.S. colony (200 pairs) in the Dry Tortugas, Florida. Forages over coastal and pelagic tropical seas.

Threats: Probably limited by suitable nesting habitat in the U.S. Human disturbance, development, and introduced mammalian predators have extirpated this species from c.50% of Caribbean breeding sites. Feral goats degrade vegetation in the British Virgin Islands. Collisions with towers and power lines, and entanglement in fishing gear have been recorded.

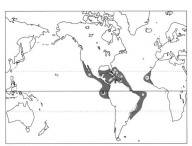

Conservation: Protected within the Dry Tortugas NP. Largest Caribbean colony on Barbuda lacks formal protection. Man 'O War Cay in Belize is protected in a reserve. Other populations outside the region also protected (e.g., in Galapagos National Park).

Actions: Legally protect remaining breeding colonies and limit disturbance there; remove invasive mammals where they occur near colonies.

REDDISH EGRET
Egretta rufescens

GLOBAL POPULATION: **10,000-20,000**
20-33% BREED IN U.S.

TREND: **INCREASING**
WATCHLIST: **HIGHEST CONCERN**

18

Distribution: Typically nests in mangroves and on vegetated islands, and forages in coastal lagoons and on tidal flats. Found from Florida to the Gulf Coast states, ranging south through Panama and east through the Caribbean. On Pacific Coast, ranges from Mexico to extreme northern South America. Roughly 2,000 pairs occur in the U.S., with the largest colonies breeding on islands in Laguna Madre, Texas.

Threats: Plume-hunting from the late 1880s to 1912 probably extirpated populations of this species from much of the U.S. (particularly Florida) and Jamaica. The Florida population recovered to 400 pairs by the 1990s. Habitat loss due to coastal development is now the most severe threat, though sea level rise due to climate change may eventually become more serious.

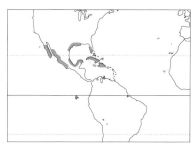

Conservation: Habitat protected in Everglades NP, Florida Keys, and Ding Darling NWR. Much of the Texas population protected within Laguna Atascosa NWR, and Audubon's Green Island Sanctuary.

Actions: Protect coastal wetland habitats; reduce greenhouse gas emissions to combat climate change.

CALIFORNIA CONDOR
Gymnogyps californianus

GLOBAL POPULATION: **C. 350**
c.93% IN U.S.

TREND: **INCREASING**
WATCHLIST: **HIGHEST CONCERN**
ESA: **ENDANGERED**

20

Distribution: Once ranged from southern British Columbia, Canada, through the western U.S. to northern Baja California, Mexico. Today, >150 occur in the wild in central and southern California, Arizona, Utah, and Mexico, with a similar number in captivity.

Threats: Historic declines caused by persecution, exploitation, and poisoning from carcasses baited to kill bears, wolves, and mountain lions. Accidental lead poisoning from scavenging on hunted carcasses and gut piles remains the biggest threat; also mortality from power line collisions, electrocutions, and shootings.

Conservation: Last remaining 22 wild birds captured in 1987; captive-breeding program began releasing birds back into the wild in 1992. Plastic markers added to power lines in Big Sur to reduce collisions. Released birds protected in Grand Canyon NP, Arizona and Utah; Tehachapi Mountains, Hopper Mountain NWR, Los Padres NF, and Big Sur, California; and in Baja California Norte, Mexico. Lead bullets banned within condor habitat in California in 2007.

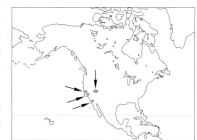

Actions: Eliminate lead ammunition where condors occur; place markers on power lines to reduce collisions; expand release efforts to reestablish populations within former range.

SWALLOW-TAILED KITE
Elanoides forficatus

GLOBAL POPULATION: **150,000**
<3% IN U.S.

TREND: **DECREASING**
WATCHLIST: **DECLINING**

16

Distribution: The *forficatus* subspecies once bred over much of the central and southern U.S., foraging over forests, swamps, marshes, and prairies. The U.S. breeding range contracted greatly from the late 19th to mid 20th Centuries. In 1990, <1,150 breeding pairs were restricted to fragmented habitat across the coastal plain from North Carolina south through Florida (>60% of population), and west along the Gulf of Mexico to eastern Texas. Populations of the *yetapa* subspecies breed from southern Mexico south into South America in humid lowland and montane forests. U.S. and Central American populations winter in South America.

Threats: Historic declines blamed on breeding habitat loss and degradation due to agriculture, logging, and shooting.

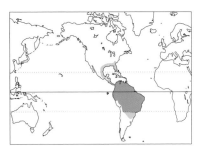

Conservation: Occurs within many protected areas, including Apalachicola NF, Okefenokee NWR, Big Cypress National Preserve, and Everglades NP, Florida.

Actions: Restore large tracts of pine and bottomland forests in the Southeast on public and private lands, and avoid cutting trees near nests; protect and monitor populations at large communal roosts used prior to migration (e.g., near Lake Okeechobee); research threats on wintering grounds.

SWAINSON'S HAWK
Buteo swainsoni

GLOBAL POPULATION: **500,000**
>50% BREED IN U.S.

TREND: **UNKNOWN**
WATCHLIST: **RARE**

15

Distribution: Breeds in open grassland, shrub-land, and agricultural land from Alaska through the Canadian prairies, then south through the western U.S. to northern Mexico. The California population has declined by 90%, and declines have been observed in Canada, but populations stable elsewhere. Migrates in flocks through Central America to winter in grasslands of Argentina.

Threats: Persecuted across Canada and northern U.S. prior to 1930s, especially in Canada and Washington. Threatened by habitat alteration due to agriculture, urban development, and reduction of nesting trees and prey (e.g., Richardson's ground squirrel). Tens of thousands killed by the pesticide monocrotophos in agricultural areas in Argentina during mid-1990s. Collisions with vehicles and fences, and power line electrocutions are additional threats.

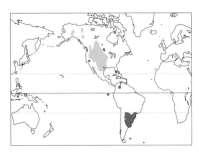

Conservation: ABC successfully advocated for a ban on monocrotophos in Argentina in 1999.

Actions: Restrict application of harmful pesticides across species' breeding range and migratory routes; include the protection and restoration of riparian woodlands in landscapes where nest sites are limited.

HAWAIIAN HAWK
Buteo solitarius

GLOBAL POPULATION: **3,085**
U.S. ENDEMIC

TREND: **STABLE**
WATCHLIST: **HIGHEST CONCERN**
ESA: **ENDANGERED**

17

Distribution: Restricted to the Island of Hawaii, where it uses a variety of habitats from the lowlands to almost 9,000 feet, including native ohia forests, exotic forests, and pastures. Sub-fossil evidence suggests that this species was widely distributed throughout the main Hawaiian Islands prior to Polynesian colonization; vagrants have been recorded on Maui, Oahu, and Kauai since 1778. Surveys on Hawaii in 2007 estimated the population to be 3,085 birds, roughly the same as in 1998.

Threats: Historically persecuted by ranchers. Harassed birds may abandon nests. Fortunately, this species seems resistant to introduced diseases that have decimated native songbirds, and nests in both native and exotic forests.

Conservation: Listed under the ESA as Endangered, but proposed for down-listing to Threatened. Protected areas include Hawaii Volcanoes NP, Mauna Kea Forest Reserve, Puu Waawaa State Wildlife Sanctuary, Hakalau Forest NWR, and Kau Forest Reserve.

Actions: Continue efforts to protect and restore native forests on Hawaii and control invasive species.

YELLOW RAIL
Coturnicops noveboracensis

GLOBAL POPULATION: **10,000-25,000**
10% BREED IN **U.S.** (99% WINTER)

TREND: **UNKNOWN**
WATCHLIST: **HIGHEST CONCERN**

18

Distribution: Breeds in wet meadows across much of Canada's boreal region (>90% of population) and in the northern U.S. from Oregon to Michigan and Maine. Winters in coastal wetlands across southeastern U.S. from North Carolina south to Florida and Texas.

Threats: Loss and degradation of wetland habitats due to agriculture; urban development; changes in fire, siltation, or hydrology regimes; invasive plants; grazing pressure; acid rain; and erosion and saltwater intrusion (along Gulf Coast). Sea level rise caused by climate change could be a significant future threat. Collisions with communication towers also recorded.

Conservation: Manitoba's Douglas Marsh protects one of the largest known breeding populations (500 pairs). In the U.S., protected areas include Klamath Marsh NWR, Oregon; Seney NWR, Michigan; and Anahuac NWR, Texas. Numerous wetland restoration projects across the southeastern U.S. could benefit this species.

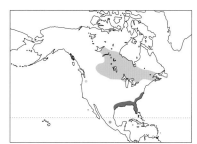

Actions: Population monitoring and habitat studies are needed to detect trends and manage habitat for this little-known species; continue wetland protection and restoration efforts across range, especially in Canada's boreal region.

BLACK RAIL
Laterallus jamaicensis

GLOBAL POPULATION: **35,000-110,000**
% IN U.S. UNKNOWN

TREND: **DECREASING**
WATCHLIST: **HIGHEST CONCERN**

18

Distribution: Occurs along dryer edges of salt and freshwater marshes over a large but disjunct range. The *jamaicensis* subspecies was formerly widespread along the Atlantic and Gulf Coasts, as well as in the Midwest, but has undergone a long-term decline. The *corturniculus* subspecies is resident in central California, along the lower Colorado River, and in Baja California, Mexico. Also in Latin America and the Caribbean.

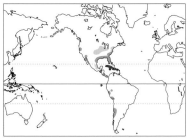

Threats: Wetland loss to agriculture and development, invasive plants and mammals, grazing, and pollution. Impact of saltmarsh management for mosquito control potentially problematic. Sea level rise caused by climate change could be a significant future threat. Collisions with towers and other structures recorded. Pesticides may also be a risk.

Conservation: Occurs in numerous U.S. protected areas including St. Johns NWR, Florida; Cedar Island NWR, North Carolina; Suisun Marsh, California; and on public lands in Chesapeake and San Francisco Bays. Wetland restoration projects across southeastern U.S. may benefit this species.

Actions: Population monitoring and habitat studies needed; continue wetland protection, management, and restoration efforts; combat sea level rise and climate change by reducing greenhouse gas emissions.

CLAPPER RAIL
Rallus longirostris

GLOBAL POPULATION: **UNKNOWN**
% IN U.S. UNKNOWN

TREND: **UNKNOWN**
WATCHLIST: **RARE**
ESA: **ENDANGERED (3 SUBSPECIES)**

14

Distribution: Mainly inhabits coastal and estuarine saltmarshes and mangroves, with 21 subspecies ranging from the northeastern U.S. to South America. Most are resident, though populations breeding in the U.S. north- and mid-Atlantic regions move to the southern Atlantic coast during winter.

Threats: Wetland loss to agriculture, salt ponds, urban development, and invasive plants. Other threats include heavy metal and pesticide contamination, introduced predators (rats, foxes, and cats), and collisions with vehicles, towers, and other structures. Sea level rise due to climate change is a future threat.

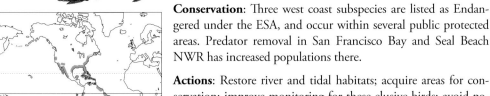

Conservation: Three west coast subspecies are listed as Endangered under the ESA, and occur within several public protected areas. Predator removal in San Francisco Bay and Seal Beach NWR has increased populations there.

Actions: Restore river and tidal habitats; acquire areas for conservation; improve monitoring for these elusive birds; avoid potential for inbreeding by using translocations for subspecies with small populations.

KING RAIL
Rallus elegans

GLOBAL POPULATION: >35,000
>75% BREED IN U.S.

TREND: **DECREASING**
WATCHLIST: **DECLINING**

16

Distribution: Inhabits mainly freshwater marshes, but also rice fields and tidal brackish marshes throughout eastern North American, from extreme southern Ontario, Canada, to eastern and central Mexico; also in Cuba. Birds breeding in northern portion of range migrate south during winter, with peak winter concentrations in southern Louisiana and the Everglades, Florida. Severe population declines recorded over much of U.S., except perhaps Louisiana, Mississippi, Georgia, and Florida.

Threats: Loss and degradation of habitat to agricultural, industrial, and urban development. Other threats include pollution and pesticide contamination; accidental mortality in muskrat traps; and collisions with vehicles, towers, barbed wire fences, and other structures. Legally but seldom hunted across much of U.S. range.

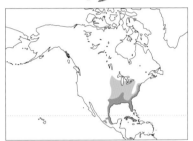

Conservation: Most high quality habitat now remains on public refuges, such as Sabine NWR in Louisiana. Strong wetland protection and pollution laws (e.g., Clean Water Act) are helpful.

Actions: Continue wetland habitat protection and restoration efforts on public and private lands; monitor to determine current range, population size, and trends; reduce mortality associated with communication tower collisions.

HAWAIIAN COOT
Fulica alai

GLOBAL POPULATION: **3,000**
U.S. ENDEMIC

TREND: **STABLE**
WATCHLIST: **HIGHEST CONCERN**
ESA: **ENDANGERED**

18

Distribution: Inhabits freshwater wetlands, brackish fishponds, flooded taro fields, reservoirs, and sewage-treatment ponds throughout the major Hawaiian Islands (80% on Kauai, Oahu, and Maui). Population stable but fluctuates; peaks following years with high rainfall.

Threats: Habitat loss due to agricultural and urban development is the most severe threat. Other threats include outbreaks of botulism; and predation by introduced mongooses, cats, and dogs. Occasionally suffers collisions with power lines and wind turbines. Pesticides and other contaminants may be additional stresses.

Conservation: Many breeding sites are protected in public reserves, including Hanalaei, Huleia, James Campbell, Pearl Harbor, Kakahai, and Kealia Pond NWRs; and Kawaiele Sanctuary, Hamakua Marsh, Kanaha Pond Sanctuary, and Aimakapa Pond within Kaloko-Honokohau National Historical Park. Kaelia Pond was temporarily dried out to mitigate a botulism outbreak in 2000.

Actions: Prevent further loss of wetlands and continue to protect, restore, or enlarge wetlands where possible; control invasive predators; minimize or mitigate risk of botulism outbreaks.

WHOOPING CRANE
Grus americana

GLOBAL POPULATION: **535**
U.S. WINTER ENDEMIC

TREND: INCREASING
WATCHLIST: HIGHEST CONCERN
ESA: ENDANGERED

20

Distribution: Once bred in marshes from northwest Canada to Illinois, and wintered along the Gulf and Atlantic Coasts, with a non-migratory population in Louisiana. Now in three wild populations: c. 250 birds migrate between Canada's Wood Buffalo NP and Aransas NWR, Texas; 100 in Wisconsin's Necedah NWR, migrating to Florida; and 30 resident in Florida's Kissimmee Prairie. Approximately 150 in captivity.

Threats: Water diversion for San Antonio is increasing salinity (apparently causing the recent deaths of 23 birds due to food shortages), and feral hogs degrade coastal wetlands; oil spills at Aransas; water draw-down on the Platte River; collisions with power lines, and occasional shootings.

Conservation: Listed as Endangered under the ESA. Conservation efforts increased the population from historic low of 16, to c. 380 wild birds today. Most key sites occur within protected areas. Captive-breeding began in 1967. Ultralight aircraft are used to jump-start migration of introduced birds.

Actions: Ensure sufficient water from the Guadalupe River reaches Aransas NWR; maintain water flows on the Platte; use captive-breeding and ultra light techniques to establish additional migratory flocks.

AMERICAN GOLDEN-PLOVER
Pluvialis dominica

GLOBAL POPULATION: **200,000**
C.50% BREED IN U.S.

TREND: DECREASING
WATCHLIST: DECLINING

15

Distribution: Nests in dry lowland tundra and sub-arctic alpine areas from western Alaska to Canada's Baffin Island and Hudson Bay. Follows an easterly migration path in fall to winter in grasslands of southern South America; many flying over the Atlantic. Uses grasslands and agricultural fields from eastern Mexico, to Texas and the central U.S. during spring migration.

Threats: Exploited by market hunters during the 1800s. Conversion of prairies to agriculture, and absence of bison, grasshoppers, and natural fires reduce habitat and food during migration. Loss of grasslands to tree plantations in Argentina and Brazil threatens wintering habitat. Oil and gas drilling in Alaska further threatens this species, as does shooting in Barbados, and exposure to pesticides.

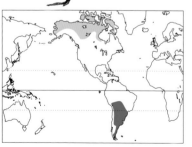

Conservation: Several important protected areas exist within its breeding, staging, and wintering range, including the Arctic NWR, and Cheyenne Bottoms WMA in the U.S., Blow River Delta in Canada, and Lagoa do Peixe NP in Brazil.

Actions: Restrict oil and gas extraction in breeding range; reduce conversion of South American grasslands to tree plantations; protect important staging areas within the U.S.

SNOWY PLOVER
Charadrius alexandrinus

GLOBAL POPULATION: **300,000-460,000**
4% BREED IN U.S.

TREND: **DECREASING**
WATCHLIST: **DECLINING**
ESA: **THREATENED (WESTERN POPULATION)**

14

Distribution: Resident on salt flats and beaches of Pacific Coast from Baja California, Mexico, north to Oregon; and along portions of the Gulf Coast. Disjunct populations occur inland in western U.S., and in Saskatchewan, Canada. Half the U.S. population breeds around the Great Salt Lake, Utah. Widely distributed worldwide.

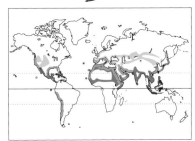

Threats: Beachfront development threatens habitat; beach raking degrades foraging areas, and may destroy nests. Vehicles on beaches disturb birds; invasive grasses and plants used in dune stabilization eliminate open habitat and encourage predators. Reduced water levels in the Great Salt Lake, and rising sea levels caused by climate change are significant threats.

Conservation: Pacific Coast population listed under the ESA. Many nesting beaches are roped-off, and predator exclosures increase breeding success. Invasive grasses removed from many Pacific beaches. Efforts are ongoing to protect Mono, Owens, and Great Salt Lakes, and the Salton Sea from water diversion projects.

Actions: Restrict coastal development; maintain water levels at inland breeding sites; remove invasive plants and control predators; restrict vehicle traffic on beaches.

WILSON'S PLOVER
Charadrius wilsonia

GLOBAL POPULATION: **>6,000**
% IN U.S. UNKNOWN

TREND: **DECREASING**
WATCHLIST: **DECLINING**

14

Distribution: Inhabits open beaches, sand dunes, salt flats, and tidal mudflats along the Atlantic and Gulf Coasts of the U.S. Also in Latin America and the Caribbean. Much of the population along the Atlantic Coast migrates south in winter as far as Brazil. Breeding range has retracted south from New Jersey to Virginia since 1940.

Threats: Coastal habitats threatened by beachfront development. Free-ranging cattle, vehicle and human traffic on beaches, and loose dogs and feral cats can disturb birds and predate nests. Sea level rise due to global warming is a potential future threat.

Conservation: Protected areas for this species include the Altamaha River Delta, Georgia; Cape Hatteras National Seashore, North Carolina; Canaveral National Seashore, Florida; and Laguna Atascosa NWR, Padre Island National Seashore, and Bolivar Flats Shorebird Sanctuary, Texas. Habitat protection efforts aimed at Piping Plovers may also benefit this species.

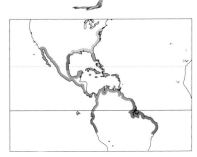

Actions: Maintain and increase beach habitat, with restricted access to vehicles, predators, cattle, and other causes of disturbance; restore dunes to maintain habitat in the face of future sea level rise.

PIPING PLOVER
Charadrius melodus

GLOBAL POPULATION: **8,000**
c.90% IN U.S.

TREND: **DECREASING**
WATCHLIST: **HIGHEST CONCERN**
ESA: **ENDANGERED (GT LAKES POP.); OTHERS THREATENED**

20

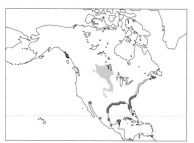

Distribution: Breeds on sandy beaches along the Atlantic Coast from Newfoundland south to North Carolina; along the shores of the Great Lakes; and on large rivers, reservoirs, and alkali lakes of the northern Great Plains. Winters along Atlantic, Caribbean, and Gulf Coasts, mainly in the U.S., but also in Mexico, Cuba, and the Bahamas.

Threats: Beachfront development and recreational use threaten habitat along coasts and the Great Lakes. Vehicles on beaches may destroy nests. Beach raking removes roosting habitat and foraging material. Beach grasses used in dune stabilization eliminate open habitat and encourage predators. River management in the Great Plains results in habitat loss and increased predation pressure.

Conservation: Listed as Endangered in the U.S. and Canada. Roped-off areas and predator exclosures around nests increase breeding success. Closing beaches to vehicle traffic has also helped.

Actions: Increase habitat at inland breeding sites; reduce nest disturbance and predation with predator exclosures, fencing, and restrictions on vehicle traffic on beaches during the breeding season.

MOUNTAIN PLOVER
Charadrius montanus

GLOBAL POPULATION: **>9,000**
>98% BREED IN U.S.

TREND: **DECREASING**
WATCHLIST: **HIGHEST CONCERN**

17

Distribution: Nests in heavily grazed short-grass prairies and dry shrub-lands from southern Canada to New Mexico and western Texas. Most thought to breed in Colorado (60%) and Wyoming (20%). Winters in central California south and east through Texas, then south into Mexico.

Threats: The most severe threats are loss and degradation of habitat to agriculture, urban development, and the absence of grazers (prairie dogs, bison, and grasshopper swarms) that historically kept grass short. Prairie dog population control still occurs on both public and private lands.

Conservation: Conservation groups and state and federal agencies have collaborated on a conservation plan for this species, and are engaging private landowners in on-the-ground conservation. The USDA avoids spraying pesticides for grasshoppers in known plover areas.

Actions: Increase and restore native grasslands throughout range (especially in Mexico) along with native herbivores; ensure that public lands in U.S. are managed to maintain a diversity of grassland habitats; increase efforts on private lands through conservation easements and Farm Bill incentives; avoid construction in plover habitat.

WANDERING TATTLER
Tringa incana

GLOBAL POPULATION: **10,000-25,000**
>50% BREED IN **U.S.**

TREND: **DECREASING**
WATCHLIST: **RARE**

15

Distribution: Nests in montane tundra, often foraging along rocky glacial streams, gravel bars, or around alpine lakes of Alaska and northwestern Canada, with small numbers nesting in nearby Russia. Winters along rocky shorelines, reefs, and jetties throughout the Pacific, from southern British Columbia south to Peru, and west to Australia, with concentrations in the Galapagos Islands and on islands in the south-central Pacific.

Threats: Threats are not well known, but potentially include habitat degradation caused by oil spills and other contaminants.

Conservation: Occurs within several coastal protected areas. Efforts to restore sea otters and kelp beds likely benefit this species by increasing onshore kelp debris used for foraging.

Actions: Continue to restore healthy rocky intertidal zones and kelp beds along Pacific Coast; reduce risk of oil spills.

ESKIMO CURLEW
Numenius borealis

GLOBAL POPULATION: **c.0**
% IN U.S. UNKNOWN

TREND: **PROBABLY EXTINCT**
WATCHLIST: **HIGHEST CONCERN**
ESA: **ENDANGERED**

20

Distribution: Likely once bred from northern Canada through Alaska, and possibly into Siberia. During the fall, it travelled east before making an over-sea flight to South America. It wintered in grasslands and along the coasts of Argentina, Brazil, Uruguay, and Chile; stopped during return migration on the Texas coast, then travelled north through the prairies of the central U.S. and Canada.

Threats: Decimated by market and sport hunters between 1850 and 1900. Further harmed by extensive loss of prairie habitat, and eradication of grasshoppers that served as a major food source. Last confirmed bird shot in Barbados in 1963, but there have since been 29 unconfirmed sightings from Alaska to Argentina. Argentine grasslands continue to be converted to agriculture and tree farms, and oil and mineral exploration could threaten any remaining breeding areas.

Conservation: Without a confirmed record in nearly 50 years, the top priority would be to find a bird, and satellite track it if it could be safely caught.

Actions: Surveys of areas used by the species in the past should be undertaken.

BRISTLE-THIGHED CURLEW
Numenius tahitiensis

GLOBAL POPULATION: **7,000**
U.S. BREEDING ENDEMIC

TREND: **UNKNOWN**
WATCHLIST: **RARE**

17

Distribution: Nests in montane tundra in the lower Yukon River and central Seward Peninsula of Alaska. Winters in coastal habitats and open areas of tropical Pacific islands, from Hawaii south to Fiji and Pitcairn.

Threats: Hunting by Polynesian settlers on the Hawaiian, Marshall, and Tahitian Islands has likely reduced the population. Oil and mineral exploration degrades habitat on the Seward Peninsula, and garbage dumps boost predatory Raven populations. Sea level rise could be significant, as many wintering islands are low-lying. The species is flightless during its winter moult, making it vulnerable to introduced predators.

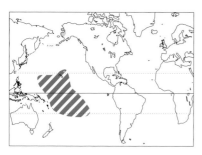

Conservation: Important wintering areas (Laysan, Midway Atoll, and Lisianski Island), are protected, and c.75% of breeding areas occur within the Yukon Delta NWR and Bering Land Bridge National Preserve in Alaska. The Togiak NWR in Alaska protects a major staging area at Bristol Bay.

Actions: Reduce hunting at remote wintering locations; continue efforts to manage habitat and remove invasive predators in Hawaii. Restrict oil and gas development from known breeding locations, and ensure private lands in Alaska are appropriately managed.

LONG-BILLED CURLEW
Numenius americanus

GLOBAL POPULATION: **20,000-50,000**
>75% BREED IN U.S.

TREND: **DECREASING**
WATCHLIST: **DECLINING**

15

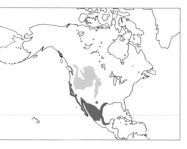

Distribution: Breeds in agricultural fields, short- and mixed-grass prairies, and grasslands over much of the Great Plains and Great Basin of the western U.S. and southwest Canada; though it once reached the tall-grass prairies and eastern grasslands. During winter and migration, it uses a variety of wetland and grassland habitats along the Pacific Coast of California, the Gulf Coast of Louisiana and Texas, and both coasts and the interior of Mexico.

Threats: Conversion of wetland and grassland habitat to agriculture, urban development, and reduction in native grazing animals (bison, prairie dogs) continue to diminish its numbers and range. Populations appear to be declining in the Great Plains, but increasing in some areas west of the Rockies.

Conservation: Occurs within many protected areas in its breeding and wintering ranges. Recent purchases by TNC at Matagorda Ranch in Montana, and by ABC and Pronatura Noreste near Saltillo in Mexico are securing key areas of habitat.

Actions: Continue habitat protection and restoration efforts for coastal and grassland habitats across range.

HUDSONIAN GODWIT
Limosa haemastica

GLOBAL POPULATION: **50,000-70,000**
C.**20 %** BREED IN **U.S.**

TREND: **UNKNOWN**
WATCHLIST: **RARE**

16

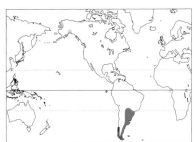

Distribution: Nests in open sedge meadows near tidal mudflats in Canada's Hudson and James Bay regions, on the Mackenzie and Anderson River Deltas, on the Seward Peninsula, and in south-central Alaska. Stages at Cook Inlet near Anchorage; on Quill and Luck Lakes in Saskatchewan; and on the upper James Bay, Ontario. Winters on coastal mudflats at three sites in Argentina and Chile. During northbound migration, many birds stage in the central U.S. and at Lago do Peixe NP in Brazil.

Threats: Oil and gas development around the Mackenzie River Delta, development and coal mining in Cook Inlet, over-abundant Snow Geese in the Hudson Bay that feed on tundra grasses, and urban development in Texas are major threats. Others include contamination from military bases and refineries in Argentina, disturbance by seaweed collectors and boat traffic in Chile, and climate change.

Conservation: Many sites are protected. Stopover habitat restoration efforts have been successful at some playa lakes.

Actions: Protect remaining breeding, wintering, and stopover sites; restrict cattle access to wetlands within Lagoa do Peixe NP.

BAR-TAILED GODWIT
Limosa lapponica

GLOBAL POPULATION: **1.1-1.2 MILLION**
<**8%** BREED IN **U.S.**

TREND: **DECREASING**
WATCHLIST: **DECLINING**

13

Distribution: Three subspecies breed in open meadows and tundra habitats near water from Alaska through Eurasia to Scandinavia. Winters in intertidal areas along the coasts of Europe, Africa, the Middle East, and Australasia. The *baueri* subspecies breeds from northeastern Siberia to western and northern Alaska, winters in New Zealand and southeastern Australia, and returns north through coastal China.

Threats: Loss of habitat along the East Asian coast is the greatest threat (e.g., due to the Saemangeum tidal barrier in South Korea), where foraging birds are also disturbed by fishing and shrimp-farming. Indigenous people harvest up to 1,900 birds a year in Alaska, and 2,000-3,000 are killed for food during migration in eastern China. Other threats include contamination, oil spills, collisions with lighted structures, and emergent diseases.

Conservation: Most breeding habitat in Alaska is secure within protected areas. Legally protected in U.S., Australia, and New Zealand.

Actions: Reduce habitat loss and disturbance at wintering and stopover sites, especially in China and South Korea; assess population trends and the impact of hunting in Alaska and China.

MARBLED GODWIT
Limosa fedoa

GLOBAL POPULATION: **173,500**
<50% BREED IN U.S.

TREND: **DECREASING**
WATCHLIST: **DECLINING**

15

Distribution: Nests in prairies of the Great Plains, with smaller numbers on James Bay and the Alaska Peninsula. Winters along the Pacific Coast from California to Panama, and along the Atlantic and Gulf Coasts from North Carolina to Florida, and Texas through Mexico. Large flocks stopover at Great Salt Lake, and winter in San Francisco Bay and western Mexico.

Threats: Loss of habitat and market hunting extirpated this species from 25% of its former range by the early 1900s; loss of grasslands and coastal wetlands continues. Invasive plants and the absence of native bison also degrade habitat.

Conservation: Many important breeding, stopover, and wintering areas in the U.S. occur on NWRs. Grassland conservation and restoration at Lostwood and Chase Lake NWRs, North Dakota; wetland restoration in San Francisco Bay, and rotational grazing systems have been beneficial.

Actions: Continue land management, protection and restoration efforts to improve habitat at breeding and wintering sites (especially in northwest Mexico); maintain and strengthen incentives for grassland conservation on private lands through the Conservation Reserve Program.

BLACK TURNSTONE
Arenaria melanocephala

GLOBAL POPULATION: **95,000**
U.S. BREEDING ENDEMIC

TREND: **UNKNOWN**
WATCHLIST: **RARE**

16

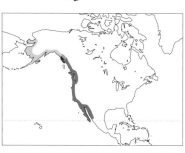

Distribution: Breeds in coastal sedge meadows in western Alaska, from Point Hope south to the Alaska Peninsula and Nunivak Island. Winters along rocky coasts from Kodiak Island in Alaska, south to the Gulf of California.

Threats: Habitat degradation and contamination from oil spills is the most serious threat, and may be linked to reduced clutch sizes. Ingestion of plastics and lead shot recorded, but effects unknown. Christmas Bird Counts indicate declines in the Pacific Northwest and on the Farallon Islands, but numbers dramatically increased (1969-1986) following restoration of kelp beds off the Palos Verdes Peninsula, California.

Conservation: Breeding areas are protected within the Bering Land Bridge National Preserve and Yukon Delta NWR. Important wintering and staging areas are uninhabited or protected, including Kachemak Bay and Northeast Montague Island in Alaska, and others along the coasts of Washington, Oregon, and California.

Actions: Respond rapidly to any oil spills; need effective cleanup plan for Montague Island.

SURFBIRD
Aphriza virgata

GLOBAL POPULATION: **70,000**
>75% BREED IN U.S.

TREND: **UNKNOWN**
WATCHLIST: **RARE**

15

Distribution: Nests in high-elevation tundra in Alaska (>75% of population) and the western Yukon of Canada. Winters in rocky intertidal habitats along the Pacific Coast from southern Alaska to Chile, with highest densities from southern Canada to Oregon. Up to 56,000 stop in Prince William Sound during spring migration.

Threats: Oil spills, especially in areas where much of the population concentrates during spring migration (e.g., Prince William Sound, Montague Island), represent the most severe potential threat to this species; climate change effects on tundra may be a long-term threat.

Conservation: Most of the known breeding areas are protected, including the Arctic, Yukon Delta, and Yukon Flats NWRs; and Denali NP. Many wintering areas are also protected.

Actions: Respond rapidly to any oil spills; need effective cleanup plan for Montague Island.

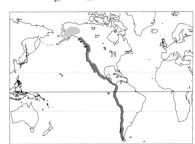

RED KNOT
Calidris canutus

GLOBAL POPULATION: **<1.1 MILLION**
10% BREED IN OR MIGRATE THROUGH U.S.

TREND: **DECREASING**
WATCHLIST: **DECLINING**
ESA: **CANDIDATE**

15

Distribution: Breeds in coastal arctic tundra, migrating and wintering along sandy beaches, mudflats, and estuaries. Three subspecies in North America (120,000 birds): *roselaari* breeds in northwestern Alaska, migrating and wintering along the Pacific Coast to South America; *rufa* breeds in arctic tundra in central Canada, winters in Argentina, Brazil, and southeastern U.S., and stages at Delaware Bay during spring migration; *islandica* breeds in eastern Canada and coastal Greenland and winters in Europe.

Threats: Overharvest of horseshoe crabs greatly reduced supplies of crab eggs in Delaware Bay, causing a rapid decline in the *rufa* subspecies from >100,000 birds to c.18,000. Other threats include human disturbance at beaches, and risk of oil spills.

Conservation: Breeding grounds and some staging areas (30% of Delaware Bay shore) are protected. Important sites in Delaware and Tierra del Fuego lack full protection. ABC has advocated for horseshoe crab harvest restrictions in Delaware Bay that are stemming declines in crab numbers.

Actions: List both *roselaari* and *rufa* subspecies under the ESA; strengthen restrictions on horseshoe crab harvests until crab numbers recover; need effective cleanup plan for Delaware Bay beaches in case of a major oil spill.

SANDERLING
Calidris alba

GLOBAL POPULATION: PERHAPS C. 600,000-700,000
C. >50% MIGRATE THROUGH U.S.

TREND: DECREASING
WATCHLIST: DECLINING

14

Distribution: Nests in high-arctic tundra near coasts in northern Alaska and central Canada (300,000 birds); also in Greenland and Eurasia. During migration and winter, inhabits sandy beaches throughout much of the world, including most of North and South America, with highest wintering densities in southern Peru and northern Chile. Important staging areas occur along coasts of Oregon and Washington, alkali lakes in Saskatchewan, and Delaware Bay.

Threats: Human disturbance at staging areas decreases foraging rates. Overharvest of horseshoe crabs greatly reduces food supplies (crab eggs) in Delaware Bay. Pesticide and heavy metal contamination and oil spills are also threats. Reduction of breeding habitat due to climate change is a future risk.

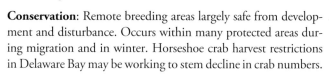

Conservation: Remote breeding areas largely safe from development and disturbance. Occurs within many protected areas during migration and in winter. Horseshoe crab harvest restrictions in Delaware Bay may be working to stem decline in crab numbers.

Actions: Reduce disturbance by people on beaches where Sanderlings migrate and winter; strengthen restrictions on horseshoe crab harvest in Delaware Bay until crab numbers recover.

SEMIPALMATED SANDPIPER
Calidris pusilla

GLOBAL POPULATION: 4 MILLION
100% MIGRATE THROUGH U.S.,
SOME BREED

TREND: DECREASING
WATCHLIST: DECLINING

14

Distribution: Nests on tundra near water in Alaska, Canada, and eastern Siberia. Uses shallow wetland habitats during migration, and winters on tidal mudflats and mangrove lagoons along the Pacific Coast from Mexico to Peru, and along the Atlantic Coast from the Caribbean to Argentina, particularly in Suriname and French Guiana. Important staging sites include Newfoundland's Bay of Fundy, Cheyenne Bottoms in Kansas, and Delaware Bay.

Threats: Birds wintering in northern South America have declined 80% since the 1980s, mainly due to overhunting. Overharvest of horseshoe crabs greatly reduces food supplies (crab eggs) in Delaware Bay. Habitat loss and catastrophic oil spills at staging areas are potential threats, as is climate change.

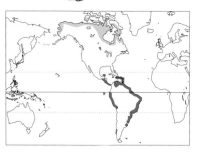

Conservation: Exploited by market hunters in the late 1800s, but has recovered from related declines. Many breeding and staging areas are either officially protected, or naturally so by their remoteness.

Actions: End unregulated hunting in South America; continue habitat protection and restoration efforts throughout range; reduce risk of oil spills; reduce Delaware Bay horseshoe crab harvests.

WESTERN SANDPIPER
Calidris mauri

GLOBAL POPULATION: **3.5 MILLION**
>75% BREED IN **U.S.**

TREND: **DECREASING**
WATCHLIST: **RARE**

15

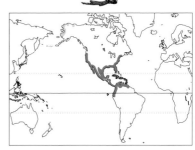

Distribution: Breeding restricted to coastal tundra of western Alaska and Siberia's Chukotski Peninsula. Winters on intertidal mudflats along the Pacific Coast from California to Peru, and on the Atlantic Coast from North Carolina to Suriname. Uses shallow wetlands and alkali lakes on passage. Almost entire population stops at Alaska's Copper River Delta during northward migration.

Threats: Wetland loss to urban development, agriculture, aquaculture, sea walls, waterways, water diversion, and invasive cordgrass are the main threats. Disturbance by people at stopover sites is also a problem. Climate change, oil spills, contamination from agricultural runoff, sewage, and emergent diseases are potential threats.

Conservation: Most breeding and some important non-breeding areas are at least partially protected: Copper River Delta, Alaska; San Francisco Bay, California; Cheyenne Bottoms, Kansas; and Laguna Atascosa NWR, Texas; in the U.S.; Fraser River Delta in British Columbia, Canada; the Mitsubishi Shorebird Reserve at Bahía Santa María, Mexico, supported by ABC; and the Parte Alta de la Bahía de Panama.

Actions: Continue habitat protection and restoration efforts throughout range; reduce risk of oil spills.

WHITE-RUMPED SANDPIPER
Calidris fuscicollis

GLOBAL POPULATION: **1.1 MILLION**
100% MIGRATE THROUGH **U.S.**, SOME BREED

TREND: **DECREASING**
WATCHLIST: **DECLINING**

14

Distribution: Nests in tundra near water from northeastern Alaska through the Canadian Arctic to southern Baffin Island. Uses a variety of shallow wetland habitats during migration, and while wintering in southern South America. Important spring staging areas include Decatur in Alabama; Cheyenne Bottoms NWR, Kansas; and Foxe River Basin in northern Canada.

Threats: Habitat loss and degradation at important staging areas are the most severe threats (including agriculture that depletes aquifers of inland wetlands), with climate change threatening to reduce habitat on the breeding grounds and disrupt migration patterns. Oil and gas development in the Arctic NWR could impact this species if it were permitted.

Conservation: Exploited by market hunters in the late 1800s, but now recovered from related declines. Many breeding and wintering areas are either secure due to their remote geography, or within protected areas (e.g., Arctic NWR).

Actions: Continue habitat protection and restoration efforts throughout range (especially on wintering grounds); ensure adequate water supplies for inland wetlands such as Cheyenne Bottoms.

ROCK SANDPIPER
Calidris ptilocnemis

GLOBAL POPULATION: >150,000
>75% BREED IN U.S.

TREND: DECREASING
WATCHLIST: HIGHEST CONCERN

17

Distribution: Breeds in coastal tundra using both rocky and soft sand or mudflat habitats along the Bering Sea and north Pacific. Four to five subspecies recognized: *ptilocnemis* breeds in the Pribilofs (25,000), "northern" *tschuktschorum* breeds in eastern Siberia and western Alaska (50,000), *couesi* in the Aleutians (75,000), and others in the Russian Commander Islands and Kamchatka Peninsula (c.5,600 birds). Northern subspecies winters from Gulf of Alaska to northern California; Pribilof subspecies winters on Alaskan peninsula (e.g., Izembek NWR); other subspecies sedentary or winter close to breeding areas.

Threats: Oil spills present a grave threat to congregations in coastal habitats. Introduced reindeer degrade habitat, and foxes predate nests, on the Aleutian Islands.

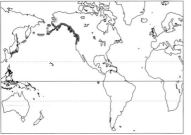

Conservation: Many breeding and wintering areas are protected on state or federal lands, including Yukon Delta, Izembek, and Alaska Maritime NWRs in the U.S.; and Beringia Regional Park, Komandorsky State Nature Reserve, and Yuzhno-Kamchatsky Sanctuary in Russia.

Actions: Respond rapidly to any oil spills; remove reindeer and predators from Aleutian Islands.

STILT SANDPIPER
Calidris himantopus

GLOBAL POPULATION: 820,000
100% MIGRATE THROUGH U.S.,
SOME BREED

TREND: UNKNOWN
WATCHLIST: RARE

15

Distribution: Nests in tundra meadows. Patchily distributed from James Bay, Canada, to northern Alaska. Main wintering areas are on ponds and lagoons in central and southern South America, though some winter in southern Florida, and along the U.S. Gulf Coast. Major migration concentrations occur at Quill Lakes in Saskatchewan, Canada; at Cheyenne Bottoms in Kansas, and in Suriname.

Threats: Over-grazing of nesting habitat by over-abundant Snow Geese in Manitoba is blamed for declines there. Loss of wetland habitats along migration routes and wintering grounds is likely the most severe threat, including wetland loss in the U.S., and coastal development at lagoons in northern South America.

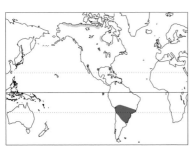

Conservation: Many breeding (e.g., Arctic NWR) and stop-over sites (e.g., Cheyenne Bottoms, Kansas; and Texas Coast NWRs) are protected.

Actions: Continue efforts to protect and restore habitat at wintering and staging areas in the U.S. and South America.

BUFF-BREASTED SANDPIPER
Tryngites subruficollis

GLOBAL POPULATION: **16,000-84,000**
100% MIGRATE THROUGH U.S., SOME BREED

TREND: **DECREASING**
WATCHLIST: **HIGHEST CONCERN**

18

Distribution: Nests in drier, high-arctic tundra from Canada's Queen Elizabeth Islands west through northern Alaska, to Chukotka, Russia. Winters in grasslands, pastures, and wetland margins in southern South America. During migration, uses grasslands, pastures, and sandbars. The Eastern Rainwater Basin of Nebraska is a primary spring stopover area.

Threats: Market hunting, conversion of grassland habitat to agriculture, and loss of grazers and natural fire regimes that kept grass short significantly reduced populations in the past. Grassland loss in non-breeding areas remains the most severe current threat. Exposure to pesticides in agricultural areas in the U.S. and South America likely common, though effects unknown.

Conservation: Many breeding and stop-over sites are protected, and many grassland restoration projects are underway, yet little conservation is directed specifically at this species.

Actions: Restrict oil extraction from known breeding areas within the Arctic NWR and National Petroleum Reserve in Alaska; continue efforts to protect and restore habitat in wintering and staging areas in the U.S. and South America; reduce pesticide use and maintain grazed pasturelands on wintering grounds.

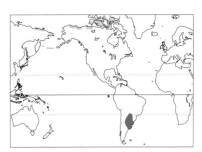

RED-LEGGED KITTIWAKE
Rissa brevirostris

GLOBAL POPULATION: **300,000**
C.>80% BREED IN U.S.

TREND: **DECREASING**
WATCHLIST: **RARE**

16

Distribution: Nests on high cliff faces on islands in the Bering Sea, with 70% on St. George and St. Paul Islands in the Pribilofs. Small populations on Buldir and Bogoslof Islands in the Aleutians, and >30,000 on Bering and Commander Islands in Russia. Forages over deep pelagic waters. Winter distribution in North Pacific poorly known. Total population has declined by 35% since 1970s.

Threats: Reasons for declines unknown, but reduction in prey caused by commercial fisheries is suspected. Introduced predators (rats, foxes) may also be problematic on breeding islands.

Conservation: Colonies in the U.S. are protected within the Alaska Maritime NWR. A 1994 international treaty closed pollock fishing in international waters between the U.S. and Russia (including around Bogoslof Island), but enforcement is insufficient.

Actions: Assess status, trends, wintering distribution, and effects of fisheries; eradicate predators at breeding sites; improve sustainability of fisheries in the Bering Sea, and consider closure of fisheries near the largest colonies in the Pribilofs.

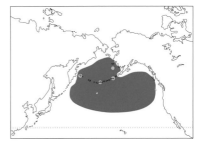

Ivory Gull
Pagophila eburnea

GLOBAL POPULATION: **15,000-25,000**
% IN U.S. UNKNOWN (RARE)

TREND: **DECREASING**
WATCHLIST: **HIGHEST CONCERN**

18

Distribution: Nests on cliffs, broken ice fields, and islands in northern Canada, Greenland, Norway, and Russia. Winters near sea ice edge in areas of North Atlantic with 70-90% ice cover, from Davis Strait to Labrador Sea, and also in the Bering and Chukchi Seas. Migrates offshore along Greenland coast. Canadian breeding population declined >70% from 1987 to 2005.

Threats: Global climate change threatens polar bear and seal populations, reducing scavenging opportunities. Pesticides, mercury, and other toxins recorded at very high levels in adults and eggs. Oil pollution also a threat. Other threats may include subsistence hunting by native people, disturbance at nesting sites, and loss of eggs to dogs at polar stations.

Conservation: Listed as Endangered in Canada in 2006. Some benefit from scavenging on seal carcasses and garbage dumps in arctic communities, but generally appears to be in serious decline.

Actions: Identify additional breeding sites in Canada; prevent human disturbance at breeding sites; reduce risk of oil spills and other sources of pollution in arctic seas.

Ross's Gull
Rhodostethia rosea

GLOBAL POPULATION: **25,000-100,000**
% IN U.S. UNKNOWN (REGULAR MIGRANT IN ALASKA)

TREND: **UNKNOWN**
WATCHLIST: **RARE**

14

Distribution: Poorly known, but breeds locally in swampy arctic estuaries, boggy tundra, and taiga. Main breeding areas are in northeastern Siberia, with small numbers (<200) breeding in northern and western Greenland, on Canadian Arctic islands, and near Churchill, Manitoba. Large numbers migrate past Point Barrow, Alaska, during the fall. Winters in open sea and at ice edges, mainly in the Arctic, with vagrants occasionally wandering far south into the U.S.

Threats: Disturbance at nests by humans recorded near Churchill, and some birds are hunted by indigenous people in Alaska. Oil development in the Beaufort Sea and across the Arctic is a potential threat. Global climate change and the resulting reduction in sea ice may be the most severe long-term threat.

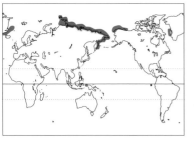

Conservation: Little effort is devoted to this species due to its remote range. Breeding areas near Churchill are legally protected.

Actions: Assess status and trends, identify additional breeding sites in North America, and map migratory routes. Reduce risk of oil spills and other sources of pollution in arctic seas.

HEERMANN'S GULL
Larus heermanni

GLOBAL POPULATION: **300,000-500,000**
c.**50%** WINTER IN **U.S.**

TREND: **INCREASING**
WATCHLIST: **RARE**

16

Distribution: Nests during the spring on nine rocky islands along the Pacific Coast and Gulf of California in Mexico, with >90% of population breeding on Isla Rasa. Migrates north along the coastline and offshore during summer and fall, reaching north to Vancouver, Canada.

Threats: Formerly, exploitation of eggs and introduced mammalian predators at main breeding colony. Environmental contamination from pesticides and oil spills, overfishing, bycatch, and climate change are potential threats. Population is vulnerable to catastrophic events at Isla Rasa.

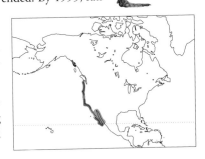

Conservation: In 1964, Isla Rasa was protected and egg harvests ended. By 1995, rats and mice were eradicated. Other breeding islands are protected within the Gulf of California Biosphere Reserve. Island Conservation works to eradicate invasive mammals from seabird islands in this region. The Monterey Bay National Marine Sanctuary protects important non-breeding areas.

Actions: Eradicate invasive mammals at smaller breeding colonies; legally protect the San Benito Islands and other breeding sites off the Pacific coast of Baja California; limit environmental contaminants and oil pollution.

YELLOW-FOOTED GULL
Larus livens

GLOBAL POPULATION: **40,000-60,000**
≤**5%** OCCUR IN **U.S.**POST-BREEDING

TREND: **INCREASING**
WATCHLIST: **RARE**

16

Distribution: Nests on rocky islands within the Gulf of California, Mexico. Occurs throughout Gulf of California and north to Salton Sea in the U.S. after breeding. Numbers at the Salton Sea have increased dramatically since the 1960s.

Threats: Exploitation of eggs, human disturbance, and introduced mammalian predators (cats, dogs, rats) are likely the main threats at breeding colonies. Environmental contamination from toxins, pesticides, and oil spills; overfishing, fisheries bycatch, and climate change are all potential threats.

Conservation: Most breeding islands (including Isla Rasa) in the Gulf of California are protected, but enforcement could be improved.

Actions: Restrict access on breeding islands to trails marked with bilingual signs to reduce disturbance to nesting birds; improve enforcement of egg collecting restrictions; continue efforts to limit environmental contaminants and oil pollution.

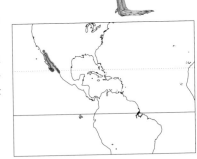

THAYER'S GULL
Larus thayeri

GLOBAL POPULATION: **10,000-12,000**
% IN U.S. UNKNOWN

TREND: UNKNOWN
WATCHLIST: RARE

15

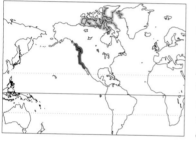

Distribution: Nests on ledges of steep rocky sea cliffs along coasts of the central Canadian Arctic and northwestern Greenland. Most winter along the northwest Pacific Coast (mainly British Columbia, Canada, to Washington, but south to central Baja California, Mexico) with smaller numbers in the Great Lakes, interior North America, and the north Atlantic Coast.

Threats: Few serious threats identified. Human disturbance and hunting by indigenous people may occur at colonies in Greenland, but most colonies are inaccessible to people. Other threats include oil and pesticide contamination. Effects of climate change not yet known, but could cause changes in arctic habitat.

Conservation: Inaccessibility of breeding colonies offers best protection.

Actions: Improve monitoring especially at remote and seldom-visited nesting colonies, and of migrating birds in Alaska; reduce risk of oil spills and pesticide contamination; reduce potential future impacts of global climate change by reducing greenhouse gas emissions.

ICELAND GULL
Larus glaucoides

GLOBAL POPULATION: **200,000-400,000**
% IN U.S. UNKNOWN

TREND: UNKNOWN
WATCHLIST: RARE

14

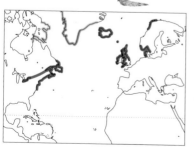

Distribution: Nests on sea cliffs on islands and coasts. Roughly 5,000 pairs of the *kumlieni* subspecies breed in the eastern Canadian Arctic (e.g., Baffin Island), whereas the more numerous *glaucoides* subspecies breeds in southwestern and eastern Greenland. Both winter mainly within arctic sea ice openings, but also along coasts in southeastern Canada, the northeastern U.S., and Europe.

Threats: Few serious threats identified. Human disturbance and hunting by indigenous people occurs at colonies in Greenland, though population effects are unknown, and most colonies are inaccessible to people. Oil and pesticide contamination and ingestion of plastics recorded in this species. Effects of climate change or reductions in sea ice not yet known, but could cause changes in arctic habitat.

Conservation: Inaccessibility of breeding colonies offers best protection.

Actions: Improve population monitoring, especially at remote nesting colonies; reduce risk of oil spills and other contamination; reduce potential future impacts of global climate change by reducing greenhouse gas emissions.

BRIDLED TERN
Onychoprion anaethetus

GLOBAL POPULATION: **600,000-1 MILLION**
<3% BREED IN U.S.

TREND: **DECREASING**
WATCHLIST: **DECLINING**

14

Distribution: Nests in sheltered crevices on small islands throughout tropical and subtropical oceans; most abundant in the Persian Gulf. Two subspecies occur in the Americas: *nelsoni* off the Pacific Coast of Mexico, Nicaragua, and Costa Rica; and *melanoptera* from the Florida Keys and Bahamas to Belize and northern Venezuela. A handful nest in the Florida Keys each year.

Threats: Human exploitation of eggs, and predation by introduced mammals are likely the most severe threats. Oil pollution, heavy metals, and entanglement in fishing line are also potential threats.

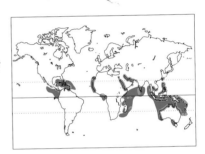

Conservation: Single Florida breeding site at Pelican Shoal is protected, as are colonies at Culebra NWR in Puerto Rico. Legal restrictions on egg collecting in Caribbean appear successful.

Actions: Stop egg collecting at remaining colonies; reduce risk of environmental contaminants.

ALEUTIAN TERN
Onychoprion aleuticus

GLOBAL POPULATION: **16,200-25,000**
c.50% BREED IN U.S.

TREND: **DECREASING**
WATCHLIST: **RARE**

15

Distribution: Nests at widely scattered locations in marshes, on flat islands, and on sandbars in estuaries along the Pacific and Bering Sea coasts of Alaska (9,000-12,000 birds), and Siberia (7,200-13,000 birds). In winter, ranges into the southwestern Pacific, with migrants passing by the Philippines, Hong Kong, and Japan.

Threats: Threats at nesting colonies include human disturbance through egg harvests, reindeer herding, grass burning in spring, and introduced foxes. High pesticide residues recorded in eggs, but effects unknown. Oil spills and other contaminants are general threats to marine ecosystems in Alaska. Low-lying islands could be vulnerable to sea level rise.

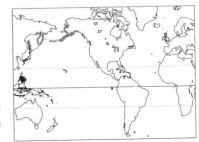

Conservation: Most U.S. breeding sites are within federally protected areas (e.g., Alaska Maritime and Yukon Delta NWRs, Bering Land Bridge National Preserve, Copper River Delta, and Glacier Bay NP); fox eradication on Aleutian Islands by FWS has led to re-colonizations.

Actions: Eradicate invasive predators on Aleutian Islands; reduce risk of oil spills and other sources of pollution to Alaskan marine ecosystems.

LEAST TERN
Sternula antillarum

GLOBAL POPULATION: **60,000-100,000**
% IN U.S. UNKNOWN, BUT LIKELY c.50%

TREND: **UNKNOWN**
WATCHLIST: **HIGHEST CONCERN**
ESA: **ENDANGERED (CA AND INTERIOR POPS)**

16

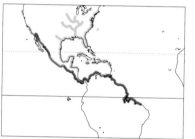

Distribution: Nests on coastal beaches and sandbars of major rivers. Winters along coasts of Mexico and the Caribbean south to South America. The *antillarum* subspecies breeds along the East and Gulf Coasts of the U.S., and in Mexico and the Caribbean; *athalassos* breeds primarily in the Mississippi and Rio Grande drainages; *brownii* breeds along the Pacific Coast from central California to Baja California.

Threats: Historic declines due to commercial hunting, egg collecting, and pesticide contamination. Continuing threats include poor water flow management on rivers, coastal development; human disturbance; and invasive or over-abundant predators.

Conservation: California and "Interior" populations listed under the ESA. Populations boosted by restricting access to nesting areas by people and predators, and the creation of sandbar nesting habitat on rivers. Dredge-material disposal along coasts, and an increase in gravel rooftops have provided additional nesting habitat. ABC is working to improve habitat on interior rivers.

Actions: Continue habitat protection, predator controls, and efforts to reduce human disturbance; reduce pesticide, heavy metal, oil, and other contamination of marine and riparian environments.

GULL-BILLED TERN
Gelochelidon nilotica

GLOBAL POPULATION: **150,000-420,000**
c.5% BREED IN U.S.

TREND: **DECREASING**
WATCHLIST: **DECLINING**

14

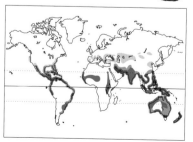

Distribution: Six subspecies distributed globally in tropical and temperate climates, but nowhere abundant: *aranea* nests on barrier islands, coastal marshes, and occasionally rooftops along the U.S. East and Gulf Coasts; *vanrossemi* nests on the Salton Sea and in San Diego Bay, California, and Isla Montague, Mexico. Winters in flooded rice fields, and along the coast from Texas to the Carolinas, and in Mexico and Central America.

Threats: Disturbance from recreational boating and coastal development are the most severe threats; others include pesticides and other toxins, and over-abundant gulls. Controlled in San Diego to reduce predation on Least Tern and Snowy Plover chicks.

Conservation: Recovered somewhat from historical exploitation for feathers. Many breeding sites protected including the Salton Sea and San Diego Bay NWR, California. The Army Corps of Engineers created the Baptiste Collette Bird Islands with dredge material, which benefit this species in Louisiana.

Actions: Protect known and potential breeding sites from human disturbance; cull gulls near colonies; improve population monitoring; measure benefits of control at Least Tern and Snowy Plover colonies.

ROSEATE TERN
Sterna dougallii

GLOBAL POPULATION: **70,000-82,000**
9% BREED IN THE **U.S.**

TREND: **DECREASING**
WATCHLIST: **DECLINING**
ESA: **EN (NE), THREATENED (SE)**

15

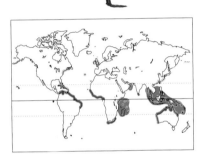

Distribution: Nests on partially vegetated beaches, or occasionally on flat gravel rooftops, from Nova Scotia to New York; also in Florida and the Caribbean. Probably winters off the coast of northern South America and Brazil. Other populations in Europe, South Africa, Arabian Sea, Indian Ocean, and Australasia.

Threats: The northeastern population was reduced to 2,000 pairs by the feather and egg trades. Has since recovered somewhat, but still fluctuates. Coastal development, over-abundant gulls, human disturbance, sea level rise, and contamination (heavy metals, pesticides) are all threats. Exploitation still occurs in some parts of the Caribbean and South America.

Conservation: U.S. colonies protected at Stewart B. McKinney NWR, Connecticut; Bird Island and Monomoy NWR, Massachusetts; Acadia NP, Maine; and Dry Tortugas NP, Florida. Provision of artificial nest boxes and gull control are beneficial.

Actions: Continue habitat protection, predator control, provision of artificial nests, and efforts to reduce human disturbance; research winter distribution and threats.

ELEGANT TERN
Thalasseus elegans

GLOBAL POPULATION: **34,000-90,000**
<10% BREED IN U.S.

TREND: **UNKNOWN**
WATCHLIST: **RARE**

17

Distribution: Nests on rocky or sandy islands: on Isla Rasa (perhaps as much as 90-97% of population) and Isla Montague in Mexico's Gulf of California; and in San Diego Bay, Los Angeles Harbor, and Bolsa Chica Ecological Reserve, California. Forages over shallow waters along California coast. Winters along coast from central Mexico south to Chile. Population trends unknown, but apparently expanding in southern California.

Threats: Extirpated from some historical breeding sites by illegal egg harvesting, human disturbance, and introduced predators. Overfishing of sardines and anchovies, climate change, and marine contamination are also threats.

Conservation: Isla Rasa is protected, and rodents have been eradicated there. Other potential breeding islands are protected within the Gulf of California Biosphere Reserve. In the U.S., it is protected within San Diego NWR. Some wintering areas also protected (e.g., Peru's Paracas NP).

Actions: Increase breeding populations outside Isla Rasa, reduce human disturbance at nesting sites, continue removal of invasive predators; improve law enforcement on protected islands in Mexico; reduce risk of pollution and contamination of marine environment throughout range.

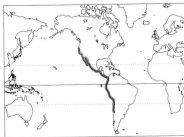

BLACK SKIMMER
Rhynchops niger

GLOBAL POPULATION: **120,00-210,000**
25-50% BREED IN U.S.

TREND: **DECREASING**
WATCHLIST: **DECLINING**

14

Distribution: The *niger* subspecies nests on barrier islands, sandbars, gravel bars, dredge spoil mounds, and in saltmarshes along most of the East and Gulf Coasts, and in southern California, with key breeding concentrations in Louisiana and Texas. Winters from North Carolina to Mexico and Central America. Forages in shallow coastal waters, estuaries, and lagoons. Two other races occur in South America.

Threats: Historically exploited for egg and feather trades. Loss of coastal habitats and human disturbance at nesting sites are biggest threats, but also contamination (heavy metals, pesticides, oil pollution) and entanglement in garbage (fishing line, six-pack-holders).

Conservation: Protected by many federal and state lands, including San Diego Bay NWR, California; and East Timbalier Island NWR, Louisiana. The Baptiste Collette Bird Islands, Louisiana, created from dredge material, may be used by >40% of U.S. population. Restricting nest disturbance through signs, ropes, and public education, benefits beach-nesting colonies.

Actions: Protect key breeding colonies.

GREAT SKUA
Stercorarius skua

GLOBAL POPULATION: **48,000**
% IN U.S. UNKNOWN (REGULAR IN WINTER IN ATLANTIC)

TREND: **UNKNOWN**
WATCHLIST: **RARE**

14

Distribution: Nests on islands in the northeastern Atlantic outside North America, from Iceland (5,400 pairs), to Scotland (7,900 pairs), and the Faeroes (250 pairs), then east to northwestern Russia. Mainly winters over marine waters off southwestern Europe, with younger birds ranging to northwestern Africa and northern South America. Small numbers winter on the Grand Banks, Newfoundland, and some venture further south along the East Coast.

Threats: Birds accumulate pesticides, mercury, and other toxins, and some die accidentally in fishing gear.

Conservation: Almost extirpated from the Faeroes and Shetlands at the end of the 19th Century due to exploitation. Populations there have since increased following strict protection, and also benefit from consuming scraps discarded by fishing boats. European populations are increasing and expanding their range eastward, but no trend information is available in North America.

Actions: Maintain protection at breeding sites.

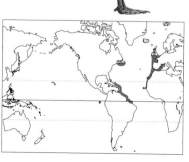

RAZORBILL
Alca torda

GLOBAL POPULATION: **1.5 MILLION**
<1% BREED IN U.S.

TREND: **INCREASING**
WATCHLIST: **RARE**

14

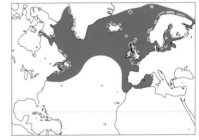

Distribution: Nests on rock ledges or within cliff crevices along coastlines of the north Atlantic from Maine to eastern Canada, Greenland, Iceland, northern Europe, and northwestern Russia. Winters in ice-free waters over the continental shelf, especially the outer Bay of Fundy and Gulf of Maine in North America. Populations are expanding in eastern North America.

Threats: Oil pollution, accidental death in gillnet fisheries, reduction of food supplies due to overfishing, and over-abundant gull populations are all threats. Sensitive to human disturbance at nesting sites. Pesticides, mercury, and other pollutants recorded in adults and eggs. Commercial hunting of adults and eggs diminished numbers historically.

Conservation: The most important breeding sites in Canada are protected. Gillnet fishery closures in Quebec and other fishery moratoriums likely reduced bycatch and may have contributed to local population increases during 1990s, but mortality continues in other fisheries.

Actions: Reduce risk of oil spills near breeding and wintering areas; minimize gillnet impacts; reduce over-fishing near colonies.

MARBLED MURRELET
Brachyramphus marmoratus

GLOBAL POPULATION: **c.500,000**
>50% LIKELY BREED IN U.S.

TREND: **DECREASING**
WATCHLIST: **DECLINING**
ESA: **THREATENED (IN PART)**

16

Distribution: Nests high in old-growth conifers of the Pacific Northwest, and on the ground in western Alaska. There is an isolated population further south in California at Half Moon Bay.

Threats: Annual population declines of 4-7% are blamed primarily on loss of old-growth forest to logging and coastal development, as well as accidental mortality in gillnets and from oil spills. Overfishing of sardines has decreased food availability.

Conservation: 3.9 million acres of Critical Habitat designated in Washington, Oregon, and California (where population is listed as Threatened under the ESA). Measures to reduce mortality in gillnets taken outside Alaska. Important breeding sites occur within Tongass and Chugach NFs and Kodiak NWR, Alaska; Olympia NP, Washington; and Carmanah-Walbran Provincial Park and Gwaii Hanas NP in British Columbia, Canada. Recent efforts to remove ESA and habitat protections to boost old-growth logging were unsuccessful.

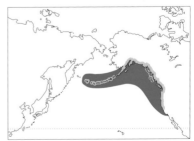

Actions: List under the ESA in Alaska; restrict logging in remaining old-growth forests within breeding range, reduce mortality in gillnets in Alaska and Canada; reduce risk of oil spills.

KITTLITZ'S MURRELET
Brachyramphus brevirostris

GLOBAL POPULATION: **7,000-35,000**
70% BREED IN U.S.

TREND: **DECREASING**
WATCHLIST: **HIGHEST CONCERN**
ESA: **CANDIDATE**

19

Distribution: Nests in boulder fields and forages in nearby offshore waters near glaciers, from southeastern Alaska through the larger Aleutian Islands, Bristol Bay, and the Seward and Lisbourne Peninsulas. Also patchily distributed along coasts in Siberia. Wintering areas are not known, but likely in ice-free waters over the continental shelf far from shore.

Threats: Rapid population declines ranging from 59-90% noted at several Alaskan sites perhaps due to glacial retreat caused by climate change, oil spills and other pollution, disturbance at foraging areas caused by frequent boat traffic, and accidental mortality in salmon gillnet fisheries. Some recent surveys suggest declines are less severe, however.

Conservation: Breeding sites are mostly protected within NPs and NWRs. Candidate for listing under the ESA.

Actions: List under the ESA; reduce disturbance from boat traffic in foraging areas by restricting boat numbers and speed limits; respond rapidly to any oil spills; address climate change to reduce greenhouse gas emissions; measure impact of fisheries; determine winter distribution.

XANTUS'S MURRELET
Synthliboramphus hypoleucus

GLOBAL POPULATION: **5,000-20,000**
c.**13-25%** BREED IN U.S.

TREND: **DECREASING**
WATCHLIST: **HIGHEST CONCERN**
ESA: **CANDIDATE**

18

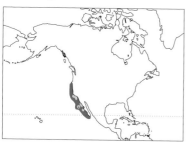

Distribution: Nests on rocky islands off southern California and Baja California, Mexico. After breeding, forages from Baja California to southern Canada, with highest concentrations off California. Most of the northern *scrippsi* population breeds on Santa Barbara Island, whereas the southern *hypoleucus* subspecies is restricted to islands off Guadalupe.

Threats: Introduced rats and cats wiped out or greatly reduced colonies on many islands. Lights on fishing boats cause collisions and attract predatory gulls. Mortality in gillnets, and oil and other contaminants are also threats; proposed development of a natural gas terminal near the Coronado Islands is an incipient threat.

Conservation: U.S. breeding islands protected within the Channel Islands NP. Rats removed from Santa Barbara and Anacapa (where birds now breed again). Invasive mammals also removed from most Mexican breeding sites (except Guadalupe and Cedros Islands).

Actions: List both subspecies under the ESA; remove invasive predators from breeding islands (e.g., Los Coronados); restrict light pollution caused by fishing boats at breeding sites; reduce risk of oil spills and other contaminants; monitor impacts of gill-net fisheries.

CRAVERI'S MURRELET
Synthliboramphus craveri

GLOBAL POPULATION: **9,000-20,000**
% IN U.S. UNKNOWN (REGULAR OFFSHORE)

TREND: **DECREASING**
WATCHLIST: **HIGHEST CONCERN**

18

Distribution: Nests in crevices on at least ten offshore rocky islands in Mexico's Gulf of California (90% on Islas Partida, Rasa, and Tiburón). Unconfirmed nesting on San Benito Island on the Pacific coast of Baja California. After breeding, forages in warm marine waters near to shore from Sonora, Mexico, north to southern California.

Threats: Introduced rats and cats on nesting islands are the most severe problem. Accidental mortality in gillnets, collisions with squid fishing boats at night, and oil pollution and other contaminants are also threats.

Conservation: Most breeding islands are protected within the Gulf of California Biosphere Reserve, or are wildlife sanctuaries. Rats and mice were removed from Isla Rasa by 1995. The Monterey Bay National Marine Sanctuary protects non-breeding areas.

Actions: Protected status for San Benito Island needed; remove invasive predators, reduce human disturbance, and halt illegal fishing at breeding islands; reduce risk of oil spills and other contaminants; continue population monitoring to asses status, trends, effects of gillnet fisheries, and predator eradication efforts.

ANCIENT MURRELET
Synthliboramphus antiquus

GLOBAL POPULATION: **1.3 MILLION**
C.**30% BREED IN U.S.**

TREND: **DECREASING**
WATCHLIST: **DECLINING**

14

Distribution: Nests in burrows on islands along the North Pacific Coast from British Columbia (Queen Charlotte Islands), Canada, and the Aleutians, to Russia, Japan, Korea, and China. Most abundant in North America, where birds winter south to central California. Forages in marine waters over continental shelf and between islands and the mainland.

Threats: Introduced rats are the most severe threat to adults and young at nesting islands, and have reduced numbers in the Aleutians by 80%, and the Queen Charlotte Islands by 50%. Also vulnerable to oil spills.

Conservation: Almost all Alaskan breeding sites protected within the Alaska Maritime NWR. Half of the Canadian nesting sites are within the Gwaii Haanas NP, and other large colonies are protected within provincial reserves. Populations recovered quickly on Aleutian Islands following fox removal, but attempts at removing raccoons in the Queen Charlotte Islands have been less successful due to re-colonization.

Actions: Control raccoons in the Queen Charlotte Islands and continue invasive predator removal in the Aleutians; reduce risk of oil spills.

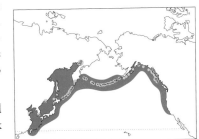

WHISKERED AUKLET
Aethia pygmaea

GLOBAL POPULATION: **116,000**
>80% BREED IN U.S.

TREND: **INCREASING**
WATCHLIST: **RARE**

17

Distribution: Nests in rocky crevices on islands in Alaska's Aleutian chain, with concentrations in the Krenitzen Islands, Islands of Four Mountains, Andreanof Islands, and on Buldir Island. Also nests in the Russian Commander and Kurile Islands. Forages in tidal rips close to shore near breeding areas year-round.

Threats: Introduced rats and foxes at breeding islands are the most severe problem. Collisions with brightly lit vessels at night also kill birds. Oil spills and climate change are also potential threats.

Conservation: Most breeding sites in the U.S. occur within the Alaskan Maritime NWR. The population is rebounding and re-colonizing islands following the removal of rats. Between 1974 and 2003, the Aleutian population grew from about 25,000 to around 116,000 birds.

Actions: Continue to remove rats and foxes from Aleutian Islands where Whiskered Auklets were extirpated or greatly reduced in numbers.

WHITE-CROWNED PIGEON
Patagionas leucocephala

GLOBAL POPULATION: **550,000**
≤5% BREED IN U.S.

TREND: **UNKNOWN**
WATCHLIST: **HIGHEST CONCERN**

17

Distribution: Nests in mangroves on islands, and forages in tropical forests of the Caribbean and Central America. Breeds locally along the U.S. Caribbean coast in southern Florida. Most U.S. birds appear to leave Florida for Cuba or the Bahamas in winter.

Threats: Mangrove and forest loss to coastal development, shrimp farms, agriculture, and charcoal production. Exploitation of nestlings for food occurs in parts of the Dominican Republic, and regulated hunting occurs in the Bahamas. Hurricanes, and the spread of rats and raccoons in the Florida Keys are also threats.

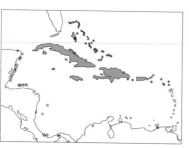

Conservation: Most nesting habitat in U.S. is protected, including in Everglades NP, and Key West and Great White Heron NWRs. Also in protected areas in Cuba and elsewhere, but enforcement often lacking.

Actions: Protect remaining mangroves and forests within range, and improve enforcement in protected areas in the Caribbean. Reduce hunting pressure where this threat is problematic. Eradicate invasive rats and raccoons from breeding islands.

GREEN PARAKEET
Aratinga holochlora

GLOBAL POPULATION: **200,000**
≤1% BREED IN U.S.

TREND: **INCREASING**
WATCHLIST: **HIGHEST CONCERN**

17

Distribution: Nominate subspecies nests in large trees and palms in urban areas in the Lower Rio Grande Valley of Texas, and in tropical forests from Tamaulipas south to Veracruz, Mexico. Origins of Texas population are unclear. Some birds escaped from the pet trade, but others likely colonized naturally. Small introduced populations also occur in southern California and Florida.

Threats: Exploitation for the pet trade and habitat loss to agriculture are the major threats in Mexico, and at least 30% of its original habitat is lost. The species is increasing in the U.S.

Conservation: The Sierra de Tamaulipas and El Cielo Biosphere Reserve are important sites in Mexico, but enforcement is lacking. In the U.S., occurs mainly in urban and residential areas near Brownsville, McAllen, San Benito, and Laredo.

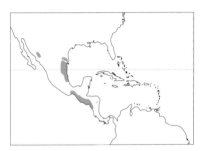

Actions: Continue land conservation efforts in the lower Rio Grande Valley of Texas and within the Mexican range. Clarify protected status in the U.S.

THICK-BILLED PARROT
Rhynchopsitta pachyrhyncha

GLOBAL POPULATION: **2,500**
0% BREED IN U.S.

TREND: **LIKELY EXTINCT IN U.S.**
WATCHLIST: **HIGHEST CONCERN**

20

Distribution: Nests in cavities of old-growth trees and snags in high-elevation coniferous forests in mountains of northwestern Mexico. Formerly occurred in southern Arizona and New Mexico.

Threats: The most severe threat is loss and degradation of its habitat (<0.1% of habitat remains pristine) due to logging of old-growth pines (that historically supplied timber products to the U.S.). Felling nest trees to capture nestlings is illegal but still occurs.

Conservation: Reintroduction attempts from 1986-1993 failed due to problems with disease, social behavior, and predators. In Mexico, this species occurs within the Ajos Bavispe National Forest and Wildlife Refuge, and Sonora Tancitaro NP; and on a 6,000 acre conservation easement at Cebadillas. With funding from ABC, Pronatura Noreste acquired a vital tract of breeding habitat at Mesa de las Guacamayas, and confirmed nesting within artificial nest boxes in 2008.

Actions: Protect remaining old-growth habitat and nest snags from logging and poachers. Long-term and large scale forestry planning is needed to maintain both productive forests and parrot habitat.

RED-CROWNED PARROT
Amazona viridigenalis

GLOBAL POPULATION: **3,000-6,500**
>50% BREED IN U.S.

TREND: **INCREASING**
WATCHLIST: **HIGHEST CONCERN**

20

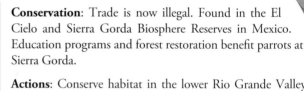

Distribution: Declining population of approximately 2,500 birds found in subtropical forests of northeastern Mexico. Some Lower Rio Grande Valley, Texas, birds are probably escapes, but others began wintering naturally from c. the 1920s, and began breeding in the 1980s. Now more in Texas than in Mexican range. Feral populations also in southern California, Florida, Hawaii, Puerto Rico, and inside Mexico.

Threats: More than 16,000 were legally or illegally exported from Mexico from 1970-1982, and more birds likely died in captivity prior to export. Export is now illegal, but continues, and poachers can destroy trees to capture nestlings. Also suffers from habitat loss for agriculture.

Conservation: Trade is now illegal. Found in the El Cielo and Sierra Gorda Biosphere Reserves in Mexico. Education programs and forest restoration benefit parrots at Sierra Gorda.

Actions: Conserve habitat in the lower Rio Grande Valley and within the Mexican range including on private lands. Outreach on illegal trade. Erect nest boxes within protected areas.

MANGROVE CUCKOO
Coccyzus minor

GLOBAL POPULATION: **200,000**
≤ 1% BREED IN U.S.

TREND: **DECREASING**
WATCHLIST: **DECLINING**

14

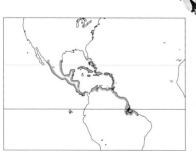

Distribution: Resident in mangroves, dense scrub, and tropical forests in coastal southern and western Florida. Also ranges south through the Caribbean, along the Atlantic and Pacific coasts of Mexico and Central America, and to the northern coast of South America.

Threats: Loss and fragmentation of mangrove and tropical forest habitat (60% lost in Florida by 1991) is the most severe threat across its range. Pesticide contamination could be a problem in areas sprayed for mosquitoes.

Conservation: In U.S., protected within Everglades and Biscayne NPs, Key Largo Hammocks State Botanical Preserve, and NWRs in the lower Florida Keys. TNC has also acquired lands within the Florida Keys. This species occurs in protected areas outside the U.S., though enforcement of park boundaries is lacking in parts of its range.

Actions: Protect remaining mangroves and forests within range, continue acquisition of habitat in Florida to connect protected areas, and improve enforcement of protected areas in the Caribbean.

FLAMMULATED OWL
Otus flammeolus

GLOBAL POPULATION: **37,000**
75% BREED IN **U.S.**

TREND: **UNKNOWN**
WATCHLIST: **RARE**

16

Distribution: Nests in cavities of dead and dying trees in open, montane ponderosa pine forest. Patchily distributed from southern British Columbia through the western U.S. to central Mexico. Northern populations thought to winter from Mexico, south to Guatemala and El Salvador.

Threats: Degradation and loss of habitat, including fire suppression that results in the loss of open areas required for foraging. The cutting of dead trees, and declines in woodpecker populations (e.g., extinction of Imperial Woodpecker in Mexico) reduce numbers of nest cavities. Spraying pesticides to combat spruce budworms may reduce insect prey for owls.

Conservation: Occurs within many NPs and NFs in the U.S. Thinning ponderosa pine is widely used to reduce fire threat and is compatible with this species' habitat needs if snags (standing dead trees) are left. Nest boxes may be used in the interim when snags are not available. Since 2003, ABC has led a project to restore ponderosa pine habitats for cavity-nesting birds on private lands in the western U.S.

Actions: Expand ponderosa pine habitat restoration efforts across species' range.

ELF OWL
Micrathene whitneyi

GLOBAL POPULATION: **190,000**
24% BREED IN **U.S.**

TREND: **UNKNOWN**
WATCHLIST: **RARE**

15

Distribution: Nests in cavities in trees, cacti, utility poles, and in nest boxes in a variety of habitats including deserts, mountain evergreen forests, riparian forests, and subtropical woodlands from the southwestern U.S. to western Mexico. U.S. populations migrate to southern Mexico during the winter; Mexican populations are non-migratory.

Threats: Habitat loss, degradation, and loss of nesting cavities are the most serious threats in the U.S., especially along the Lower Colorado River and around expanding urban areas (e.g., Phoenix, Tucson). Invasive salt cedar has replaced many riparian cottonwood-willow forests (e.g., 99% along the Lower Colorado River).

Conservation: Occurs within many western parks and wildlife refuges, including along the Lower Colorado River; Coronado NF, Arizona; and Big Bend NP, and several protected areas in the lower Rio Grande Valley, Texas.

Actions: Continue to restore and protect riparian forest habitats throughout the southwestern U.S., including removing salt cedar and other invasive species; expand nest box use.

SPOTTED OWL
Strix occidentalis

GLOBAL POPULATION: **15,000**
C.**70%** BREED IN U.S.

TREND: **DECREASING**
WATCHLIST: **HIGHEST CONCERN**
ESA: **THREATENED ("NORTHERN" AND "MEXICAN" SUBSPECIES)**

16

Distribution: Resident in old-growth conifer forests; as well as in mature oak, canyon, and riparian forests in southern range. Three subspecies: *caurina* ("Northern") in the Pacific Northwest; *occidentalis* from the Sierra Nevada to northern Baja California; and *lucida* ("Mexican") in Utah, Colorado, Arizona, New Mexico, western Texas, and central Mexico.

Threats: Loss of old-growth habitat to logging, also, in some areas, catastrophic forest fires due to past fire suppression; competition with expanding populations of Barred Owls.

Conservation: Among the most controversial of all species in U.S. history. Northern and Mexican populations listed under the ESA in the early 1990s, and also legally protected in Canada and Mexico. Despite protection, and presence in protected areas, populations continue to decline, especially in Canada, Washington, and southern California, due to logging.

Actions: Prevent loss of remaining old-growth forest due to logging; control Barred Owls which have spread into the species' range due to anthropogenic habitat fragmentation that favors Barred over Spotted Owls.

SHORT-EARED OWL
Asio flammeus

GLOBAL POPULATION: **2.4 MILLION**
C.**10%** BREED IN U.S.

TREND: **DECREASING**
WATCHLIST: **DECLINING**

13

Distribution: Nests on ground in open habitats (tundra, grasslands, marshes, agricultural lands, coastal dunes) throughout Eurasia and North America. Additional subspecies in the Caribbean, South America, and Hawaii. Alaskan and Canadian populations winter in southern Canada and the U.S. Abundance linked to population cycles of rodent prey. Caribbean population possibly expanding into the Florida Keys.

Threats: Loss and fragmentation of grassland, marsh, and coastal habitats due to agriculture, over-grazing, and urban and coastal development. Also invasive predators; potentially West Nile virus, pesticides.

Conservation: Occurs within NPs and NWRs throughout the U.S. Has benefited from grasslands created on reclaimed mine sites in the East, as well as from habitat protection on Conservation Reserve Program lands.

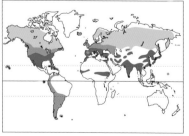

Actions: Maintain large tracts (>250 acres) of open habitat throughout range, but especially in the northeastern U.S. Maintain conservation incentives for private lands through the Conservation Reserve Program, and conservation easements. Monitor arctic populations not covered by Breeding Bird Survey. Determine causes of "sick owl syndrome" on Kauai.

ANTILLEAN NIGHTHAWK
Chordeiles gundlachii

GLOBAL POPULATION: **200,000**
≤1% BREED IN U.S.

TREND: **UNKNOWN**
WATCHLIST: **RARE**

14

Distribution: Nests on bare ground (or flat gravel rooftops) and forages over open areas, second growth scrub, hammock and pine forests; breeds in the lower Florida Keys, Bahamas, and Greater Antilles. Non-breeding range is unknown, but likely in South America.

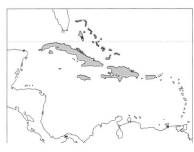

Threats: Little information on threats, but potentially affected by pesticide spraying targeting mosquitoes; formerly hunted for food in Jamaica. Invasive predators (rats, cats, mongooses) on nesting islands are potentially a threat, though effects are undocumented.

Conservation: Probably benefited from vegetation clearance in the Florida Keys, and occurs at airports (Marathon and Key West) and other urban sites (e.g., near community college on Stock Island). Several state and federal protected areas (e.g., National Key Deer Refuge) protect habitat for this species in the Florida Keys.

Actions: Research effects of pesticides and predators; identify and map winter range.

BLACK SWIFT
Cypseloides niger

GLOBAL POPULATION: **150,000**
C.20% BREED IN U.S.

TREND: **DECREASING**
WATCHLIST: **DECLINING**

14

Distribution: Nests behind waterfalls and in sea caves, breeding locally from southeastern Alaska south to Costa Rica; also in the Caribbean. Continental populations are migratory, probably wintering in Central and South America.

Threats: Poorly known, but may include reduction of prey through exposure to pesticides, and reduced water flow at nesting sites due to water diversion projects or climate change. Many Rocky Mountain nesting sites are behind waterfalls fed by rapidly shrinking glaciers and seasonal snowfields.

Conservation: Occurs in several protected areas including Misty Fiords NM and Tongass NF, Alaska; Wenatchee and Willamette NFs, Oregon; Olympic NP, Washington; Glacier NP, Montana; and Rocky Mountain NP, Colorado. ABC-led surveys in Montana, Oregon, Washington, and southeast Alaska during 2003-2005 raised the number of known nesting sites from five to >25 in those states.

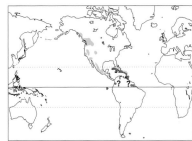

Actions: Locate and monitor nesting sites, especially outside the U.S. Ensure nesting sites at waterfalls are not disturbed by hikers or rock-climbers on U.S. public lands. Reduce pesticides in environment. Identify and protect key sites in non-breeding range.

BLUE-THROATED HUMMINGBIRD
Lampornis clemenciae

GLOBAL POPULATION: **2 MILLION** TREND: **INCREASING**
≤**5% BREED IN U.S.** WATCHLIST: **RARE**

14

Distribution: Breeds in mountain pine-oak forests and riparian (e.g., sycamore) woodlands, forest edge, and second growth, in Arizona, New Mexico, Texas, and Mexico. Populations in the U.S. and northern Mexico are mostly migratory, wintering further south, but some do winter in the U.S., particularly where feeding stations are available. Winter frequency may be increasing in U.S., perhaps due to habitat recovery and feeders.

Threats: Habitat loss and degradation due to logging, over-grazing of riparian forests, mining, and catastrophic fires caused by prior fire suppression potentially threaten this species, especially in Mexico.

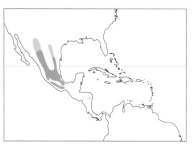

Conservation: Texas populations protected within Guadalupe Mountains and Big Bend NPs, some habitat in Arizona protected within Coronado NF and other preserves (e.g., TNC's Ramsey Canyon). Populations appear to thrive in close proximity to people and bird feeders, as long as most habitat remains intact.

Actions: Improve population monitoring to assess status and trends, especially in Mexico.

COSTA'S HUMMINGBIRD
Calypte costae

GLOBAL POPULATION: **3.6 MILLION** TREND: **STABLE**
50% BREED IN U.S. WATCHLIST: **RARE**

14

Distribution: Breeds in desert scrub and washes (Sonoran and Mojave Deserts), and xeric coastal scrub and chaparral; resident from southern California and Arizona to northwestern Mexico and Baja California. Migratory populations breed in the eastern and more arid portions of range; winters south to Jalisco, Mexico. Population appears to be stable in most areas (though fluctuates in Arizona) and perhaps expanding its non-breeding distribution in California.

Threats: Loss of coastal habitat and deserts to urban development (e.g., coastal California, Tucson, Phoenix) agriculture, and over-grazing. Planting buffelgrass for cattle forage increases fire frequency, which can damage nesting trees in deserts, whereas fire suppression degrades chaparral habitat. Ornamental flowers likely favor dominant competitors such as Anna's Hummingbird.

Conservation: Occurs in Joshua Tree NP, California; and Saguaro NP and Cabeza Prieta NWR, Arizona: and in some other protected areas.

Actions: Increase habitat protection and restoration efforts, including controlling fire-prone invasive grasses in deserts.

CALLIOPE HUMMINGBIRD
Stellula calliope

GLOBAL POPULATION: **1 MILLION**
C.70% BREED IN U.S.

TREND: **STABLE**
WATCHLIST: **RARE**

14

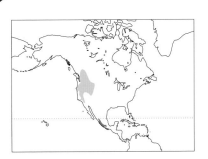

Distribution: Breeds in early successional shrub-land (eight years old), riparian woodland, and in openings in montane forest from central British Columbia and southwestern Alberta, Canada, south through the western U.S. to northern Baja California, Mexico. Migrates to wintering grounds in dry thorn forest and humid pine-oak forest in southwestern and south-central Mexico. Vagrants occasionally occur in the southeastern U.S. during winter.

Threats: Few threats are known. Grazing in riparian habitats reduces structural complexity

Conservation: Occurs in many U.S. NPs and NFs, including in the Sierra Nevada, Rocky Mountain NP, and elsewhere.

Actions: Determine whether this species benefits from early successional habitat created after logging, or from riparian restoration that enhances structural diversity.

ALLEN'S HUMMINGBIRD
Selasphorus sasin

GLOBAL POPULATION: **530,000**
U.S. BREEDING ENDEMIC

TREND: **DECREASING**
WATCHLIST: **RARE**

15

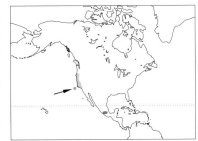

Distribution: Breeds along a narrow coastal strip or "fog belt" in scrub and open forest, from southern Oregon to southern California, then migrates to a small area in central Mexico. The slightly larger *sedentarius* subspecies is resident in dense chaparral and riparian forests near Los Angeles and on the Channel Islands, and is expanding.

Threats: Loss of habitat to development within its highly restricted range is the major threat to the migratory population. Feeders and exotic plants likely facilitate expanding *sedentarius* populations, though these may favor the dominant Anna's Hummingbird to the detriment of migratory populations. Climate change could alter flowering patterns and reduce breeding range.

Conservation: The migratory population occurs within several protected areas in California, including Arcata Marsh, Humboldt Bay, Bodega Bay, Point Reyes National Seashore, Golden Gate National Recreation Area, and Vandenberg Air Force Base, although remaining unprotected habitat is threatened by development. Most habitat for *sedentarius* is protected within the Channel Islands NP.

Actions: Protect and restore coastal scrub and riparian habitats in California. Assess winter-limiting factors in Mexico.

ELEGANT TROGON
Trogon elegans

GLOBAL POPULATION: **200,000**
≤1% BREED IN U.S.

TREND: **FLUCTUATING**
WATCHLIST: **RARE**

15

Distribution: Nests in tree cavities in forest from southeastern Arizona through much of Mexico, and Central America. In U.S., small numbers breed in pine-oak and riparian forests in the mountains of southeastern Arizona, adjacent New Mexico, and rarely along the Rio Grande Valley, Texas. Populations appear to fluctuate in the U.S., where they are at the northern range limit.

Threats: Degradation of habitat in riparian forests due to development or water draw-down is a potential threat. Disturbance by birders is a concern, though one study found no effect on nesting success.

Conservation: Most habitat in the U.S. is protected on public lands (e.g., Coronado NF, Fort Huachuca). Use of tape-recordings is not allowed at most sites popularly visited by birders searching for this species.

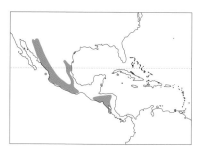

Actions: Maintain restrictions on vehicle use and tapes at Fort Huachuca to limit disturbance near nests. Ensure sufficient water flows in riparian areas for sycamores to grow. Protect habitat in Mexico, where vast majority of population occurs.

LEWIS'S WOODPECKER
Melanerpes lewis

GLOBAL POPULATION: **130,000**
C. 80% BREED IN U.S.

TREND: **DECREASING**
WATCHLIST: **HIGHEST CONCERN**

16

Distribution: Occurs locally in the western U.S. and southern British Columbia, Canada, breeding mainly in open ponderosa pine forests in mountains (especially burned forests), but also using open cottonwoods, aspen and oak woodlands (especially in winter), and pinyon-juniper forest. Northern populations migrate south during winter, sometimes as far as northern Baja California, Mexico.

Threats: Habitat loss and degradation due to fire suppression in ponderosa pine and oak. Cutting dead trees reduces suitable sites for nesting or storing acorns. Riparian cottonwoods and aspens fail to regenerate where over-grazing occurs, and may be replaced by unsuitable vegetation. Pesticide spraying may reduce food supplies on breeding grounds or contaminate birds in orchards and other agricultural areas during winter.

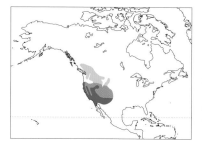

Conservation: Occurs within many NPs and NFs in the U.S. Since 2003, ABC has worked to improve ponderosa pine habitat for this and other cavity-nesting birds on private lands in the western U.S.

Actions: Expand ponderosa pine habitat restoration efforts in the western U.S.

RED-HEADED WOODPECKER
Melanerpes erythrocephalus

GLOBAL POPULATION: **2.5 MILLION**
C.**95% BREED IN U.S.**

TREND: **DECREASING**
WATCHLIST: **DECLINING**

13

Distribution: Breeds in deciduous forest and pine savanna, usually with open understory, from southern Canada through most of the eastern U.S. Highest densities in oak savanna of Midwest. Migrates out of northern 20% of range during winter, and expands into southwest Texas. Declined range-wide by 63% from 1966-2005.

Threats: Habitat loss and degradation, especially the loss of dead snags and mature trees with dead branches, is the most severe threat. Over 98% of Midwestern oak-savannas, 95% of long-leaf pine savannas, and 75% of bottomland swamps along the Mississippi are now gone.

Conservation: Occurs in many NWRs and other protected areas. Work to protect or restore prime habitat within this species' range, including prescribed burns and water management that create dead snags helps this species. Beaver dams also create dead snags and forest openings.

Actions: Increase habitat through the protection and restoration of forests, particularly in large tracts where natural disturbance processes (floods and fires) are maintained. Continue to restore pine and oak savannas on public and private lands using prescribed burns.

WILLIAMSON'S SAPSUCKER
Sphyrapicus thyroideus

GLOBAL POPULATION: **310,000**
C.**90% BREED IN U.S.**

TREND: **STABLE**
WATCHLIST: **RARE**

14

Distribution: Breeds in high elevation (usually above 5,000 feet) conifer and mixed-conifer (spruce-fir, Douglas fir, lodgepole pine, ponderosa pine, western larch) and aspen forests in western North America from south-central British Columbia south through the U.S. to northern Baja California, Mexico. Northern populations spend the winter in pine-oak, oak-juniper, and riparian woodlands in California, Arizona, New Mexico, and Mexico.

Threats: Degradation and loss of habitat likely threaten this species: fire suppression reduces the open structure of pine forests, and cutting dead trees reduces availability of snags for nesting.

Conservation: Occurs within many NPs and NFs in the U.S. Since 2003, ABC has led a project to restore ponderosa pine habitats for cavity-nesting birds that benefit this and other species on private lands in the western U.S.

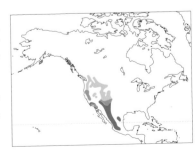

Actions: Continue and expand habitat restoration efforts in western U.S. to maintain groups of large snags, large trees, and natural fire frequencies.

NUTTALL'S WOODPECKER
Picoides nuttallii

GLOBAL POPULATION: **290,000**
C.95% BREED IN U.S.

TREND: **INCREASING**
WATCHLIST: **RARE**

14

Distribution: Resident in oak woodlands in California and extreme northern Baja California, Mexico, but also uses riparian woodlands (in the southern portion of its range), and locally, pinyon pines (e.g., in the San Bernardino Mountains).

Threats: Loss and degradation of habitat due to urban and agricultural development are the most severe threats within its relatively small range. In recent years, an introduced fungal pathogen (*Phytophthora*) has caused sudden oak death syndrome and killed tens of thousands of oaks in California. In the near-term, woodpeckers will likely benefit from these dead trees, but the disease will ultimately degrade and reduce habitat for woodpeckers and other birds, as dead trees are not replaced by maturing younger trees.

Conservation: Occurs within several protected areas on federal, state, and private lands, including Camp Pendleton, San Bernardino NF, Vandenberg Air Force Base, Morro Bay, Point Reyes National Seashore, the Sacramento NWR complex, and Kern River Valley.

Actions: Increase acreage managed for healthy oak woodlands on public and private lands.

ARIZONA WOODPECKER
Picoides arizonae

GLOBAL POPULATION: **200,000**
≤5% BREED IN U.S.

TREND: **UNKNOWN**
WATCHLIST: **RARE**

15

Distribution: Resident in pine-oak and riparian (walnut and sycamore) woodlands in mountains from southeastern Arizona and adjacent New Mexico, south through the Sierra Madre Occidental into central Mexico. In the U.S., some birds may move down-slope during winter.

Threats: Habitat loss and degradation due to logging, mining, and catastrophic fires caused by prior fire suppression potentially threaten this species, especially in Mexico. Riparian woodlands are degraded by over-grazing and the drawdown of groundwater.

Conservation: In U.S., most of its habitat is protected on public lands (e.g., Coronado NF).

Actions: Increase habitat protection and management efforts in Mexico.

RED-COCKADED WOODPECKER
Picoides borealis

GLOBAL POPULATION: **20,000**
U.S. ENDEMIC

TREND: **INCREASING**
WATCHLIST: **HIGHEST CONCERN**
ESA: **ENDANGERED**

19

Distribution: Breeds patchily in regularly burned pine savannas across the southeast U.S., occurring in family groups that use 80-100 year-old trees. Mainly in longleaf pine, but also in other types of pine forest near the northern edge of its range. Populations declined through 1980s, but most have since stabilized or increased.

Threats: Habitat loss, fragmentation, and degradation due to logging, fire suppression, conversion to short rotation plantations, clearance for agriculture, and urban development, are the most severe threats. Longleaf pine forest once occupied 60-90 million acres, but only 90,000 acres remain, with less than 5,000 acres being old-growth.

Conservation: Removing hardwoods and using controlled burns has helped this species. Especially large populations occur at Apalachicola NF, Florida; and Croatan NF, North Carolina. Safe Harbor agreements have protected >500 woodpecker groups on private lands.

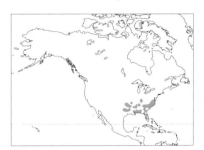

Actions: Increase acreage of habitat managed for this species on public and private lands, including use of controlled burns, artificial cavities, land purchases, conservation easements, and Safe Harbor agreements.

WHITE-HEADED WOODPECKER
Picoides albolarvatus

GLOBAL POPULATION: **72,000**
>95% BREED IN U.S.

TREND: **UNKNOWN**
WATCHLIST: **RARE**

16

Distribution: Nests in cavities in dead or dying trees from southern California north to western Nevada, Oregon, and western Idaho; and from Washington to the Okanagan Valley of British Columbia, Canada. Abundance appears to decrease north of California. Inhabits pine-oak forests in the Sierra Nevada, and ponderosa pine forests elsewhere.

Threats: The most severe threat is habitat loss, fragmentation, and degradation due to clear-cutting, fire suppression (favoring firs over pines), removal of large dead snags, and management of forests as even-aged stands. Chronic human disturbance near nest sites is an additional threat for the southern California *gravirostris* subspecies. Sierra Nevada populations appear stable, while populations elsewhere in ponderosa pine forests appear to be declining.

Conservation: Occurs within many NPs and NFs in the U.S., particularly in the Sierra Nevada where it appears to be most common. Since 2003, ABC has worked to restore ponderosa pine habitats for this and other cavity-nesting birds on private lands in the western U.S.

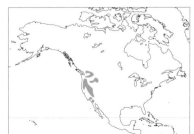

Actions: Continue and expand ponderosa pine habitat restoration efforts in western U.S.

GILDED FLICKER
Colaptes chrysoides

GLOBAL POPULATION: **1.1 MILLION**
25% BREED IN U.S.

TREND: **DECREASING**
WATCHLIST: **RARE**

16

Distribution: Resident in the Sonoran Desert and in riparian forests from Arizona and the Lower Colorado River of southeastern California, south through Baja California, Mexico, to Sonora and Sinaloa.

Threats: Suspected declines in Mexico due to conversion of desert to irrigated farmland. Water withdrawal and invasive salt cedar degrade riparian forests along the Lower Colorado River. Though significant population declines of Northern Flickers are, in part, blamed on the loss of nest sites to European Starlings, starlings have not yet been found to negatively affect Gilded Flicker, but more research is needed.

Conservation: Occurs on many federally protected lands along the Lower Colorado River and in Arizona, including Cabeza Prieta NWR, Organ Pipe Cactus NM, San Pedro Riparian NCA, and Saguaro NP.

Actions: Encourage retention of dead snags for nesting, e.g. in suburbs of Tucson and Phoenix; assess effects of starlings on nesting success; control salt cedar.

IVORY-BILLED WOODPECKER
Campephilus principalis

GLOBAL POPULATION: **UNKNOWN**
% IN U.S. UNKNOWN

TREND: **LIKELY EXTINCT**
WATCHLIST: **HIGHEST CONCERN**
ESA: **ENDANGERED**

20

Distribution: Two subspecies previously occupied disjunct regions, with *principalis* in mature lowland forests of the southeastern U.S., from eastern Texas to southeastern North Carolina, and southern Illinois south throughout Florida; and *bairdii* in both lowland hardwood and upland pine forests in Cuba. Last unquestioned records were 1944 for U.S., and 1956 for Cuba. Reports through 2008 in the U.S., and 1998 in Cuba provide hope that the species may persist, but definitive confirmation is lacking.

Threats: Historically, habitat loss and fragmentation through logging, conversion to agriculture, and suppression of natural disturbances (floods and fires) combined with hunting and trophy collection.

Conservation: Efforts to save the last known breeding population in the Singer Tract of northeastern Louisiana failed, and listing under the ESA did not occur until 1967. Intensive searches following a recent reported sighting have not located any birds.

Actions: Continue search efforts, especially in Cuba.

OLIVE-SIDED FLYCATCHER
Contopus cooperi

GLOBAL POPULATION: **1.2 MILLION**
C.**40% BREED IN U.S.**

TREND: **DECREASING**
WATCHLIST: **DECLINING**

14

Distribution: Breeds in large tracts of forest, often near natural openings, throughout the boreal region and south locally into the Appalachians; also in mountains of the western U.S. south to northern Mexico. Winters from southern Mexico to northern and western South America.

Threats: Clear-cutting and fire suppression degrade habitat. Pesticide application may impact food supplies. Other threats include habitat loss to energy development projects (dams, tar sands), and acid rain (eastern U.S.).

Conservation: Responds positively to post-fire conditions in forest. Occurs in protected areas in Canada, the U.S., and Latin America. New commitments by Quebec and Ontario in 2008 to protect large areas of boreal forest will help this species. Peru established the 5.1 million acre Cordillera Azul NP in 2001. ABC, ProAves, ECOAN, and the Jocotoco Foundation protect habitat for this species in private reserves in the Andes.

Actions: Continue efforts to protect large tracts of breeding and wintering habitat, and maintain natural fires and snags on breeding grounds. Improve management and enforcement of protected areas on wintering grounds.

WILLOW FLYCATCHER
Empidonax traillii

GLOBAL POPULATION: **3.3 MILLION**
C.**90% BREED IN U.S.**

TREND: **DECREASING**
WATCHLIST: **DECLINING**
ESA: **ENDANGERED ("SOUTHWESTERN" SUBSPECIES)**

14

Distribution: Breeds in a variety of shrubby habitats, often near water (and restricted to riparian areas in much of the arid west), across the western, Midwestern, and northeastern U.S, and into southern Canada. Winters in similar habitats from southern Mexico to northern South America.

Threats: Riparian habitats in arid western U.S. threatened by dams, channelization, urbanization, water drawdown, and over-grazing. Cowbird parasitism is problematic in California and Arizona.

Conservation: The southwestern *extimus* subspecies is listed as Endangered under the ESA, and has disappeared from much of its former range, but the population has not declined further since listing. Cowbird control is used to help breeding birds, mainly in California. Occurs within many protected areas throughout its range, with *extimus* found at Vandenberg Air Force Base, Camp Pendleton, and San Diego NWR in California; and at several NWRs on the Lower Colorado River in Arizona (e.g., Havasu, Bill Williams River, Imperial).

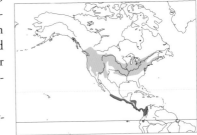

Actions: Protect and restore riparian habitats, and maintain cowbird control efforts for the *extimus* subspecies.

THICK-BILLED KINGBIRD
Tyrannus crassirostris

GLOBAL POPULATION: 2 MILLION
≤1% BREED IN U.S.

TREND: UNKNOWN
WATCHLIST: RARE

14

Distribution: Breeds in canyons and riparian woodlands (with sycamores or cottonwood-willow forests in the U.S.), from extreme southeastern Arizona and southwestern New Mexico (from the Guadalupe Mountains to the Baboquivaris), through Mexico to Oaxaca. U.S. and northern Mexican populations migrate south in winter, through Mexico south to Chiapas.

Threats: Threats to riparian woodlands (e.g., urbanization, water withdrawal, over-grazing, invasive salt cedar) in the U.S. could harm this species; threats in Mexico are not well known.

Conservation: In the U.S., occurs on publicly and privately protected lands including the San Pedro Riparian NCA, TNC's Patagonia-Sonoita Creek Preserve (and elsewhere in the Patagonia area), Buenos Aires NWR (Arivaca Creek), and Coronado NF (e.g., Pajarito Wilderness).

Actions: Continue to protect and restore riparian woodlands in southeastern Arizona.

BELL'S VIREO
Vireo bellii

GLOBAL POPULATION: 1.5 MILLION
75% BREED IN U.S.

TREND: DECREASING
WATCHLIST: HIGHEST CONCERN
ESA: ENDANGERED ("LEAST" SUBSPECIES)

17

Distribution: Four subspecies breed in dense early successional habitats in the central and southwestern U.S., and northern Mexico, wintering in similar habitats in western Mexico south to Nicaragua. Endangered *pusillus* ("Least"), the subspecies of most significant conservation concern, breeds in riparian habitats in southwestern California and northern Mexico, and winters in southern Baja California.

Threats: Cowbird nest parasitism, combined with habitat loss (90% of riparian systems present in 1850 are now lost), fragmentation, and degradation by over-grazing and invasive plants (e.g., salt cedar) are the most severe threats. "Least" territories located near urban or agricultural areas have lower reproductive success, perhaps due to increased levels of predation and parasitism. Some populations in riparian habitat in Arizona face similar threats to those in California.

Conservation: "Least" population has grown from 300 to 2,000 pairs since listing due to habitat restoration and cowbird control. It has also begun re-colonizing parts of its historic range, with breeding recently recorded in the San Joaquin River NWR.

Actions: Maintain and enhance existing riparian habitat, manage cowbirds and invasive plants; reduce predation rates.

BLACK-CAPPED VIREO
Vireo atricapilla

GLOBAL POPULATION: **8,000**
60% BREED IN U.S.

TREND: **DECREASING**
WATCHLIST: **HIGHEST CONCERN**
ESA: **ENDANGERED**

 20

Distribution: Breeds in low oak scrub, between two and 15 years after natural fire, from central Texas (Edwards Plateau) to Coahuila, Mexico, with small populations in Oklahoma. Winter range in western Mexico poorly known.

Threats: Conversion to agriculture, urbanization, and fire suppression have reduced available habitat by >90%, and also benefited Brown-headed Cowbirds, which parasitize vireo nests. Other threats include nest predation by invasive fire ants and feral cats.

Conservation: Occurs in the Witchita Mountains and Balcones Canyonlands NWRs, on Fort Hood (2,500+ adults), Kerr Wildlife Management Area, and Kickapoo Cavern State Park. A combination of prescribed burns, and cowbird trapping (reducing parasitism rates from 90% to 6%) has helped the species at Fort Hood. Safe Harbor agreements assist private landowners to manage their lands to benefit this species.

Actions: Increase the size of Balcones Canyonlands NWR to the proposed 46,000 acres. Restore scrub habitat using prescribed burns, and control cowbirds in breeding areas on both public and private lands. Assess threats to winter habitat.

GRAY VIREO
Vireo vicinior

GLOBAL POPULATION: **410,000**
90% BREED IN U.S.

TREND: **UNKNOWN**
WATCHLIST: **RARE**

16

Distribution: Breeds in juniper woodlands (sometimes mixed with pinyon or oak) and chaparral scrub-lands across the southwestern U.S. into northern Mexico. Winters in desert scrub, particularly where elephant trees (*Busera microphylla*) are present, or near streams; ranging from southern Baja California and southwestern Arizona through Sonora, Mexico; and in Big Bend NP, Texas, and adjacent Mexico.

Threats: Habitat loss and degradation (due to clearance for cattle pasture, pinyon firewood collection, fire suppression, and urban development) is the most severe threat. From 1950-1964, three-million acres of pinyon-juniper habitat were converted to cattle pasture in the U.S. Such habitat alterations also favor Brown-headed Cowbirds, which sometimes parasitize the nests of this species.

Conservation: Occurs within several protected areas in the U.S. (Canyons of the Ancients and Colorado NMs, Colorado; Zion and Arches NPs, Utah; and Big Bend NP and Kickapoo Cavern State Park, Texas) and in Mexico (El Pinacate, and El Vizcaino Biosphere Reserves).

Actions: Increase and restore juniper-scrub habitat by reducing grazing pressure and using prescribed burns.

FLORIDA SCRUB-JAY
Aphelocoma coerulescens

GLOBAL POPULATION: **6,500**
U.S. ENDEMIC

TREND: **DECREASING**
WATCHLIST: **HIGHEST CONCERN**
ESA: **THREATENED**

20

Distribution: Resident in fire-dependent oak scrub (five to 15 years post-burn) on dry sandy soils in peninsular Florida. The population was estimated to total 10,000 across 42 fragmented subpopulations in 1993, but many of these have since experienced rapid declines.

Threats: Habitat loss, fragmentation, and degradation due to fire suppression and conversion to agriculture and urban development. Birds in urban environments suffer predation rates (e.g., by cats) that exceed reproduction. Only 10-15% of the original habitat remains, much on private land where it continues to be degraded by fire suppression. Disease outbreaks are a periodic threat.

Conservation: This species depends on active management using prescribed burns or removal of vegetation where fire risk to nearby houses is high. Ocala NF, Merritt Island and Lake Wales Ridge NWRs, and Archbold Biological Station are important protected areas.

Goals: Protect and manage remaining natural scrub on private lands using Safe Harbor agreements and Farm Bill incentives. Manage public lands using controlled burns to increase suitable habitat. Upgrade ESA status to Endangered.

ISLAND SCRUB-JAY
Aphelocoma insularis

GLOBAL POPULATION: **9,000**
U.S. ENDEMIC

TREND: **STABLE**
WATCHLIST: **RARE**

17

Distribution: Endemic to Santa Cruz Island, California, where it is resident in woodland and chaparral dominated by oaks and Bishop pine, which cover less than 40% of the 60,000-acre island. It likely occupied other nearby Channel Islands historically.

Threats: This single island population is vulnerable to invasive species and emergent diseases, though currently the population appears stable. Island habitats were historically degraded by cattle, sheep, pigs (that competed with jays for acorns) and other invasive species. Over 200 introduced plant species also occur on the island.

Conservation: The entirety of Santa Cruz Island is protected, 76% by TNC, and the remainder within the Channel Islands NP. Both TNC and the NP Service have active programs to control or eradicate invasive species, removing cattle and sheep by 2000, pigs by 2006, and currently eradicating invasive plants. Following these eradications, native vegetation is recovering.

Goals: Continue efforts to restore native habitats, eradicate invasive species, prevent accidental introductions of additional invasive species, and prevent accidental fires.

PINYON JAY
Gymnorhinus cyanocephalus

GLOBAL POPULATION: **4.1 MILLION**
>95% BREED IN U.S.

TREND: **DECREASING**
WATCHLIST: **DECLINING**

15

Distribution: Occurs in large flocks in the western U.S., with a small local population in northern Baja California, Mexico. Mainly in pinyon-juniper woodland, but also in sagebrush, scrub-oak, chaparral, and locally in ponderosa and Jeffrey pine forests. Its seed-caching behavior is important for the regeneration of pinyon woodlands.

Threats: Habitat loss and degradation due to clearance for cattle grazing, fire suppression, and urban development are blamed for a substantial population decline from 1966-2005. Cattle grazing reduces fire frequency, which in turn increases vegetation density, ultimately leading to more severe fires that degrade habitat. Mortality due to West Nile virus also recorded.

Conservation: Occurs in several protected areas, including El Malpais NM, New Mexico; Grand Canyon NP, Arizona and Utah; Great Basin NP and Desert NWR, Nevada; and Mojave Desert National Preserve, California.

Goals: Increase and restore pinyon-juniper on public lands by reducing grazing pressure and using prescribed burns. Improve fire management in other pine habitats in southern California to reduce risk of catastrophic fires.

YELLOW-BILLED MAGPIE
Pica nuttalli

GLOBAL POPULATION: **180,000**
U.S. ENDEMIC

TREND: **DECREASING**
WATCHLIST: **RARE**

15

Distribution: Resident in California's oak savannas from the Central Valley to the foothills of the western Sierra Nevada, and along the coast from southeast San Francisco Bay to Santa Barbara and Ventura counties.

Threats: In 2004, West Nile virus arrived in California and may have killed nearly half the population from 2004-2006. Historical threats include habitat conversion to housing and agriculture. An introduced fungal pathogen (*Phytophthora*) causes sudden oak death syndrome and may affect this species, as acorns are part of its diet. The species has also been accidentally killed by poisons targeting ground squirrels.

Conservation: Occurs within several protected areas including East Park Reservoir, the Sacramento NWR complex, Consumnes River Preserve, Morro Bay, the San Luis NWR complex, and Great Valley Grasslands State Park.

Actions: Increase acreage of healthy oak woodlands; restrict use of poisons near breeding locations.

HAWAIIAN CROW
Corvus hawaiiensis

GLOBAL POPULATION: **60**
U.S. ENDEMIC

TREND: **EXTINCT IN THE WILD**
WATCHLIST: **HIGHEST CONCERN** 20
ESA: **ENDANGERED**

Distribution: Common in ohia-koa forests on Hawaii Island before the 1890s; fossil evidence suggests that crows were also once more widespread in the islands. Not reported from the wild since 2002, when only a single pair survived. Almost 60 individuals now remain in captive-breeding facilities on Maui and Hawaii.

Threats: Most habitat lost to agriculture and logging, or degraded by feral grazing animals, fire, and invasive plants. Farmers shot crows from the 1890s until the 1980s. They also likely suffered predation by Hawaiian Hawks and introduced rats, mongooses, and cats. Mosquito-transmitted diseases killed young birds, and made survivors susceptible to predation.

Conservation: Legally protected in Hawaii in 1931. Captive-breeding began in 1973. In the 1990s, 27 were released into the wild, but 21 died, and the surviving six birds were returned to captivity for protection in 1999.

Goals: Create new wild populations from captive stock and reduce threats to released birds, including controlling introduced predators and mosquitoes, reducing predation by native Hawaiian Hawks, and restoring habitat. Develop strategy to cope with West Nile virus if it reaches Hawaii.

ELEPAIO
Chasiempis sandwichensis

GLOBAL POPULATION: **237,000**
U.S. ENDEMIC

TREND: **DECREASING**
WATCHLIST: **HIGHEST CONCERN** 19
ESA: **ENDANGERED (OAHU SUBSPECIES)**

Distribution: Inhabits a variety of native and exotic forests in the Hawaiian Islands, preferring forests along rivers. Distinct subspecies occur on Kauai (*sclateri*) and Oahu (*ibidis*), with three subspecies on Hawaii (*sandwichensis*, *bryani*, and *ridgwayi*). The Oahu population numbers <2,000 birds.

Threats: Habitat loss historically, but many areas of suitable habitat are unoccupied due to nest predation by invasive rats, avian diseases spread by mosquitoes, or habitat fragmentation due to its typically low dispersal (<1 mile from the nest). Oahu subspecies occupies only 4% of historic range; fires ignited by military training threaten habitat at Schofield Barracks and Makua Reservation on the island.

Conservation: On Oahu, rat control is ongoing at several sites where populations are stable, though they decline rapidly without it (but could double in 19 years if both rats and disease could be controlled). On Mauna Kea (Hawaii), habitat restoration is aided by the removal of feral sheep and goats.

Actions: Expand rat control programs, continue disease research, and protect remaining forests from development and fire.

MEXICAN CHICKADEE
Poecile sclateri

GLOBAL POPULATION: **2 MILLION**
≤1% BREED IN **U.S.**

TREND: **UNKNOWN**
WATCHLIST: **RARE**

14

Distribution: Nests in tree cavities in coniferous and pine-oak forests of the Sierra Madre Occidental, Sierra Madre del Sur, and Sierra Volcánica Tranversal mountain ranges in Mexico. Small isolated populations occur in central Mexico, and in the upper elevations of the Chiricahua Mountains of Arizona (>300 pairs) and Animas Mountains of New Mexico (200-300 pairs). Birds may move down-slope during winter.

Threats: Habitat loss and degradation due to logging, fire suppression, and over-grazing are problems in Mexico.

Conservation: Habitat in the U.S. is largely protected within Coronado NF, Arizona; and Gray Ranch, New Mexico. The species occurs within several protected areas in Mexico.

Actions: Increase habitat protection and management efforts in Mexico; investigate the use of artificial nest boxes to boost U.S. populations.

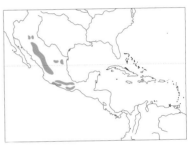

OAK TITMOUSE
Baeolophus inornatus

GLOBAL POPULATION: **900,000**
>95% BREED IN **U.S.**

TREND: **UNKNOWN**
WATCHLIST: **RARE**

14

Distribution: Resident cavity-nester in oak and pine-oak woodlands from southern Oregon south through California to Baja California, Mexico.

Threats: Loss and degradation of habitat for urban development, pasture, and agriculture, as well as fire suppression, over-grazing, and fuel-wood harvesting are the most severe threats. California's original oak woodlands have been reduced by more than 25%. In recent years, an introduced fungal pathogen (*Phytophthora*) "sudden oak death syndrome" has killed tens of thousands of oaks. Declines have also been noted following the advent of West Nile virus.

Conservation: Occurs within several protected areas on federal, state, and private lands including Camp Pendleton; San Bernardino, Los Padres, and Shasta-Trinity NFs; Vandenberg Air Force Base, Morro Bay, and Point Reyes National Seashore, California.

Actions: Increase acreage managed for healthy oak woodlands on public and private lands, through habitat restoration and educational outreach to private landowners.

MILLERBIRD
Acrocephalus familiaris

GLOBAL POPULATION: **800**
U.S. ENDEMIC

TREND: **FLUCTUATING**
WATCHLIST: **HIGHEST CONCERN** **18**
ESA: **ENDANGERED**

Distribution: Currently restricted to <100 acres of shrubby and grassy vegetation on the 156-acre island of Nihoa, Hawaii. c.1,500 Millerbirds once occurred on Laysan, where they became extinct by 1923. Nominate Laysan and Nihoa (*kingi*) subspecies have sometimes been considered separate species.

Threats: Extinction on Laysan was caused by elimination of vegetation by rabbits introduced in 1903. Numbers on Nihoa are likely limited by habitat area and food supplies that fluctuate with precipitation. Introduction of invasive species and avian diseases are the most serious potential threats. Outbreaks of the alien grasshopper *Schistocerca nitens* periodically defoliate vegetation and degrade habitat on Nihoa.

Conservation: Nihoa is uninhabited and protected as part of the Hawaiian Islands NWR. Restricted access helps guard against the accidental introduction of invasive species and fire. Work supported by ABC aims to establish a second population by translocating hatch-year birds from Nihoa during years when Nihoa population is high.

Actions: Establish a second wild population; manage Nihoa to prevent the accidental introduction of invasive species.

CALIFORNIA GNATCATCHER
Polioptila californica

GLOBAL POPULATION: **77,000**
8% BREED IN U.S.

TREND: **DECREASING**
WATCHLIST: **RARE** **14**
ESA: **THREATENED ("COASTAL") SUBSPECIES**

Distribution: Resident in coastal sagebrush scrub in southern California (3,000 pairs) south through Baja California, Mexico (including on some offshore islands), from sea level to 4,500 feet.

Threats: 70-90% of its habitat in southern California has been lost to housing development, including 33% from 1993-2001. Habitat loss is also problematic in this species' Mexican range, where introduced buffelgrass also increases fire frequency to the detriment of native scrub. Nest parasitism by Brown-headed Cowbirds is problematic in fragmented habitats.

Conservation: Northern subspecies listed as Threatened under the ESA in 1993 and >513,000 acres of Critical Habitat (83% on private lands) has been designated. California has a mitigation program that allows landowners to develop land when they set aside habitat for conservation; >89,000 acres of coastal scrub have so far been protected this way. Occurs within several protected areas including California's Camp Pendleton (600 pairs) and San Diego NWR complex (more than 500 pairs). Cowbird trapping occurs in many occupied areas.

Actions: Restore habitat and control invasive plants; evaluate the success of mitigation program.

KAMAO
Myadestes myadestinus

GLOBAL POPULATION: **UNKNOWN**
U.S. ENDEMIC

TREND: **LIKELY EXTINCT**
WATCHLIST: **HIGHEST CONCERN** | 20
ESA: **ENDANGERED**

Distribution: Inhabited forests throughout Kauai during the 1880s; disappeared from the lowlands by the 1920s; and was largely restricted to, but rare in, the Alakai Wilderness by the 1960s. Last sighting was in 1989 with last unconfirmed reports in 1993, despite intensive searches in 1995 and 1997.

Threats: Diseases (avian pox and malaria) are likely the causes of its rapid decline, but invasive predators (rats), competitors (birds, predatory insects), feral pigs, goats, and invasive plants may all have contributed. Hurricane Iniki in 1993 damaged forest and reduced other bird populations on Kauai, possibly killing the last of this species.

Conservation: Conservation actions targeting Puaiohi on Kauai, including locally controlling rats, may help any surviving Kamao.

Actions: Conduct surveys to locate any surviving populations, and if found, quickly assess management options, including controlling invasive predators at site and captive-breeding.

OLOMAO
Myadestes lanaiensis

GLOBAL POPULATION: **UNKNOWN**
U.S. ENDEMIC

TREND: **LIKELY EXTINCT**
WATCHLIST: **HIGHEST CONCERN** | 20
ESA: **ENDANGERED**

Distribution: Once inhabited forests throughout the islands of Lanai and Molokai. If considered the same species as the extinct Amaui, then it once occurred on Oahu and Maui as well. Last recorded in the highlands of Molokai in 1988. Surveys have since failed to locate this species.

Threats: Disease likely greatest cause of decline, but habitat degradation by invasive feral pigs, sheep, goats, and deer, as well as predation by introduced rats may all have contributed. Lanai populations disappeared by 1931, following construction of Lanai City and the introduction of avian diseases.

Conservation: Ungulate and weed control programs are underway in the Puu Alii Natural Area Reserve and TNC's Kamakou Preserve on Molokai, which might benefit any potential surviving Olomao.

Goals: Conduct surveys to locate any surviving populations, especially on the remote Olokui Plateau. If found, assess management options, including controlling invasive predators at site and captive-breeding. Meanwhile, eradicating feral pigs and controlling rats will benefit other native birds as well as any potential survivors of this species.

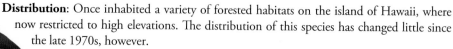

Omao
Myadestes obscurus

GLOBAL POPULATION: **170,000**
U.S. ENDEMIC

TREND: **STABLE**
WATCHLIST: **HIGHEST CONCERN**

14

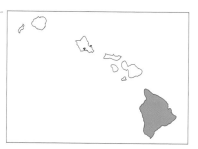

Distribution: Once inhabited a variety of forested habitats on the island of Hawaii, where now restricted to high elevations. The distribution of this species has changed little since the late 1970s, however.

Threats: Introduced diseases (pox, malaria) and predators (rats), combined with forest loss to logging, agriculture, cattle pasture, and urban development, are the primary threats. Cattle and goats degrade forest understory, and feral pigs facilitate the spread of disease-carrying mosquitoes. May be developing resistance to introduced diseases, but still largely restricted to high elevations where disease prevalence is low.

Conservation: Occurs within several protected areas including Hawaii Volcanoes NP and Hakalau Forest NWR. Efforts to fence out and remove pigs, sheep, goats, and cattle from protected areas has helped vegetation recover, but invasive rats remain problematic.

Actions: Protect and restore high-elevation forests on Hawaii, eradicate pigs and other invasive mammals within refuges, and use fences to prevent pigs and grazing animals from re-entering cleared areas. Investigate the potential for large scale rat control.

Puaiohi
Myadestes palmeri

GLOBAL POPULATION: **350**
U.S. ENDEMIC

TREND: **INCREASING**
WATCHLIST: **HIGHEST CONCERN**
ESA: **ENDANGERED**

18

Distribution: Now restricted to ten square miles of the Alakai Wilderness Preserve above 3,450 feet, on the Hawaiian island of Kauai.

Threats: Introduced avian diseases (pox, malaria) limit range at lower reaches of stream drainages; 48% of nest failures in one study were caused by rat predation. Introduced mammals (feral pigs, deer, goats) degrade understory. Invasive insects and introduced birds may compete for food. Climate change threatens to facilitate the spread of avian diseases upslope into the last refuge of many native birds on Kauai.

Conservation: Remaining populations occur within the Alakai Wilderness Preserve. Since 1995, a captive-breeding program has hatched >200 chicks and released 176, with released birds successfully breeding in the wild. Rat control benefits breeding success, but bait stations are logistically difficult to maintain in much of the Alakai.

Actions: Continue captive-breeding and release programs; control invasive species; increase the total population to 1,000 birds over the next 30 years.

BICKNELL'S THRUSH
Catharus bicknelli

GLOBAL POPULATION: **40,000**
>**90%** BREED IN U.S.

TREND: **UNKNOWN**
WATCHLIST: **HIGHEST CONCERN**

18

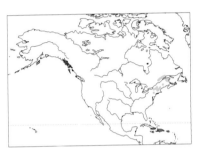

Distribution: Breeds in dense stands of fir on mountain tops and coasts from New York to Maine in the U.S., and from southeastern Quebec to Nova Scotia, Canada. Winters in the Caribbean, especially in moist broadleaf forest in the Dominican Republic.

Threats: Habitat destruction on wintering grounds (>90% lost on Hispaniola, >80% lost in Cuba, 75% lost in Jamaica); acid rain also damages breeding habitat and reduces food supplies.

Conservation: Nests in the High Peaks Wilderness Area of Adirondack Park, New York; and in White Mountain NF, New Hampshire; in the U.S., and in Canada's Cape Breton Highlands and Forillon NPs. Winters in the Dominican Republic's Sierra de Bahoruco NP (where ABC is working to expand and secure park boundaries).

Actions: Consider for ESA listing; protect and restore habitat on wintering grounds, reduce air pollution; ensure availability of large tracts of habitat in New Brunswick through forest management; improve population monitoring at breeding and wintering sites.

WOOD THRUSH
Hylocichla mustelina

GLOBAL POPULATION: **14 MILLION**
>**90%** BREED IN U.S.

TREND: **DECREASING**
WATCHLIST: **DECLINING**

14

Distribution: Nests in deciduous and mixed forests in the eastern U.S. into southeastern Canada, with highest abundance in the Appalachian Mountains. Winters in lowland tropical forests from southeastern Mexico to Panama.

Threats: Habitat loss and fragmentation in both breeding and wintering areas is likely the most serious threat. Nests in forest fragments or close to forest edges suffer greater predation rates. Birds unable to find territories in optimal habitat during winter (due to habitat loss) suffer lower survival rates. Habitat fragmentation facilitates nest parasitism by Brown-headed Cowbirds. Acid rain reduces snail densities required for food by breeding birds. Collisions with towers and glass windows during migration, and exposure to pesticides during winter are additional sources of mortality.

Conservation: Occurs within numerous protected areas on federal and state lands, including Adirondack Park, New York; Great Smoky Mountains NP, North Carolina; and many large NFs (Ozark, Ouachita, Monongahela, George Washington, Jefferson, Allegheny, Green Mountain).

Actions: Protect large tracts of mature forests in both breeding and wintering areas. Reduce threats from collisions, cats, pesticides, and acid rain.

VARIED THRUSH
Ixoreus naevius

GLOBAL POPULATION: **26 MILLION**
C.**35% BREED IN U.S.**

TREND: **DECREASING**
WatchList: **DECLINING**

14

Distribution: Nests in montane coniferous forests from Alaska south through western Canada to northern California and the northern Rockies, with highest abundance in the Cascades and northern Pacific coastal forests. Winters in similar habitats; uses non-forested areas at lower elevations during harsh winters. During winter irruptions, vagrant individuals can wander far to the east.

Threats: Loss and fragmentation of old-growth forests, particularly those dominated by cedar and hemlock, is likely the main threat, and birds suffer higher rates of nest predation near forest edges.

Conservation: This species benefits from old-growth forest protection for "Northern" Spotted Owls, and forest management that promotes diverse older forests. Occurs in many protected areas including Olympic NP, Washington; Glacier NP, Montana; Mount Hood and Willamette NFs, Oregon; and Tongass NF, Yukon Flats NWR, and Bering Land Bridge National Preserve, Alaska.

Actions: Prevent loss of remaining old-growth forest habitat to logging.

WRENTIT
Chamaea fasciata

GLOBAL POPULATION: **1.5 MILLION**
>**90% BREED IN U.S.**

TREND: **UNKNOWN**
WatchList: **RARE**

14

Distribution: Resident in coastal scrub and chaparral, but also uses thickets in urban parks and yards, oak woodland, and riparian and old-growth forest with dense understory, from Oregon south through California to northern Baja California, Mexico. Habitat for this species improves following deforestation; increased fire frequencies and warming climate apparently facilitate localized range expansions.

Threats: Habitat loss, fragmentation, and shrub removal for fire control due to urban development, and invasive plants (broom, eucalyptus) within this species' restricted range are the most severe threats. Loss of natural fire regimes that maintain healthy shrub ecosystems continues to reduce, isolate, and extirpate coastal Wrentit populations. Suffers predation by domestic cats, and nest parasitism by Brown-headed Cowbirds.

Conservation: Occurs in many protected areas, including Camp Pendleton, Los Padres and San Bernardino NFs, Sespe Condor Sanctuary, Morro Bay State Park, Point Reyes National Seashore, Redwoods NP, and Palomar Mountain State Park, among others.

Actions: Restore and protect fire regimes that maintain coastal scrub and chaparral ecosystems.

BENDIRE'S THRASHER
Toxostoma bendirei

GLOBAL POPULATION: **170,000**
75% BREED IN U.S.

TREND: **DECREASING**
WATCHLIST: **HIGHEST CONCERN**

17

Distribution: Breeds locally in open desert grasslands, desert scrub, and sagebrush with scattered junipers (at higher elevations) in the Sonoran and Mojave Deserts of the southwestern U.S. and northwestern Mexico. Northern populations retreat south during the winter.

Threats: Poorly understood, but likely include habitat loss and degradation due to urban development (e.g., near Tucson), agriculture (e.g., along the Gila River), invasive buffelgrass, off-road vehicle use, the harvesting of Joshua trees and yuccas, and over-grazing.

Conservation: The BLM manages the 25 million acre California Desert Conservation Area, which includes 3.5 million acres of wilderness. Much of this area is closed to mining, road construction, and off-road vehicles. This species also breeds within the Mojave Desert National Preserve, and Death Valley and Joshua Tree NPs, California; and occurs year-round in Cabeza Prieta and Buenos Aires NWRs, Organ Pipe Cactus NM, and Coronado NF, Arizona.

Goals: Research needed to clarify winter range, population size and trends, potential impacts or benefits of cattle grazing; identify habitat management techniques that might benefit the species.

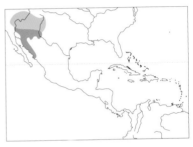

CALIFORNIA THRASHER
Toxostoma redivivum

GLOBAL POPULATION: **220,000**
90% BREED IN U.S.

TREND: **DECREASING**
WATCHLIST: **RARE**

16

Distribution: Resident in a variety of habitats along the coast and foothills of California south to northern Baja California, Mexico, with highest densities in chaparral; also occurs in coastal sage-scrub, riparian and oak woodlands, pine-juniper scrub, and urban parks.

Threats: Habitat loss, fragmentation, and degradation due to urban development and agriculture are the most severe threats; predation by feral cats; historically poisoned by pesticides applied to citrus orchards.

Conservation: Abundance is maximized within habitat patches >50 acres. Occurs in many protected areas including Camp Pendleton, Los Padres and San Bernardino NFs, Joshua Tree NP, and Butterbredt Spring Wildlife Sanctuary.

Actions: Continue efforts to restore, manage, and protect larger tracts of chaparral habitat in California.

LE CONTE'S THRASHER
Toxostoma lecontei

GLOBAL POPULATION: **190,000**
75% BREED IN U.S.

TREND: **UNKNOWN**
WATCHLIST: **RARE**

15

Distribution: Resident at low population densities in sparsely vegetated areas in the Mojave and Sonoran Deserts of the southwestern U.S., and adjacent northwestern Mexico.

Threats: Habitat loss due to agricultural development in lower Gila, Salt, San Joaquin, Imperial, and Coachella river valleys of Arizona and California; and to urban development near Phoenix, Las Vegas, and to the north of Los Angeles. Off-road vehicle use, fires, and over-grazing can also degrade habitat by destroying bushes and leaf litter.

Conservation: Little conservation effort focused directly on this species has occurred, but large areas of desert habitat are protected including the Carrizo Plain Natural Area and NM, Butterbredt Springs Wildlife Sanctuary, Mojave National Preserve, Joshua Tree NP, and Anza-Borrego Desert State Park, California; and Cabeza Prieta NWR, Arizona.

Goals: Develop habitat restoration techniques; research dispersal barriers, effect of competition with other thrasher species, limiting factors, and effects of drought.

SPRAGUE'S PIPIT
Anthus spragueii

GLOBAL POPULATION: **870,000**
C.20% BREED IN U.S.

TREND: **DECREASING**
WATCHLIST: **DECLINING**

15

Distribution: Breeds in large tracts (>350 acres) of open native short-grass prairie in the northern Great Plains of Canada and the U.S. Winters in open grassland, pasture, and fallow agricultural fields in the southern central U.S., south through central and eastern Mexico.

Threats: Continued loss and degradation of native grassland through conversion to agriculture, over-grazing, fire suppression, absence of bison, and invasive plants (e.g., leafy spurge) responsible for a 79% population decline from 1966-2005.

Conservation: Prairie restoration using prescribed burns and light cattle grazing has boosted numbers at Lostwood NWR. Habitat restoration activities are underway at several other NWRs where this species breeds (e.g., Chase Lake NWR, North Dakota) or winters (e.g., Attwater's Prairie-Chicken NWR, Texas). In Mexico, ABC and TNC are working with local partners to purchase ranches and protect grasslands (Saltillo grasslands, and Cuatro Ciénegas Valley).

Goals: This species needs coordinated efforts in Canada, the U.S., and Mexico, including habitat management recommendations for public lands and incentives for private lands.

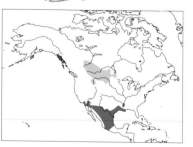

BACHMAN'S WARBLER
Vermivora bachmanii

GLOBAL POPULATION: **UNKNOWN**
U.S. BREEDING ENDEMIC

TREND: **LIKELY EXTINCT**
WATCHLIST: **HIGHEST CONCERN**
ESA: **ENDANGERED**

20

Distribution: Nested in the dense understory of canebreaks and blackberry thickets in the southeastern U.S.; wintered in Cuba and the Isle of Pines, and passed through south Florida during migration. Last confirmed in South Carolina's I'On Swamp in 1962, though subsequent reports keep faint hope alive.

Threats: Habitat loss on both breeding and wintering grounds likely caused populations to disappear. Canebreaks in bottomlands were often first to be cleared for agriculture. 80-85% of forests cleared in Cuba for agriculture, replacing some of the most potentially important wintering areas.

Conservation: Listed under the ESA in 1973, but an extensive search from 1975-1979 in South Carolina, Missouri, and Arkansas failed to detect survivors. Bottomland forest restoration is ongoing in the Mississippi Alluvial Valley and elsewhere; several protected areas could provide habitat for any surviving birds, including Francis Marion NF and Congaree Swamp NP in the U.S., and the Cienega de Zapata Biosphere Reserve in Cuba.

Actions: Conduct searches to follow up on any reported sightings to document existence.

BLUE-WINGED WARBLER
Vermivora pinus

GLOBAL POPULATION: **390,000**
>90% BREED IN U.S.

TREND: **DECREASING**
WATCHLIST: **RARE**

15

Distribution: Breeds in early successional habitats, ranging from the Midwest, east to New England and the Appalachians, and north to Ontario, Canada. In winter, joins mixed species flocks in tropical forests from Mexico to Panama.

Threats: Local population declines in the U.S. due to loss of breeding habitat as it matures into forest, or is converted to urban development. Hybridization with Golden-winged Warbler is also an issue. Loss of wintering habitat (e.g., conversion of shade coffee to sun coffee, forest clearing) may also be a factor. Predation by feral cats, collisions with communication towers, and nest parasitism by Brown-headed Cowbirds also recorded.

Conservation: Occurs on state and federal lands, including Ozark, Allegheny, George Washington, and Jefferson NFs.

Actions: Research survival on wintering grounds to test whether birds do better in undisturbed forests than in secondary forest. Maintain early successional breeding habitat through mechanical removal of vegetation, prescribed burns, and management of power line corridors. Continue forest restoration efforts and the maintenance of shade trees in coffee plantations on wintering grounds.

GOLDEN-WINGED WARBLER
Vermivora chrysoptera

GLOBAL POPULATION: **210,000**
C.**80% BREED IN U.S.**

TREND: **DECREASING**
WATCHLIST: **HIGHEST CONCERN**

17

Distribution: Breeds in early successional forest from the southern Appalachians to the Great Lakes and parts of New England, then into the upper Midwest and southern Canada where its greatest densities occur. The population is declining across the southeastern portion of its range, but stable in the upper Midwest. Winters in canopy of semi-open forest, from Guatemala to northern South America.

Threats: Loss of breeding habitat as shrub-lands mature into forest or are converted to housing developments, deforestation on wintering grounds, and hybridization with Blue-winged Warblers. Also suffers nest parasitism by Brown-headed Cowbirds.

Conservation: Occurs on public lands, including many NFs. ABC is working with private landowners to maintain early successional habitats for this species in Maryland and Pennsylvania.

Actions: Consider for listing under the ESA; identify and enhance critical breeding sites where hybridization with Blue-winged Warblers is minimal. Maintain early successional conditions using timber harvest or prescribed burning. Study potential benefits of cowbird control.

VIRGINIA'S WARBLER
Vermivora virginiae

GLOBAL POPULATION: **410,000**
U.S. BREEDING ENDEMIC

TREND: **DECREASING**
WATCHLIST: **RARE**

16

Distribution: Patchily distributed in pinyon-juniper, oak woodland, and other dry scrubby habitats in mountains of the western U.S, with highest abundance in Arizona, New Mexico, and Colorado. Winters in thorn scrub and tropical deciduous forests in the mountains of southwestern Mexico.

Threats: Loss of habitat in both breeding and wintering areas due to clearance for pasture, firewood collection, and housing development. Over-grazing removes ground cover required for nesting, and poorly managed fire regimes can negatively affect this species. Nest parasitism by Brown-headed Cowbirds increases in fragmented habitats. Lower reproductive success in drier years when birds may be displaced into areas with more predators.

Conservation: Little conservation effort is directed at this species, though it is protected on federal lands including Rocky Mountain NP, Colorado; Grand Staircase-Escalante NM, Utah; and several NFs (e.g., Kaibab, Coconino, Tonto, Apache-Sitgreaves, Gila), as well as within Mexico's Sierra de Manantlán Biosphere Reserve.

Actions: Increase habitat managed to benefit this species; decrease over-grazing on public lands.

COLIMA WARBLER
Vermivora crissalis

GLOBAL POPULATION: **25,000**
≤1% BREED IN U.S.

TREND: **UNKNOWN**
WATCHLIST: **RARE**

17

Distribution: Nests on the ground in oak and pine-oak woodlands with grassy ground cover from 5,000-10,500 feet elevation from the Chisos Mountains, Texas, south through the Sierra Madre Oriental mountains to San Luis Potosi, Mexico. In winter it joins mixed species flocks in brushy habitats at similar elevations in the mountains along the Pacific slope of southwestern Mexico.

Threats: In Mexico, habitat loss due to logging, land clearance for agriculture and pasture threaten this species, with perhaps 40% of pine-oak forests lost.

Conservation: Little conservation effort is directed at this species, though the entire U.S. breeding range is protected within Big Bend NP, Texas, where the population is also monitored. Occurs within additional protected areas in Mexico.

Actions: Increase research in Mexico to monitor population status and trends, study habitat use during migration and winter, and the impacts of habitat alterations on survival and reproductive success.

LUCY'S WARBLER
Vermivora luciae

GLOBAL POPULATION: **1.2 MILLION**
80% BREED IN U.S.

TREND: **STABLE**
WATCHLIST: **RARE**

15

Distribution: Nests in tree cavities and behind loose bark in mesquite, desert washes, and riparian woodlands in the southwestern U.S. and adjacent Mexico. Most winter in similar habitats along the Pacific coast of Mexico, with smaller numbers wintering in mesquite and salt cedar in the Big Bend region of Texas.

Threats: Cutting of mesquite, loss and degradation of riparian habitats due to groundwater pumping (to supply drinking water to urban areas, agricultural lands, and cattle ranches), dams changing natural flood regimes, over-grazing, and urban development have reduced or extirpated some populations in the U.S. (e.g., locally along the Gila and lower Colorado rivers).

Conservation: Occurs within many protected areas on federal lands and TNC reserves along the Colorado, San Pedro, and Gila Rivers, including Havasu NWR, California; and Cibola and Imperial NWRs, Grand Canyon NP, and the San Pedro Riparian and Gila Box NCAs, Arizona.

Actions: Protect and restore riparian habitat in the southwestern U.S., including restoring flood regimes, restricting cattle-grazing from riparian areas, and habitat protection.

GOLDEN-CHEEKED WARBLER
Dendroica chrysoparia

GLOBAL POPULATION: **21,000**
U.S. BREEDING ENDEMIC

TREND: **DECREASING**
WATCHLIST: **HIGHEST CONCERN** **20**
ESA: **ENDANGERED**

Distribution: Breeds in mature juniper-oak woodland in limestone hills and canyons of central Texas, primarily in the Edwards Plateau. Winters in pine-oak woodland in mountains (3,600-9,800 feet) from Chiapas, Mexico, to Nicaragua.

Threats: Conversion of breeding habitat to pasture, agriculture, and sprawl; over-browsing by deer; oak wilt fungus; predation by invasive fire ants; and Brown-headed Cowbird parasitism. Logging and agricultural expansion affect habitat on wintering grounds.

Conservation: Occurs in several protected areas, with the largest populations at Fort Hood (5,400 singing males) and Balcones Canyonlands NWR (800 singing males). Cow-

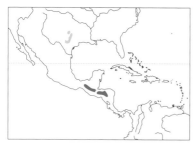

bird control programs are successful where implemented. Habitat on private lands has been protected under Safe Harbor agreements and other incentive programs. Occurs within several protected areas in Mexico, Guatemala, and Honduras during winter.

Actions: Increase Balcones Canyonlands NWR to the proposed 46,000 acres. Continue using fire to manage habitat while balancing the needs of this species against those of the Black-capped Vireo on publicly-owned lands; control deer populations and cowbird parasitism.

HERMIT WARBLER
Dendroica occidentalis

GLOBAL POPULATION: **2.4 MILLION**
U.S. BREEDING ENDEMIC

TREND: **STABLE**
WATCHLIST: **RARE**

15

Distribution: Nests high in the canopy of coniferous forests in the Coast, Cascade, and Sierra Nevada mountains of Washington, Oregon, and California. During winter, joins mixed species flocks in pine-oak and pine forests from Mexico to Nicaragua. Small numbers also winter along the southern California coast.

Threats: Extensive logging reduces closed canopy structure, and hence Hermit Warbler populations. In the Cascades, Townsend's Warblers are replacing this species through competition and there is also some hybridization.

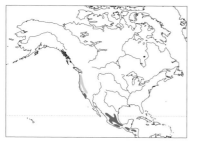

Conservation: Breeds in many protected areas including Olympic, Mount Hood, Willamette, Umpqua, Klamath, Shasta-Trinity, Tahoe, Eldorado, Stanislaus, and Sierra NFs; and Yosemite, Olympic, Kings Canyon, and Sequoia NPs, and during migration in Coronado NF.

Actions: Implement forestry practices (e.g., longer cutting rotations) to reduce edge, and maintain canopy closure; halt logging of old-growth.

GRACE'S WARBLER
Dendroica graciae

GLOBAL POPULATION: **2 MILLION**
50% BREED IN U.S.

TREND: **DECREASING**
WATCHLIST: **RARE**

15

Distribution: Nests in mature park-like stands of ponderosa pine, and pine-oak woodland in the southwestern U.S., south through Mexico and Central America to northern Nicaragua. In winter, birds from the U.S. and northern Mexico move south.

Threats: Habitat loss and degradation due to logging; fire suppression; conversion to agriculture (in Mexico); and locally, urban development, are the most severe threats. Range appears to be moving northward, perhaps due to climate change.

Conservation: Occurs in several U.S. NFs (Gila, Cibola, Cocinino, Kaibab, and Coronado), and also in protected areas outside the U.S. (e.g., the Sierra Gorda Biosphere Reserve, Mexico). Abundance increases in forests thinned to mimic naturally open habitats.

Actions: This is one of the least studied warblers in North America and it could benefit from research on the effects of fire management and other forestry practices on its population.

KIRTLAND'S WARBLER
Dendroica kirtlandii

GLOBAL POPULATION: **>3,000**
>99% BREED IN U.S.

TREND: **INCREASING**
WATCHLIST: **HIGHEST CONCERN**
ESA: **ENDANGERED**

20

Distribution: Breeds in large (>500 acres) six to 20 year old stands of jack pine in Michigan, with small numbers now also in Wisconsin and Ontario. Winters in scrub and pine habitats in the Bahamas (particularly Eleuthera), and the Turks and Caicos Islands.

Threats: Breeding habitat is dependent on natural wildfire to restart forest succession, but habitat fragmentation and fire suppression historically reduced habitat availability. Cowbirds once parasitized 70% of nests, but this has been reduced to 3% with control measures.

Conservation: Breeding habitat managed and cowbirds controlled at Kirtland's Warbler Management Area (150,000 acres of federal and state lands) near Mio, Michigan. >700,000 acres are protected within NPs in the Bahamas, where a research program on the wintering distribution and ecology is ongoing. Conservation efforts have helped the Michigan population grow from 200 to 1,700 singing males since the 1980s.

Actions: Increase breeding habitat through managed burns, clearcutting, and planting jack pines. Continue to control cowbirds at breeding areas. Determine management needs in wintering range.

PRAIRIE WARBLER
Dendroica discolor

GLOBAL POPULATION: **1.4 MILLION**
C.**99% BREED IN U.S.**

TREND: **DECREASING**
WATCHLIST: **DECLINING**

14

Distribution: Breeds in a variety of eastern early successional habitats, including coastal barrens, abandoned farmland, and power line clearings (with isolated populations in Ontario, Michigan, Iowa, and Kansas); winters in similar habitats in southern Florida and the Caribbean. Smaller *paludicola* subspecies resident in mangroves in south Florida.

Threats: Range expanded following early eastern deforestation, but decreased in recent decades as breeding habitat lost to forest maturation, fire suppression, and urban development. Pesticide spraying for mosquitoes is a threat to *paludicola*. Also susceptible to collisions, cowbird parasitism, and habitat loss in wintering areas.

Conservation: Benefits from management that maintains open shrub-land habitats, including prescribed burns and reclamation of strip mines. Occurs within numerous protected areas such as Everglades NP, Florida (both subspecies); Ouachita NF, Arkansas; Fort Campbell, Kentucky; Great Dismal Swamp NWR, Virginia; and on state lands in the pine barrens of New Jersey.

Actions: Assess effects of pesticides on mangrove breeders. Develop management guidelines to help maintain breeding habitat.

BAY-BREASTED WARBLER
Dendroica castanea

GLOBAL POPULATION: **3.1 MILLION**
≤**10% BREED IN U.S.**

TREND: **DECREASING**
WATCHLIST: **DECLINING**

15

Distribution: >90% of the population breeds in mature (>60 years) boreal forest in Canada, with <10% in the northern U.S. Breeding densities peak where spruce budworm outbreaks occur. Migrates through the eastern U.S., and winters in forested habitats and shade coffee in Costa Rica, Panama, Colombia, and Venezuela.

Threats: Forestry practices that favor young even-aged forests or trees resistant to budworm (e.g., jack pine, black spruce) over older forests reduce habitat and food supplies. Pesticide spraying (fenitrothion) for budworms disrupts brain function in this species. Also vulnerable to winter habitat loss and collisions during migration.

Conservation: Ontario and Quebec's 2008 announcements that they will protect large tracts of boreal forest will benefit this species. In the U.S., breeds within several protected areas, including Adirondack Park, New York; Superior NF, Minnesota; and Hiawatha NF, Michigan.

Actions: Continue habitat protection and restoration efforts in breeding and winter ranges, including the reduction of pesticide spraying. Assess population impacts of pesticides used to control budworm outbreaks.

CERULEAN WARBLER
Dendroica cerulea

GLOBAL POPULATION: **560,000**
>90% BREED IN U.S.

TREND: **DECREASING**
WATCHLIST: **DECLINING**

16

Distribution: Nests in broken canopy of large tracts of mature deciduous forest along rivers, and on mountain slopes and ridges, in eastern North America. In winter, inhabits moist mid-elevation forest (1,600-6,000 feet) in the Andes from Venezuela to Peru (also some records from Bolivia). Central American highlands provide important staging areas during spring migration.

Threats: Habitat destruction and fragmentation due to resource extraction (e.g., timber, coal) in the breeding range, and to sun-coffee and illicit coca production in the Andes. Also subject to collisions and other threats facing long distance migrants.

Conservation: Occurs within protected areas in the U.S. (Royal Blue WMA, Montezuma NWR; Ozark, Mark Twain, and Shawnee NFs) and in its winter range. ABC and its partners are protecting habitat in the Andes and Appalachia, and addressing collisions range-wide.

Actions: Consider for listing under the ESA. Reduce habitat loss throughout range and implement timber management that maintains complex canopy structure required for breeding. Maintain habitat in wintering and staging areas, including shade coffee. Address collisions and other threats facing migrants.

PROTHONOTARY WARBLER
Protonotaria citrea

GLOBAL POPULATION: **1.8 MILLION**
c.100% BREED IN U.S.

TREND: **DECREASING**
WATCHLIST: **DECLINING**

15

Distribution: Nests in tree cavities in large tracts (>250 acres) of mature swamp forest in the eastern U.S., with <100 pairs in southern Ontario, Canada. Winters mainly in mangroves and adjacent forests from Mexico to northern South America.

Threats: Loss of bottomland forests; mangroves cleared for shrimp farms, coastal development, agriculture, and charcoal (for e.g., Colombia lost 64% from 1956-1995). Also vulnerable to collisions and pesticide exposure.

Conservation: Laws prohibit cutting of mangroves in Costa Rica and Venezuela, but are weakly enforced. Forest cover has greatly increased in the Mississippi Alluvial Valley since 1950. Nest boxes have been used to increase densities locally. Occurs within many protected areas such as Great Dismal Swamp NWR, Virginia; White River NWR, Arkansas; and in Latin America in Colombia's Cienega Grande de Santa Marta sanctuary, Venezuela's Morocoy NP, and Costa Rica's Humedal Nacional Terraba-Sierpe.

Actions: Restore bottomland forests in U.S., increase protection and restoration efforts for mangroves in Latin America.

SWAINSON'S WARBLER
Limnothlypis swainsonii

GLOBAL POPULATION: **84,000**
U.S. BREEDING ENDEMIC

TREND: **INCREASING**
WATCHLIST: **RARE**

14

Distribution: Nests near the ground in dense vegetation such as canebreaks along upper edges of bottom-land swamp forests of the southeast, as well as locally in rhododendron thickets in the Appalachians and Arkansas' Boston Mountains. Winters in a variety of forested habitats, including mangroves, in the Yucatan Peninsula of Mexico; and in Belize, Jamaica, and Cuba.

Threats: Historic loss of 90% of southeast bottomland forests, particularly the loss of canebreaks, combined with habitat loss on wintering grounds were likely the most severe threats. Residential development in West Virginia has also reduced suitable habitat. Vulnerable to collisions during migration.

Conservation: Through protection and restoration efforts, forest cover has greatly increased in the Mississippi Alluvial Valley since 1950. Occurs within many protected areas in the U.S. (e.g., Great Dismal Swamp and White River NWRs) and in its wintering range (e.g., Jamaica's Blue and John Crow Mountains NP).

Actions: Continue restoration of canebreaks and bottomland forests in U.S., increase protection and restoration efforts for mangroves and forests in winter range.

KENTUCKY WARBLER
Oporornis formosus

GLOBAL POPULATION: **1.1 MILLION**
U.S. BREEDING ENDEMIC

TREND: **DECREASING**
WATCHLIST: **DECLINING**

14

Distribution: Nests near the ground in dense understory vegetation of deciduous forest in the eastern U.S., especially in bottomland hardwood forests. Winters in tropical lowland and foothill forest from southeastern Mexico to northern South America.

Threats: Habitat loss on both breeding and wintering grounds due to conversion to agriculture, pasture, and urban development; browsing of understory by over-abundant white-tailed deer. Vulnerable to nest parasitism and nest predation in habitat fragments, and to collisions during migration.

Conservation: Occurs within many protected areas in the U.S. (e.g., Tensas River NWR, Daniel Boone NF) and in Latin America (e.g., Mexico's Calakmul Biosphere Reserve, Guatemala's Maya Biosphere Reserve, Belize's Rio Bravo Conservation Area, and Panama's Darien NP). Loss of 90% of bottomland forest in the southeast U.S. must have greatly reduced populations over the last century, but through protection and restoration efforts, forest cover has since greatly increased, for e.g. in the Mississippi Alluvial Valley.

Actions: Protect and restore habitat on breeding and wintering grounds, including management for dense understory that may include deer population control.

CANADA WARBLER
Wilsonia canadensis

GLOBAL POPULATION: **1.4 MILLION**
16% BREED IN **U.S.**

TREND: **DECREASING**
WATCHLIST: **DECLINING**

14

Distribution: Nests on or near the ground in forest with dense understory across much of Canada's boreal region, and through the upper Midwest and New England to the Appalachians. In winter, occurs in forests within and to the east of the Andes from Venezuela to Peru.

Threats: Reduction of understory on breeding grounds due to forest maturation or browsing by over-abundant deer, as well as deforestation on wintering grounds are likely the most severe threats. Vulnerable to collisions during migration, nest parasitism by cowbirds, and likely to pesticide spraying for spruce budworm.

Conservation: Recent protection of large tracts of boreal forest will benefit this species. Over one-million acres of private lands are protected under forestry agreements and easements in Maine. Breeds in several protected areas in the U.S., including Adirondack Park, New York; and Great Smoky Mountains NP, North Carolina. ABC and its partners are working to maintain shade coffee systems, and restore degraded areas through silvipasture practices in the Andes.

Actions: Protect habitat in both breeding and winter ranges, including the reduction of pesticide spraying in both areas.

RED-FACED WARBLER
Cardellina rubrifrons

GLOBAL POPULATION: **430,000**
25% BREED IN **U.S.**

TREND: **UNKNOWN**
WATCHLIST: **RARE**

15

Distribution: Nests on the ground in pine and pine-oak forests in mountains (6,500-9,200 feet elevation) from Arizona and New Mexico south through the Sierra Madre Occidental into central Mexico. Winters in similar habitats in mountains from Mexico though Central America to Honduras. Migratory status at southern end of range is uncertain.

Threats: Disappears from forests following selective logging. Range appears to be expanding northward into Colorado, perhaps due to climate change.

Conservation: In the U.S., occurs on public lands including Coconino NF (Mogollon Rim) and Coronado NF (Madera Canyon, Chiricahua Mountains), Arizona.

Goals: Much like Grace's Warbler, this species could benefit from research both inside and outside the U.S. on reproductive success, wintering areas used by U.S. breeders, and the effects of fire management and other forestry practices on populations.

ABERT'S TOWHEE
Pipilo aberti

GLOBAL POPULATION: **230,000**
90% BREED IN U.S.

TREND: **UNKNOWN**
WATCHLIST: **RARE**

14

Distribution: Resident in dense mesquite and riparian woodland in the Sonoran Desert of Arizona, extending locally into neighboring states and Mexico.

Threats: Only 5-10% of Arizona's riparian woodlands remain. Habitat degraded due to clearance for agriculture, over-grazing by cattle, dams, channelization, and water withdrawals for drinking supplies and cattle ranches. Vulnerable to nest parasitism by Brown-headed Cowbirds. Occupies exotic vegetation along ditches in Tucson and Phoenix, as well as invasive salt cedar, but these habitats are not optimal.

Conservation: Occurs within several protected areas in California and Arizona, including NWRs along the lower Colorado River (Havasu, Bill Williams River, Cibola, and Imperial), and in Cabeza Prieta NWR and Organ Pipe Cactus NM, Buenos Aires NWR, TNC's Sonoita Creek Preserve, and San Pedro Riparian NCA. Densities nearly doubled in five years after cattle were removed from the San Pedro. Benefits from conservation of "Southwestern" Willow Flycatcher.

Actions: Restore desert riparian habitats, including reducing grazing pressure and ensuring sufficient water flows.

RUFOUS-WINGED SPARROW
Aimophila carpalis

GLOBAL POPULATION: **74,000**
12% BREED IN U.S.

TREND: **FLUCTUATES**
WATCHLIST: **RARE**

15

Distribution: Resident in thorn-brush, bunchgrass, and scrub in the Sonoran Desert of northwestern Mexico, from Sinaloa through Sonora, north into south-central Arizona. Arizona populations fluctuate, with breeding closely tied to rainfall.

Threats: Habitat loss and degradation due to over-grazing by cattle and land clearance for agriculture; urban development near Tucson.

Conservation: Little conservation attention directed at this species, but occurs on public lands and within protected areas, including Buenos Aires NWR, Organ Pipe Cactus NM, and Coronado NF.

Actions: Research potential effects of climate change on rainfall patterns and reproductive success. Develop and implement a conservation plan for this species with recommendations for habitat management on public lands.

BACHMAN'S SPARROW
Aimophila aestivalis

GLOBAL POPULATION: **250,000**
U.S. ENDEMIC

TREND: **DECREASING**
WATCHLIST: **HIGHEST CONCERN**

17

Distribution: Nests and forages near ground in open mature pine savannas of the southeastern U.S. Winters in the same habitat, but with northern populations moving south.

Threats: Habitat loss, fragmentation, and degradation due to logging of old-growth forests, fire suppression, clearance for agriculture, and urban development.

Conservation: Safe Harbor agreements and other conservation measures for the endangered Red-cockaded Woodpecker have also benefitted this species, as their habitat requirements are similar. Removing hardwoods and using controlled burns at three to six year intervals creates useable habitat. Occurs in many protected areas across the southeast including Apalachicola, Osceola, Angelina, Frances Marion, and Croatan NFs; Withlacoochee State Forest, Fred C. Babcock WMA, Fort Bragg, Weymouth Woods State Park, and TNC's Piney Grove Reserve.

JUV.

Actions: Implement a range-wide conservation plan for this species that includes incentives for habitat protection, including prescribed burns on public and private lands.

FIVE-STRIPED SPARROW
Aimophila quinquestriata

GLOBAL POPULATION: **200,000**
<1% BREED IN U.S.

TREND: **DECREASING**
WATCHLIST: **RARE**

16

Distribution: Resident in grassy desert scrub and tropical deciduous woodland in canyons and hillsides of the Sierra Madre Occidental mountains of northwestern Mexico; 60-70 breed in southern Arizona.

Threats: Threats are not well understood and potentially include disturbance by birders using playback (in Arizona), cattle grazing, mining, and nest parasitism by Brown-headed Cowbirds.

Conservation: Little conservation attention is devoted to this species, but most U.S. breeders occur in remote canyons within Coronado NF (e.g., California Gulch).

Actions: Implement a conservation plan with recommendations for habitat management on public lands.

BREWER'S SPARROW
Spizella breweri

GLOBAL POPULATION: **16** MILLION
>**75% BREED IN U.S.**

TREND: **DECREASING**
WATCHLIST: **DECLINING**

13

Distribution: Subspecies *breweri* breeds in sagebrush across the western U.S. and adjacent southern Canada, wintering from the southwestern U.S. to central Mexico; darker and grayer *taverneri* ("Timberline Sparrow") breeds in alpine scrub in southeastern Alaska and western Canada, likely overlapping with *breweri* in winter.

Threats: Destruction and fragmentation of sagebrush caused by agricultural expansion, over-grazing, altered fire regimes, invasive plants, and energy development. Frequent fires in some areas of sagebrush facilitate habitat conversion to invasive cheatgrass, whereas fire suppression allows growth of pinyon-juniper woodland.

Conservation: Protected areas include Malhuer NWR, Oregon; Ruby Lake NWR, Nevada; and Gray's Lake NWR, Idaho; wintering areas protected within TNC's Gray Ranch, New Mexico; Cabeza Prieta NWR and Organ Pipe Cactus NM, Arizona; and Mexico's Gran Desierto de Altar Biosphere Reserve. >50% of remaining sagebrush habitat in the U.S. is on BLM land.

Actions: Incorporate this species' needs into BLM land management plans and leases. Identify wintering areas for *taverneri*. Determine how efforts to manage habitat for sage-grouse can benefit this species and others dependent upon sagebrush.

BLACK-CHINNED SPARROW
Spizella atrogularis

GLOBAL POPULATION: **390,000**
80% BREED IN U.S.

TREND: **DECREASING**
WATCHLIST: **HIGHEST CONCERN**

16

Distribution: Breeds in open arid brush-land and chaparral in the southwest U.S. south through central Mexico, from sea level to 8,800 feet. Some northern populations migrate down-slope to deserts during winter; inhabiting thorn scrub, mesquite, and cactus. Three of four subspecies breed within the U.S.

Threats: Habitat degradation due to mining, off-road vehicles, fire suppression, and over-grazing (especially in wintering range) threaten this species. Population trends are unknown in Mexico, declining in California, and generally stable elsewhere, peaking after wet winters.

Conservation: Occurs within several protected areas including Guadalupe Mountains NP and Davis Mountains, Texas; TNC's Gray Ranch, New Mexico; Coronado NF, Arizona; and Vandenberg Air Force Base, California.

Actions: Restore California chaparral ecosystems, including using prescribed burns. Reduce over-grazing in grasslands within winter range. Assess causes and determine significance of irruptions in California, and the potential effects of climate change.

Sage Sparrow
Amphispiza belli

Global population: **4.3 million**
90% breed in U.S.

Trend: **Decreasing**
WatchList: **Declining**
ESA: **Threatened (San Clemente subspecies)**

14

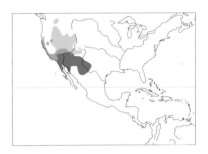

Distribution: Dark subspecies: *belli*, *clementeae*, and *cinerea*, are resident along the coast of California, on San Clemente Island, and in Baja California, Mexico; *canescens* breeds in southern and central California; the paler *nevadensis* breeds in the Great Basin, wintering in the southwest U.S. and northwestern Mexico.

Threats: Destruction and degradation of sagebrush caused by agricultural expansion, over-grazing, altered fire regimes, invasive cheatgrass, and energy development. Habitat fragmentation may increase cowbird nest parasitism. Coastal development reduces chaparral, and over-grazing by feral pigs, goats, and cattle degrades vegetation on San Clemente Island.

Conservation: Sagebrush protection and restoration benefits this species. The San Clemente Island population is listed under the ESA, and likely benefits from restoration efforts for the "San Clemente" Loggerhead Shrike, including the restriction of military target shelling.

Actions: Incorporate this species' needs into BLM land management plans. Restore natural fire frequencies to sagebrush and chaparral (reduce in Great Basin, increase in California). Control cheatgrass. Remove cattle and goats from San Clemente.

Lark Bunting
Calamospiza melanocorys

Global population: **27 million**
c.**80%** breed in U.S.

Trend: **Decreasing**
WatchList: **Declining**

13

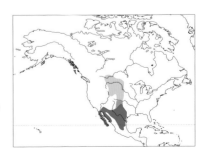

Distribution: Breeds in short- and mixed-grass prairies, agricultural lands, and sagebrush in the Great Plains of the U.S. and Canada. Distribution varies annually according to rainfall. Winters in arid scrub-lands, grasslands, playas, prairie-dog towns, agricultural lands, and cattle feed lots in the southwest U.S. through central Mexico.

Threats: Habitat loss, fragmentation, and degradation due to conversion of grasslands to agriculture; over-grazing by cattle in short-grass prairies, absence of natural fire regimes and herbivores (bison, prairie-dogs). Also vulnerable to pesticide poisoning (e.g., diazinon used to control grasshoppers), and drowning in algae-covered stock tanks.

Conservation: Many breeding and wintering areas are protected on public lands, including many NWRs (e.g., Lostwood, Salyer), National Grasslands (Comanche, Cimarron, Fort Pierre), and BLM lands. The Conservation Reserve Program provides incentives for habitat protection on many acres of private lands.

Actions: Strengthen Conservation Reserve Program and other incentives to protect and restore native grasslands; delay mowing of hayfields until after breeding, and avoid over-grazing short-grass prairies; develop stock tank mitigation methods; reduce pesticide use.

BAIRD'S SPARROW
Ammodramus bairdii

GLOBAL POPULATION: **1.2 MILLION**
C.**20% BREED IN U.S.**

TREND: **DECREASING**
WATCHLIST: **HIGHEST CONCERN**   **17**

Distribution: Breeds in large (>155 acres) mixed-grass and short-grass prairies in Canada (79% of population) and in the northern Great Plains of the U.S. Winters in Chihuahuan desert grasslands of northern Mexico and adjacent areas of Arizona, New Mexico, and Texas.

Threats: Continued loss and degradation of native grasslands due to conversion to crops, over-grazing, fire suppression that facilitates growth of shrubs, absence of bison, and invasive plants (e.g., leafy spurge, western snowberry). Also vulnerable to elevated levels of nest parasitism by cowbirds in habitat fragments, and to early mowing in agricultural fields.

Conservation: Habitat management has increased local populations in Saskatchewan, and restoration activities are underway at several NWRs. Breeding densities in North Dakota peak two to four years after controlled burns. Occurs on protected public lands such as Little Missouri NG, North Dakota; Bowdoin and Medicine Lake NWRs, Montana; and Las Cienegas NCA, Arizona.

Actions: Provide incentives for habitat restoration on private lands.

HENSLOW'S SPARROW
Ammodramus henslowii

GLOBAL POPULATION: **79,000**
>**95% BREED IN U.S.**

TREND: **DECREASING**
WATCHLIST: **HIGHEST CONCERN**   **17**

Distribution: Breeds in tall-grass prairies, on reclaimed strip mines, and in hay fields of the Midwest, and locally in grasslands in the eastern U.S. and southern Ontario. Winters in pine savannas and prairies of southeastern U.S.

Threats: Severe loss of breeding (c.90% loss of tall-grass prairie since 1850) and wintering habitat (>95% loss of southeastern pines and prairies) to row crops and pasture, fire suppression, and urban development. Also vulnerable to pesticide exposure.

Conservation: Conservation Reserve Program provides incentives for habitat protection on private lands and helps this species. Recent spikes in agricultural commodity prices have caused landowners to drop out of this program though. Benefits from habitat restoration for Greater Prairie-Chicken and Red-cockaded Woodpecker, as well as reclamation of strip mines in the East. Protected sites include Big Oaks NWR, Indiana; Fort Riley, Kansas; and Fort Campbell, Kentucky; wintering areas include Apalachicola NF, Florida; and Croatan NF, North Carolina.

Actions: Develop and implement a range-wide conservation plan, including rotational grazing and prescribed burns; strengthen Conservation Reserve Program. Manage selected reclaimed strip mines for this species.

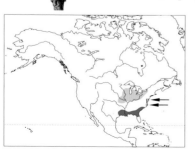

LE CONTE'S SPARROW
Ammodramus leconteii

GLOBAL POPULATION: **2.9 MILLION**
C.10% BREED IN U.S. (WINTER ENDEMIC)

TREND: **DECREASING**
WATCHLIST: **DECLINING**

14

Distribution: Breeds in wet prairies, marshy meadows, and damp hayfields from central and eastern Canada south to the extreme north-central U.S. Winters in similar wet grassland habitats and coastal prairies in the southern U.S. Commoner in the western portions of its range.

Threats: Continued loss and degradation of native grasslands due to conversion to crops, over-grazing, and fire suppression. Early haying or mowing can reduce reproductive success in hayfields.

Conservation: Benefits from habitat restoration on lands in the Conservation Reserve Program and from prescribed burns, with densities peaking two years after fire in some areas. Many NWRs and NFs

harbor breeding and wintering populations, including Medicine Lake, Lostwood, Upper Souris, and Seney NWRs; and Apalachicola and Sam Houston NFs, as well as many refuges and barrier islands along the Texas coast.

Actions: Continue habitat protection and restoration efforts, especially within wintering range; use rotational grazing and prescribed burns to manage grasslands.

NELSON'S SPARROW
Ammodramus nelsoni

GLOBAL POPULATION: **510,000**
C.8% BREED IN U.S. (WINTER ENDEMIC)

TREND: **INCREASING**
WATCHLIST: **RARE**

14

subvirgatus

Distribution: Subspecies *nelsoni* breeds in freshwater marshes in the northern Great Plains of Canada and north-central U.S.; *alterus* in brackish marshes of Hudson Bay; *subvirgatus* in saltmarshes of the Canadian Maritime provinces south to Maine. Winters in coastal marshes from New York to Texas.

Threats: Loss, degradation, and fragmentation of marsh habitat due to diking, draining, coastal development, conversion to agriculture, and invasive *Phragmites*. Climate change is predicted to raise sea levels, further reducing and fragmenting coastal habitats. Also, potentially vulnerable to contaminants (e.g., high mercury levels found in blood).

nelsoni

Conservation: U.S. habitat protected on many public lands, including Lostwood NWR, North Dakota; Agassiz NWR and Chippewa NF, Minnesota; Assateague Island National Seashore, Maryland; South Carolina's ACE Basin, and many coastal NWRs in Texas.

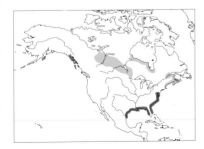

Actions: Develop and implement a range-wide conservation plan that includes protecting large remaining tracts of pristine saltmarsh (especially for *subvirgatus*); combat effects of climate change.

SALTMARSH SPARROW
Ammodramus caudacutus

GLOBAL POPULATION: **250,000**
U.S. ENDEMIC

TREND: **DECREASING**
WATCHLIST: **HIGHEST CONCERN**

18

Distribution: Inhabits coastal saltmarshes, breeding from southern Maine to Virginia, and wintering along the Atlantic coast from North Carolina to the Florida Panhandle.

Threats: Loss, degradation, and fragmentation of saltmarsh habitat due to diking, draining, coastal development, and invasive *Phragmites*. Climate change is predicted to increase storms and raise sea levels, further threatening coastal marshes. Potentially vulnerable to contaminants such as mercury.

Conservation: Occurs within many local, state, and federal protected areas including Monomoy NWR, Massachusetts; Blackwater NWR and Fishing Bay WMA, Maryland; and Cumberland Island National Seashore, Georgia. Many NGOs and government agencies are working on coastal wetland restoration and protection projects.

Actions: Develop and implement a range-wide conservation plan that includes protecting remaining large tracts of pristine saltmarsh; research effects of habitat restoration, and combat effects of climate change.

SEASIDE SPARROW
Ammodramus maritimus

GLOBAL POPULATION: **110,000**
U.S. ENDEMIC

TREND: **DECREASING**
WATCHLIST: **HIGHEST CONCERN**
ESA: **ENDANGERED ("CAPE SABLE" SUBSPECIES)**

17

Distribution: Found in coastal marshes from New Hampshire to Texas. Resident "Cape Sable" *mirabilis* restricted to brackish marshes of the Everglades; "Dusky" *nigrescens* of Florida became extinct in 1980s.

Threats: Loss, degradation, and fragmentation of saltmarshes due to diking, draining, coastal development, and invasive *Phragmites*. Insecticide spraying to control mosquitoes likely contributed to the extinction of the "Dusky" subspecies. Sea level rise could further limit habitat.

Conservation: "Cape Sable" race listed under the ESA and intensely managed, with 3,184 individuals largely within Everglades NP and adjacent Big Cypress National Preserve in 2007, down from 6,656 before Hurricane Andrew struck in 1992. Efforts to restore water flows, and controlled burns with five to ten year intervals should benefit "Cape Sable" populations. Atlantic and Gulf Coast populations each occur within multiple coastal refuges (e.g., Bombay Hook NWR, Gulf Islands National Seashore, Texas coastal refuges).

Actions: Protect large remaining tracts of undiked saltmarshes; research effects of habitat restoration; combat effects of climate change. Restore natural water flows and fire regimes to the Everglades and other marshlands.

Smith's Longspur
Calcarius pictus

GLOBAL POPULATION: <75,000
C.8% BREED IN U.S. (WINTER ENDEMIC)

TREND: UNKNOWN
WatchList: Rare

14

Distribution: Nests in wet meadows with groups of conifers in the transition zone between tundra and forest in Alaska and northern Canada. Winters across grasslands, pastures, airports, and agricultural fields of the south-central U.S.

Threats: Threats to breeding habitat are likely minimal, though climate change could affect habitat in the future. Other longspurs are vulnerable to collisions with towers during migration so this species could potentially also be affected.

Conservation: In the U.S., breeds within the National Petroleum Reserve, and Arctic and Yukon Flats NWRs.

Actions: Develop and implement a conservation plan for this species. Identify important wintering and staging areas.

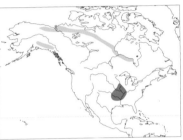

Chestnut-collared Longspur
Calcarius ornatus

GLOBAL POPULATION: 5.6 MILLION
75% BREED IN U.S.

TREND: DECREASING
WatchList: Declining

14

Distribution: Nests in large tracts (>115 acres) of moderately- to heavily-grazed or recently burned short-grass and mixed-grass prairie, and occasionally in pastures and on airstrips, in the northern Great Plains of the U.S. and Canada. Disappeared from much of Kansas, Nebraska, Minnesota, and North Dakota. Winters in grasslands and deserts of the central and southwest U.S. and central Mexico.

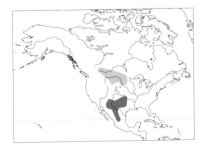

Threats: Conversion of native prairies to cropland combined with fire suppression, absence of native grazers (bison, pronghorn, prairie dogs), invasive plants, and urban and oil and gas development have greatly reduced and degraded habitat for this species. Also vulnerable to pesticide exposure.

Conservation: Occurs on many public lands, including BLM lands, National Grasslands (e.g., Pawnee, Thunder Basin, Cimarron), and NWRs (Benton Lake, Bowdoin, Medicine Lake, Upper Souris).

Actions: Increase habitat for this species using prescribed burns, rotational grazing, restoring native grazers, and mowing mixed-grass prairies. Research effects of different habitat management treatments and fragmentation caused by oil and gas development.

McKAY'S BUNTING
Plectrophenax hyperboreus

GLOBAL POPULATION: **31,200**
U.S. ENDEMIC

TREND: **STABLE**
WATCHLIST: **RARE**

16

Distribution: Endemic to Alaska. Breeds only in rocky tundra on Hall and St. Matthew Islands in the Bering Sea, and winters along Bering Sea coastal marshes and on shingle beaches of western Alaska.

Threats: Currently secure, but vulnerable to the introduction of invasive predators or reindeer that could damage vegetation on breeding islands. Reindeer introduced to St. Matthew in 1944 destroyed much of the native vegetation before they died out. Bunting breeding densities are currently still lower on St. Matthew than on Hall. Potential threats also include oil development, contaminants, and the effects of climate change.

Conservation: Hall and St. Matthew Islands are protected within the Alaska Maritime NWR. Birds winter within the Yukon Delta and Togiak NWRs.

Actions: Regular population monitoring to assess status and trends, and research on breeding success and potential effects of climate change.

VARIED BUNTING
Passerina versicolor

GLOBAL POPULATION: **2 MILLION**
≤5% IN U.S.

TREND: **DECREASING**
WATCHLIST: **DECLINING**

14

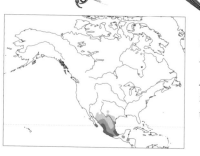

Distribution: Inhabits arid scrub, thorn forest, desert riparian areas, and other scrubby woodlands from Guatemala north through much of Mexico, and across the U.S. border into Texas, New Mexico, and Arizona, with northern populations retreating to central Mexico during winter.

Threats: Captured for caged bird trade in Mexico, where other threats likely include over-grazing and habitat loss, but are poorly known. Mining, urban development, road building, and border security measures threaten to reduce habitat in the U.S.

Conservation: The exclusion of cattle from Guadalupe Canyon, New Mexico, increased bunting territories during the 1990s. In U.S., protected sites include Coronado NF, Arizona; and Kickapoo Cavern State Park, Devils River State Natural Area, and TNC's Dolan Falls Preserve, Texas.

Actions: Increase efforts to regulate the trade in wild songbirds in Mexico. Improve population monitoring to assess status and trends, and identify additional important areas to protect.

PAINTED BUNTING
Passerina ciris

GLOBAL POPULATION: **4.5 MILLION**
80% BREED IN **U.S.**

TREND: **DECREASING**
WATCHLIST: **DECLINING**

14

Distribution: Two disjunct populations: eastern birds breed in swampy thickets, forest edges, coastal scrub and maritime hammocks along the U.S. south Atlantic coast, and winter in southern Florida, Cuba, and the Bahamas; western birds breed in early-successional scrub and riparian thickets in the south-central U.S. and adjacent Mexico, and winter from Mexico through Central America to Panama.

Threats: Historically trapped for caged bird trade in the U.S., and thousands are still trapped on wintering grounds. Loss of habitat (especially along coasts in Southeast, and riparian areas in Southwest). Also vulnerable to collisions during migration and nest parasitism by cowbirds.

Conservation: Breeds in several protected areas, including Cumberland Island National Seashore and the ACE Basin in the East; and Wichita Mountains and Balcones Canyonlands NWRs, King Ranch, and various coastal refuges in Texas.

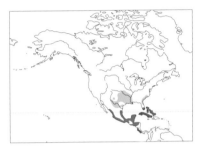

Actions: Stop trapping for the caged bird trade throughout wintering range (especially in Cuba and Mexico) and regulate international trade under CITES. Protect remaining coastal habitat and increase breeding habitat through management.

TRICOLORED BLACKBIRD
Agelaius tricolor

GLOBAL POPULATION: **395,000**
99% BREED IN **U.S.**

TREND: **UNKNOWN**
WATCHLIST: **HIGHEST CONCERN**

18

Distribution: Breeds in huge colonies in cattail marshes and agricultural fields near water; requires suitable foraging habitat close to colonies. Occurs year-round in California's Central Valley and in southern California.

Threats: Habitat loss and degradation caused by human activities. Over 90% of California's Central Valley wetlands have been destroyed. Tens of thousands of nests can be destroyed by harvesting at the wrong time. Vulnerable to pesticide exposure, and intentionally poisoned as a pest until the 1960s. Climate change is predicted to reduce water availability in its range and could further degrade habitat.

Conservation: Programs to compensate farmers for delaying harvests have saved thousands of nests from being destroyed, but are expensive. Breeding colonies occur in several protected areas including NWRs (Sacramento, Merced, San Luis, Kern), Wind Wolves Preserve, Cosumnes River Preserve, Suisun Marsh, and East Park Reservoir. The protection of 240,000 acres of the Tejon Ranch in 2008 will benefit this species.

Actions: Restore and manage habitat, including incentives for private landowners to maintain colonies on their property; protect breeding sites; reduce colony disturbance by predators.

RUSTY BLACKBIRD
Euphagus carolinus

GLOBAL POPULATION: **2 MILLION**
C.**15% BREED IN U.S. (WINTER ENDEMIC)**

TREND: **DECREASING**
WATCHLIST: **DECLINING**

13

Distribution: Breeds near wetlands in conifer forests from Alaska though Canada to the northern fringes of Michigan, New York, Vermont, New Hampshire, and Maine. Winters in bottomland forests and swamps, and sometimes with other blackbird species in agricultural areas of the southeast U.S.

Threats: Causes for the species' decline are unclear, but may include decreased food availability in the East, and the drying of wetlands in the West. Lethal control of blackbird flocks as pests also kills this species. Extensive loss of wintering and breeding habitat has occurred historically.

Conservation: Recent protection of large tracts of Canadian boreal forest may help this species. Occurs in many other protected areas, including Canadian NPs (Wood Buffalo, Wapusk, La Mauricie), and U.S. NWRs (Yukon Delta, White River, Cache River, Mingo, Pocosin Lakes, ACE Basin, and Tennessee).

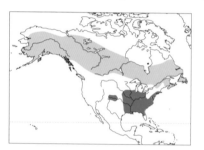

Actions: Determine causes of decline, and develop appropriate response. Assess impact of blackbird poisoning programs.

AUDUBON'S ORIOLE
Icterus graduacauda

GLOBAL POPULATION: **200,000**
% IN U.S.: **≤5%**

TREND: **UNKNOWN**
WATCHLIST: **RARE**

16

Distribution: Resident in a range of habitats from thorn and riparian forests in south Texas to pine-oak, tropical deciduous, and cloud forests in Mexico. Mainly in larger tracts of habitat, though occasionally strays into suburban yards.

Threats: Habitat fragmentation, clearance for agriculture, and urban development, facilitate nest parasitism by cowbirds. This has led to the disappearance of the species from much of the Lower Rio Grande Valley, including Laguna Atacosa and Santa Ana NWRs.

Conservation: Little conservation action is directed specifically at this species, though habitat restoration efforts in the Lower Rio Grande Valley could help. The oriole remains regular in Kennedy County on large ranches such as King Ranch, and also visits feeders for nectar, fruit, and peanut butter.

Actions: Prevent fragmentation and restore habitat along the Rio Grande Valley where nests are less likely to be parasitized; control cowbird numbers within protected areas where this species occurs; encourage maintenance of shade coffee plantations in Mexico.

BLACK ROSY-FINCH
Leucosticte atrata

GLOBAL POPULATION: **20,000**
U.S. ENDEMIC

TREND: UNKNOWN
WATCHLIST: RARE

16

Distribution: Nests in rock crevices, caves, and on cliffs of snow-capped mountains in the Great Basin and Rocky Mountains (Idaho, Montana, Oregon, Wyoming, Nevada, and Utah). Winter range expands down slope into arid valleys of the interior western U.S.

Threats: Climate change is predicted to reduce alpine habitat as the tree line moves upward. Winter flocks gathered by roadsides are potentially vulnerable to collisions with vehicles.

Conservation: Breeding habitat is protected on many public lands, including Great Basin, Yellowstone, and Grand Teton NPs; and Wallowa-Whitman, Custer, Shoshone, and Bridger-Teton NFs.

Actions: Improve population monitoring to assess validity of observed winter population declines, and determine factors that regulate population. Combat climate change.

BROWN-CAPPED ROSY-FINCH
Leucosticte australis

GLOBAL POPULATION: **45,000**
U.S. ENDEMIC

TREND: UNKNOWN
WATCHLIST: RARE

16

Distribution: Nests in rocky crevices, caves, and on cliffs near snow fields and glaciers of the southern Rocky Mountains, from southern Wyoming through Colorado to northern New Mexico. Winter range expands down slope into arid valleys in Colorado and New Mexico.

Threats: Climate change is predicted to reduce alpine habitat as tree line moves up slope, especially at southern end of range. Winter flocks gathered by roadsides are potentially vulnerable to collisions with vehicles.

Conservation: Breeding habitat is protected on public lands, including Rocky Mountain NP and Arapaho NF, Colorado. May use abandoned mine buildings above the tree line for nesting, and bird-feeders in urban areas during winter.

Actions: Improve population monitoring to assess validity of observed winter population declines, and determine factors that regulate population. Combat climate change.

LAWRENCE'S GOLDFINCH
Spinus lawrencei

GLOBAL POPULATION: **150,000**
90% BREED IN U.S.

TREND: **STABLE**
WATCHLIST: **RARE**

15

Distribution: Breeds in oak woodlands and chaparral, near open fields and water, in California west of the Sierra Nevadas, and in northern Baja California, Mexico. Breeding sites vary greatly year-to-year, complicating efforts to assess population status and trends. Usually winters in the southern portion of its breeding range, but also irrupts eastward through the arid lands of Arizona, New Mexico, northern Mexico, and rarely to western Texas.

Threats: Habitat loss and degradation due to fire suppression, conversion to agriculture, and urban development. An introduced fungal pathogen causes "sudden oak death syndrome" and may affect this species' habitat.

Conservation: Agriculture, grazing, and other human activities that disturb the landscape may benefit plants upon which this species feeds. Occurs within many protected areas, including East Park Reservoir, South Fork Kern River Valley, Los Padres and San Bernardino NFs, Camp Pendleton, and Johsua Tree NP.

Actions: Determine validity and causes of observed breeding population declines. Further research is also needed to better understand winter irruption behavior.

LAYSAN FINCH
Telespiza cantans

GLOBAL POPULATION: **c.11,500**
U.S. ENDEMIC

TREND: **FLUCTUATES**
WATCHLIST: **HIGHEST CONCERN**
ESA: **ENDANGERED**

18

Distribution: Inhabits <523 acres of bunchgrass on Laysan Island (with small numbers on Pearl and Hermes Atoll from a 1967 translocation), in the northwestern Hawaiian Islands. Population numbers fluctuate with weather conditions.

Threats: The Laysan population declined to c.100 from 1903-1923 after introduced rabbits all but eradicated Laysan's vegetation, but rebounded after rabbits were eradicated. Vulnerable to non-native diseases, droughts, hurricanes, fires, and to sea level rise caused by climate change. The arrival of rats on Midway during World War II wiped out an introduced population of finches there.

Conservation: Laysan Island is protected within the Papahānaumokuākea Marine NM, with restricted access. Native bunchgrass is doing well following eradication of an invasive grass during the 1990s. Invasive golden crownbeard benefits Laysan Finches, but harms nesting seabirds and must be removed.

Actions: Continue monitoring on occupied islands, maintain protocols to prevent introducing additional invasive species and diseases; establish new populations on other suitable islands (Lisianski, Midway) following eradication of rodents and mosquitoes; breed in captivity as a backup.

Nihoa Finch
Telespiza ultima

GLOBAL POPULATION: c.3,000
U.S. ENDEMIC

TREND: FLUCTUATES
WatchList: HIGHEST CONCERN **18**
ESA: ENDANGERED

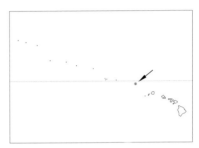

Distribution: Inhabits <100 acres of grasses and shrubs on the 156-acre Nihoa Island in the northwestern Hawaiian chain.

Threats: Vulnerable to non-native diseases (avian malaria and pox) which could be brought to the island by introduced vagrant birds (e.g., House Finch, Northern Cardinal) from the main Hawaiian Islands. Outbreaks of the alien grasshopper *Schistocerca nitens* periodically defoliate vegetation and degrade habitat on Nihoa. Also vulnerable to droughts, hurricanes, and fires.

Conservation: Nihoa is uninhabited and protected within the Papahānaumokuākea Marine NM. Annual population counts take place when weather allows landing on Nihoa. Restricted access to the island helps guard against the accidental introduction of invasive species and risk of fire. Attempts to introduce this species to French Frigate Shoals in 1967 were unsuccessful.

Actions: Continue population monitoring, maintain strict protocols to prevent introducing invasive species and diseases, and establish new populations on Kure Atoll and other islands (e.g., Midway, Niihau, Kahoolawe) following eradication of rodents and mosquitoes; breed in captivity as a backup.

Ou
Psittirostra psittacea

GLOBAL POPULATION: UNKNOWN
U.S. ENDEMIC

TREND: LIKELY EXTINCT
WatchList: HIGHEST CONCERN **20**
ESA: ENDANGERED

Distribution: Formerly among the commonest Hawaiian birds. It preferred ohia forests with ieie vines (*Freycinetia arborea*). Disappeared from all islands, except Kauai and Hawaii by 1931. Population estimates included three birds on Kauai and c.400 on Hawaii during the 1970s. Last reported from Hawaii in 1987 and Kauai in 1989; unconfirmed reports since 1995 from both islands.

Threats: The Ou's nomadic behavior increased exposure risk to introduced mosquito-borne diseases (malaria, pox). Forest clearance for agriculture, habitat degradation by feral pigs and grazing animals, and competition for ieie fruits from (and predation by) introduced rats contributed to declines. The last remnant populations may have been wiped out by habitat damage caused by Hurricane Iniki on Kauai in 1992, and by lava flows on Hawaii.

Conservation: Conservation efforts were generally too little too late for this species. Forest restoration (including ieie vines) and invasive species control at protected areas benefit other species and any Ou possibly remaining.

Actions: Conduct surveys in unsearched areas of potential habitat in an attempt to locate any remaining birds.

PALILA
Loxioides bailleui

GLOBAL POPULATION: **2,200**
U.S. ENDEMIC

TREND: **DECREASING**
WATCHLIST: **HIGHEST CONCERN**　**18**
ESA: **ENDANGERED**

Distribution: Occurs in dry mamane forests on the island of Hawaii. Approximately 95% of the population occurs in a 12 square mile area on the southwestern slope of Mauna Kea (6,600-9,400 feet).

Threats: Habitat loss to cattle ranches, degradation by feral sheep and goats, predation by rats and cats, introduced wasps parasitize caterpillars used to feed nestlings, and invasive plants heighten fire risk. Malaria and other mosquito-borne diseases likely restrict Palila from re-colonizing lower elevations.

Conservation: 46,000 sheep, goats, and other ungulates have been removed from the Mauna Kea Forest Reserve, but some sheep, cats, and rats remain. A captive-breeding program was launched in 1996, and 15 birds released in 2003; more recent translocations have also taken place. ABC is working to ensure the completion of habitat fencing.

Actions: Restore habitat; continue forest restoration efforts; construct fencing to enclose the key area for the species; control predators; remove ungulates and invasive weeds from inside the fence; install dip tanks to help fight fires. Continue to establish new populations and monitor their success.

MAUI PARROTBILL
Pseudonestor xanthophrys

GLOBAL POPULATION: **c.500**
U.S. ENDEMIC

TREND: **STABLE**
WATCHLIST: **HIGHEST CONCERN**　**18**
ESA: **ENDANGERED**

Distribution: Currently restricted to 19 square miles of wet forest (4,000-7,700 feet) on the northern and eastern slopes of Maui's Haleakala Volcano.

Threats: Mosquito-borne diseases limit habitat use at elevations below 4,500 feet. Habitat loss and degradation by feral pigs, goats, deer, and nest predation by rats further limit the population. Invasive wasps parasitize native insects, reducing food supplies. Parrotbills lay one-egg clutches and frequent storms cause high rates of nest loss.

Conservation: All currently occupied habitat is managed by the East Maui Watershed Partnership, and fencing and ungulate control efforts are restoring forests. A portion of the current range (Hanawi Natural Area) is fenced, ungulate free, and rat controlled. A captive-breeding program started in 1997 may eventually bolster the population, or establish new populations. Recent surveys suggest a possible increase in restored areas.

Actions: Establish a second population in eastern Maui with the objective of doubling the population, accompanied by fencing, ungulate removal, and restoration of degraded forests; continue captive-breeding program.

OAHU AMAKIHI
Hemignathus flavus

GLOBAL POPULATION: **40,000**
U.S. ENDEMIC

TREND: INCREASING
WATCHLIST: HIGHEST CONCERN

17

Distribution: Inhabits a variety of native and exotic forests in the Waianae and Koolau Mountains on the Hawaiian island of Oahu, but originally occurred throughout the island. Recent surveys suggest population increases and a range expansion to lower elevations.

Threats: Habitat loss, introduced mosquito-borne diseases, invasive predators, feral pigs and grazing animals likely all harm the Oahu Amakihi. This species appears to be evolving resistance to avian malaria, however.

Conservation: Little conservation action is directed at this species, but it likely benefits from efforts such as reforestation and other habitat restoration projects directed towards the Oahu subspecies of the Elepaio.

Actions: Improve population monitoring to assess status and trends. Continue researching disease resistance and competition with introduced birds. Continue habitat protection and restoration efforts.

KAUAI AMAKIHI
Hemignathus kauaiensis

GLOBAL POPULATION: **75,000**
U.S. ENDEMIC

TREND: INCREASING
WATCHLIST: HIGHEST CONCERN

16

Distribution: Inhabits native ohia-koa forests above 2,000 feet on Kauai, where it forages for insects under tree bark. It remains one of the most common native birds, though it once also occurred in the lowlands.

Threats: Forest clearance for agriculture and urban development, the introduction of invasive predators, mosquito-borne diseases, and feral pigs and other exotic ungulates likely contributed to historic population declines. This species appears to be developing resistance to avian malaria, however.

Conservation: Occurs within several protected areas, including the Alakai Wilderness, Kokee State Park, Waimea Canyon State Park, and Hono O Na Pali Natural Area Reserve. It has benefited from its ability to feed on the nectar of an introduced vine, the banana poka.

Actions: Control or eradicate invasive species (especially within the Alakai Wilderness), monitor population trends, and conduct research on disease resistance.

NUKUPUU
Hemignathus lucidus

GLOBAL POPULATION: **UNKNOWN**
U.S. ENDEMIC

TREND: **LIKELY EXTINCT**
WATCHLIST: **HIGHEST CONCERN**
ESA: **ENDANGERED**

 20

Distribution: Three subspecies once inhabited dry to wet ohia and koa forests: on Kauai *hanapepe*, on Oahu *lucidus*, and on both Maui and Molokai *affinis*, each possibly representing separate species.

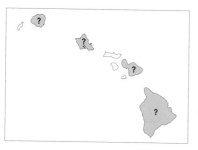

Last reported on Kauai from Kokee State Park and Alakai Wilderness Preserve, but surveys from 1981-2000 failed to observe the species; unconfirmed sightings were reported there until 1996, however. On Maui, a single bird was last observed each year from 1994-1996 on the northeast slope of Haleakala, but more recent surveys have failed to locate any there.

Threats: Most severe threats were introduced mosquito-borne diseases (malaria, pox); but also forest loss, degradation by feral grazers, and invasive predators and competitors.

Conservation: Conservation efforts were generally too little and too late for this species. Forest restoration and invasive species control at protected areas benefit other species and any Nukupuu that may possibly remain.

Actions: Locate any remaining individuals and implement immediate conservation measures.

AKIAPOLAAU
Hemignathus munroi

GLOBAL POPULATION: **1,200**
U.S. ENDEMIC

TREND: **DECREASING**
WATCHLIST: **HIGHEST CONCERN**
ESA: **ENDANGERED**

20

Distribution: Endemic to the island of Hawaii, where it inhabits mesic to wet montane koa and ohia forest. Disappeared from mamane-naio forests of Mauna Kea after 2000. Restricted from suitable habitat below 4,000 feet due to mosquitoes.

Threats: Most severe threats are introduced mosquito-borne diseases (malaria, pox), but it is also threatened by habitat loss and degradation by feral pigs and cattle, and by other invasive species that predate nests or compete for food.

Conservation: Most occur within Hakalau Forest NWR and Kau Forest Reserve, with smaller numbers in Olaa-Kilauea Management Area, Kapapala Forest Reserve, Kipuka Ainahou Nene Sanctuary, and other protected areas in Hawaii's forested uplands. Removal of sheep from Mauna Kea, and cattle from the Kapapala Forest Reserve, has helped forest regeneration. Habitat restoration within Hakalau Forest NWR, includes planting 350,000 koa seedlings (and 30,000 other native plants) since 1989, fencing, and feral pig removal.

Actions: Control or eradicate invasive species, including controlling pigs and draining cattle stock ponds to reduce mosquitoes; research broad-scale rat control options; continue forest restoration.

ANIANIAU
Magumma parva

GLOBAL POPULATION: **44,359**
U.S. ENDEMIC

TREND: **INCREASING**
WATCHLIST: **HIGHEST CONCERN**

17

Distribution: Once occurred throughout native forests of Kauai, but now largely restricted to upper elevations.

Threats: Primarily threatened by invasive species, including mosquito-borne diseases (malaria, pox), habitat degradation by feral pigs and deer, predation by invasive mammals (rats, cats), and competition for food with introduced insects (yellow jackets, ants). Mosquitoes have spread to all parts of Kauai, though their numbers, and malaria infection rates are currently limited by low highland temperatures. Climate change may allow transmission to accelerate, however.

Conservation: Most habitat is legally protected in the Alakai Wilderness Preserve and the adjacent Kokee State Park, but control of feral ungulates is difficult due to the terrain.

Actions: Manage habitat in the Alakai Wilderness; fence key areas and control or eradicate invasive species (pigs, deer, predators); identify any disease-resistant populations that might exist for potential captive-breeding programs.

AKIKIKI
Oreomystis bairdi

GLOBAL POPULATION: **1,312**
U.S. ENDEMIC

TREND: **DECREASING**
WATCHLIST: **HIGHEST CONCERN**
ESA: **ENDANGERED**

20

Distribution: Historically common at all elevations on the island of Kauai, but currently restricted to 14 square miles of ohia forest, mainly in the Alakai Wilderness Preserve above 4,000 feet elevation. From 1970-2000, its range contracted >50% and its population declined >80%.

Threats: Habitat degradation by feral pigs and deer, and predation by rats and cats. Mosquito-borne diseases restrict this species to the highest elevations. Mosquitoes have spread to all parts of Kauai, though their numbers, and malaria infection rates are currently limited by low highland temperatures. Climate change could accelerate infection rates, however. Competition with exotic birds, as well as invasive predatory insects also threaten this species.

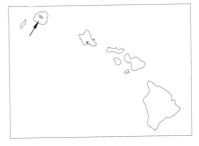

Conservation: Most habitat is legally protected in the Alakai Wilderness Preserve and the adjacent Kokee State Park, but control of feral ungulates is difficult due to the terrain, and so far ineffective. Following a petition initiated by ABC, FWS listed this species under the ESA.

Actions: Evaluate captive-breeding program; manage habitat in the Alakai Wilderness and control or eradicate invasive species (deer, pigs, predators).

HAWAII CREEPER
Oreomystis mana

GLOBAL POPULATION: **6,300**
U.S. ENDEMIC

TREND: **DECREASING**
WATCHLIST: **HIGHEST CONCERN** 20
ESA: **ENDANGERED**

Distribution: Forages along trunks in undisturbed, wet ohia and koa forests (and occasionally in dry mamane forests), from 5,000-6,230 feet on the island of Hawaii. Most of the population is now restricted to the windward slope of Mauna Kea around Hakalau Forest NWR.

Threats: Mosquito-borne diseases (malaria, pox) make habitat below 5,000 feet unsuitable. This species suffers low reproductive success, possibly due to nest predation by rats. Habitat loss and degradation by feral pigs and cattle are further threats.

Conservation: Habitat restoration efforts within Hakalau Forest NWR include the planting of 350,000 koa seedlings (and 30,000 other native plants) since 1989, along with fencing, and feral pig removal. This species has been successfully propagated in captivity, and six were released at Kipuka 21 in 2007 to establish a new population.

Goals: Remove ungulates from recovery areas, especially Kau Forest Reserve which holds the second largest population (about 2,200 birds), and continue habitat protection and restoration. Reduce mosquito breeding sites by draining ponds built for cattle; expand rodent control.

OAHU ALAUAHIO
Paroreomyza maculata

GLOBAL POPULATION: **Unknown**
U.S. ENDEMIC

TREND: **LIKELY EXTINCT**
WATCHLIST: **HIGHEST CONCERN** 20
ESA: **ENDANGERED**

Distribution: Inhabited mixed ohia and koa forests from 1,000-2,000 feet (but not mountain summits), on the island of Oahu. Last well documented observations of this species were of two birds in 1985 on the Poamoho Trail.

Threats: Habitat loss, mosquito-borne diseases, and nest predation by invasive rats all likely contributed to declines. 59% of Oahu is occupied by urban development or agriculture, especially in the lowlands. No portions of Oahu are high enough in elevation to provide refuge from mosquito-vectored diseases, and disease likely eliminated this species from its remaining habitat. A road project through some of the last occupied habitat may also have contributed to the species' demise.

Conservation: Habitat is protected in the Oahu Forest NWR in the Koolau Mountains. The recent rediscovery of small Iiwi populations on Oahu, and the lack of recent systematic surveys, lend some hope that the Oahu Alauahio may still persist.

Actions: Conduct systematic surveys of this species on Oahu to search for any surviving birds.

MAUI ALAUAHIO
Paroreomyza montana

GLOBAL POPULATION: **15,000**
U.S. ENDEMIC

TREND: **DECREASING**
WATCHLIST: **HIGHEST CONCERN**

18

Distribution: Currently restricted to native (ohia, mamane) and exotic pine forests on the slopes of Haleakala Volcano on Maui, but formerly more widespread on Maui, Lanai, and Molokai. Extinct Lanai subspecies was last seen in 1937.

Threats: Introduced mosquito-borne diseases (pox, malaria) likely limit populations from using suitable habitat at elevations below 5,250 feet, and in west Maui. Habitat loss and degradation by feral pigs, goats, and deer; fires, and nest predation by rats further limit populations.

Conservation: Little conservation attention is devoted to this species, but it likely benefits from forest restoration directed at other birds (Maui Parrotbill, Akohekohe). It occurs within the Waikamoi Preserve, Hanawi Natural Area, Polipoli State Park, and at Kahikinui on the south side of Maui.

Actions: Develop and implement a conservation plan for this species, including improved population monitoring to assess status and trends, and potentially by translocating birds to establish new populations within Maui. Continue restoration and reforestation efforts where this species occurs at high elevations, including the eradication of invasive predators and fencing out ungulates.

AKEKEE
Loxops caeruleirostris

GLOBAL POPULATION: **3,500**
U.S. ENDEMIC

TREND: **DECREASING**
WATCHLIST: **HIGHEST CONCERN**
ESA: **ENDANGERED**

20

Distribution: This species was restricted to 34 square miles of forest on the island of Kauai up to 2000, but has since disappeared from much of this tiny range, and declined from 7,800 birds in 2000 to 3,500 birds in 2007.

Threats: Invasive species, including mosquito-borne diseases (malaria and pox), habitat degradation by feral ungulates (pigs, goats, deer), predation by invasive mammals (rats and cats), and competition for food with introduced insects (yellow jackets, ants). Mosquitoes have spread across Kauai, and climate change threatens to increase disease transmission at high elevations.

Conservation: Most habitat is protected in the Alakai Wilderness Preserve and adjacent Kokee State Park, but the control of feral ungulates is difficult due to the terrain, and so far ineffective. Following a petition initiated by ABC, FWS listed this species under the ESA.

Actions: Develop a recovery plan; continue efforts to manage habitat in the Alakai Wilderness including fencing and controlling or eradicating invasive species. Expand captive-breeding programs to include the Akekee.

AKEPA
Loxops coccineus

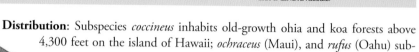

GLOBAL POPULATION: **14,000**
U.S. ENDEMIC

TREND: **STABLE**
WATCHLIST: **HIGHEST CONCERN**
ESA: **ENDANGERED**

20

Distribution: Subspecies *coccineus* inhabits old-growth ohia and koa forests above 4,300 feet on the island of Hawaii; *ochraceus* (Maui), and *rufus* (Oahu) subspecies are presumed extinct, with the last record on Maui in 1988.

Threats: Mosquito-borne diseases make habitat at lower elevations unsuitable. Habitat loss, fragmentation, and degradation by feral pigs and cattle likely limit both foraging opportunities and cavity nest sites. Invasive rats are also problematic.

Conservation: Highest densities occur within Kau Forest Reserve and Hakalau Forest NWR. Habitat restoration efforts in Hakalau have included planting 350,000 koa seedlings (and 30,000 other native plants) since 1989, fencing, and feral pig removal. Akepa will nest in artificial cavities, and have been successfully propagated in captivity; 12 were released at Kipuka 21 in 2007 to establish a new population.

Actions: Remove feral ungulates and other invasive species; expand artificial nest site program until regenerating ohia trees provide sufficient natural cavities. Research small populations at low elevations that might have resistance to mosquito-borne diseases, and could be useful for future captive-breeding and release programs.

IIWI
Vestiaria coccinea

GLOBAL POPULATION: **350,000**
U.S. ENDEMIC

TREND: **STABLE OR DECREASING**
WATCHLIST: **HIGHEST CONCERN**

16

Distribution: Currently inhabits native and non-native forests (1,000-9,500 feet) with highest densities from 4,200-6,200 feet. Once occurred in ohia-koa forests on all major Hawaiian islands down to sea level, but now extinct on Lanai, and only tiny (<100) relict populations survive on Oahu and Molokai.

Threats: Birds travel up and down slope in search of nectar, and may become exposed to mosquito-borne diseases at lower elevations. Although they depend on nectar-bearing trees, they also use understory plants which are reduced by introduced grazing animals. Also threatened by other invasive species, and by extensive forest clearance at lower elevations.

Conservation: Forest restoration and the control of invasives benefit this species, particularly fencing, and other efforts targeting feral pigs which create wet depressions where mosquitoes can breed. Occurs within many protected areas, including Kauai's Alakai Wilderness, Molokai's Kamakou Preserve, Maui's East Maui Watershed; on Hawaii, the Kau Forest Reserve and Hakalau Forest NWR are especially important. Also held in captivity.

Actions: Control or eradicate invasive species within Hawaiian forests; restore forest habitats; research disease resistance.

AKOHEKOHE
Palmeria dolei

GLOBAL POPULATION: **3,800**
U.S. ENDEMIC

TREND: **STABLE**
WATCHLIST: **HIGHEST CONCERN**
ESA: **ENDANGERED**

18

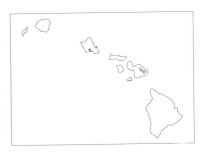

Distribution: Once distributed throughout the forests of Maui and Molokai, but currently restricted to 58 square miles of mesic to wet ohia forest on the northeastern slope of Haleakala volcano (mainly 5,000-6,600 feet) on Maui. Last recorded on Molokai in 1907.

Threats: Introduced mosquito-borne diseases likely render low elevation habitat unsuitable, and infect birds that move downslope in search of blooming ohia trees. Feral ungulates (pigs, deer, goats) degrade forest habitat. Forest clearing for pastures might limit upper elevational range. Introduced rats are abundant in current habitat and predate nests.

Conservation: Habitat is protected within the East Maui Watershed. The Hanawi Natural Area is fenced, ungulate free, and rat controlled. A captive-breeding program initiated in 1997 was not successful due to incompatible pairing of birds and their aggressive nature.

Actions: Continue habitat protection and restoration efforts, including use of fencing and ungulate removal. Widespread rodent control might help alleviate nest predation. If unoccupied and suitable mosquito-free habitat is identified, translocation might be used to establish a second population.

POO-ULI
Melamprosops phaeosoma

GLOBAL POPULATION: **Unknown**
U.S. ENDEMIC

TREND: **LIKELY EXTINCT**
WATCHLIST: **HIGHEST CONCERN**
ESA: **ENDANGERED**

20

Distribution: First discovered in 1973, and only known in life from a 3,200 acre tract of montane forest (at 5,000-7,000 feet) dominated by ohia, on the northeastern slope of Haleakala Volcano, on the island of Maui.

Threats: Restricted from possibly suitable habitat below 5,000 feet by mosquito-borne diseases. Introduced rats are abundant within its known range and likely predated nests. Feral pigs degrade habitat, and invasive garlic snails reduced native land snails, which were likely an important food source for the Poo-uli.

Conservation: Surveys from 1997-2000 located three birds within Hanawi Natural Area Reserve, which is fenced, ungulate free, and rat control is ongoing. The East Maui Watershed Partnership has also fenced areas downhill from Hanawi. Attempts at uniting two birds to form a breeding pair through translocation in 2002 were unsuccessful. The last known bird was captured, but died in 2004 from malaria likely acquired previously in the wild.

Actions: Conduct surveys in unsearched areas of potential habitat.

C H A P T E R   2

HABITATS

Habitats, the physical environments inhabited by living organisms, are fundamental to their survival. In the case of birds, habitats provide cover from predators; breeding, wintering, and migration stopover sites; and places to forage and roost. All of the habitats used by a bird play a role in its survival, and the loss or degradation of any one of them can potentially have a population-level impact. It is little surprise, then, that habitat loss is the greatest threat to birds.

The tremendous variation in birds extends beyond their physical appearance to their evolutionary strategies, and therefore their habitat requirements. Some have evolved to be habitat specialists, whereas others are generalists. Some require only one habitat throughout their entire lives, while others use—or have adapted to—a broad range of them. A bird's habitat requirements can change over the course of a day (e.g., feeding habitat versus roosting habitat), a year (e.g., nesting, versus migratory stopover, versus wintering habitat), its life-cycle (e.g., pre-breeding age versus breeding age), or as circumstances demand.

Habitats are determined by the complex interplay of geology, the seasonality of sunlight, temperature, latitude, altitude, rainfall, fire, and the organisms that live there. Changes to any one of these factors can shift the habitat towards a new equilibrium. There are few habitats on Earth that have not been affected by humans. In many cases this impact is clear and often extreme. In others it is more subtle. As we modify and reshape the land around us, we damage or destroy natural habitat for birds. The habitats we create instead—farms, suburbs, cities—are more often than not poor substitutes for the habitats they supplant, supporting a lower diversity of more adaptable (and therefore more widespread) bird species. As human populations expand into more and more natural habitats, some bird populations will inevitably decline. Anthropogenic changes provide myriad conservation challenges, but these challenges must be met if we are to retain the abundance and diversity of birds, and prevent species extinctions.

Habitats and their bird communities vary regionally, sometimes with little overlap of species between regions. To conserve the full range of bird species, we need to include many different habitat types in our conservation plans. In this book, we have identified eleven natural "Birdscapes"—nine terrestrial (Hawaiian Forests, Arctic Tundra, Northern Forests, Western Forests, Eastern Forests, Southern Forests, Grasslands, Western Arid Lands, and Southern Arid Lands), and two aquatic: Wetlands and Marine. Additionally, many birds inhabit the Human Birdscape—unnatural or semi-natural habitats dominated by human development—adding a twelfth Birdscape. The nine terrestrial Birdscapes are depicted on the map opposite (Human and Wetland Birdcapes are omitted since they overlap natural habitats).

The eleven natural Birdscapes detailed in this book comprise 38 Bird Conservation Regions (BCRs). Each BCR is defined by the similarity of

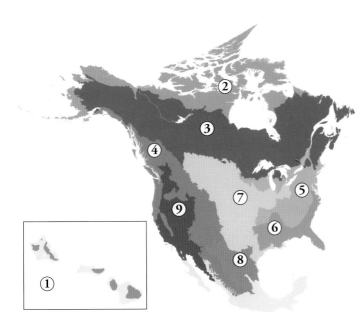

BIRDSCAPES

1. Hawaiian Uplands
2. Arctic Tundra
3. Northern Forests
4. Western Forests
5. Eastern Forests
6. Southern Forests
7. Grasslands
8. Southern Arid Lands
9. Western Arid Lands

NOT DEPICTED

10. Wetlands (see p.228)
11. Marine (see p. 246)
12. Human Birdscape (see p. 270)

habitats and bird communities within it and by collaborative conservation opportunities. BCRs were developed through the North American Bird Conservation Initiative based on earlier work by PIF, within a framework developed by the Commission on Environmental Cooperation. They provide a geographic matrix within which bird conservationists can develop habitat goals and deliver collaborative conservation programs (see p. 428 and p. 435).

Bird conservationists also identify the most important sites on which to focus their efforts. These sites are called Important Bird Areas (IBAs). IBAs provide us with a map of the highest priority places for bird conservation. IBA's can be identified on a regional, national, or global level, and are defined as the places that contain the most significant concentrations of resident or migratory birds, numbers of Endangered or WatchList species, or important populations of birds that are restricted to rare habitats.

In this chapter, we outline the habitats, birds, threats, and recommended conservation actions for each of the twelve Birdscapes. Following each Birdscape account we describe a number of ABC-designated Global IBAs that are representative of the habitats, bird populations, and threats for that Birdscape. For the full list of Global IBAs, see *The American Bird Conservancy Guide to the 500 Most Important Bird Areas in the United States*, or visit www.abcbirds.org. Several state Audubon chapters have also launched IBA programs. Information about these programs is usually available on their websites.

Like the species accounts, each Birdscape and IBA account is headed by a "fact bar" that contains a summary of its key metrics. For each Birdscape we list the BCRs that comprise it and its total area. We list the states that the Birdscape covers, and the number of WatchList species that are found there. We also provide a summary of the condition and status for each of the major bird habitat types within each Birdscape. Habitat analysis was conducted by ABC staff experts familiar with each region. For each IBA we indicate in which state, Birdscape, and BCR it is found, its area, and the number of WatchList species that occur there. Additional information is provided in the text accounts that follow below the fact bars.

HAWAIIAN UPLANDS

AREA: **5,860 SQ. MILES**
STATE: **HI**
BCR: **67**

WATCHLIST SPECIES
RED: **31**
YELLOW: **2**

Hawaii is sometimes known as "the bird extinction capital of the world". The islands once supported 113 bird species found nowhere else, including flightless geese, ibis, rails, and 59 species of honeycreepers. Since the arrival of humans, however, 71 bird species have been lost. While very significant challenges still exist, bird conservation can also point to some successes, though few native forest bird species can yet be regarded as secure.

Habitats: The windward slopes of the wettest Hawaiian Islands can receive more than 20 feet of rain annually. Here, moist tropical forest is dominated by ohia-lehua and koa trees, supporting endemic honeycreepers and other land-birds. Drier forests of mamane and naio grow on leeward slopes, with grasslands and lava fields in open areas.

Birds: Hawaiian forests support more endemic land-birds, and a higher concentration of endangered species than are found anywhere else in the U.S. Introduced birds have replaced native species at most lower elevations, and nearly 70% of all birds are exotic. Among the native forest species that survive are 13 WatchListed honeycreepers: the Palila, Maui Parrotbill, Oahu Amakihi, Kauai Amakihi, Anianiau, Akiapolaau, Akikiki, Hawaii Creeper,

ILLUSTRATION: C. VEST

Birds: 1. Apapane; 2. Elepaio; 3. Hawaiian Hawk; 4. Hawaiian Crow; 5. Omao; 6. Akiapoalaau; 7. Iiwi; 8. Palila. **Vegetation**: 9. koa; 10. ohia-lehua; 11. mamane. **Threats**: 12. mosquito-borne diseases; 13. invasive mongoose; 14. rat; 15. feral pig; 16. habitat loss due to agriculture; 17. coastal development; 18. collisions with communication towers; 19. introduction of new invasives such as the brown tree snake by boats, or 20. jetliners.

Hawaiian Upland Habitats	Condition	Threat Level	Flagship Bird
Wet ohia-lehua-koa forest	Poor	Critical	Iiwi
Dry mamane-naio forest	Poor	Critical	Palila
Hawaiian grasslands	Fair	High	Hawaiian Goose

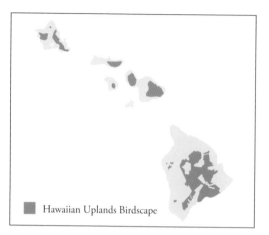

Hawaiian Uplands Birdscape

Maui Alauahio, Akekee, Akepa, Iiwi, and Akohekohe; two WatchListed solitaire thrushes, the Puaiohi and Omao; the Hawaiian Hawk, the Elepaio, and two seabirds: the Newell's Shearwater and Hawaiian Petrel (some of which nest above the tree line). The Hawaiian Crow is currently found only in captivity. The Hawaiian Goose and the Hawaiian race of the Short-eared Owl are found in open grassy areas.

Threats: Hawaiian forests and their birds face a range of serious threats. Past clearance for farmland has eliminated most lowland native forest. Exotic diseases, such as avian malaria and pox, are spread by introduced mosquitoes. Most honeycreepers are now restricted to the highest elevations where there are fewer mosquitoes and cooler temperatures limit the spread of disease. Climate change is predicted to allow malaria transmission to increase even in these last refuges though. Introduced trees and grazing animals (e.g., pigs, sheep, and goats) degrade habitat; pigs also disperse the seeds of invasive plants, and create wallows where mosquitoes breed. Invasive predators, including mongooses, feral cats, and rats, kill birds and predate nests.

Conservation: Most threatened Hawaiian species are listed under the ESA. A recent proposal by

ABC to list the Akikiki and Akekee has resulted in FWS proposing to protect these birds under a new, broad-scale ecosystem approach, which focuses on Critical Habitat to protect suites of species simultaneously. Despite their identical legal status, listed species on the mainland receive 15 times more recovery funding than Hawaiian species, an imbalance reflected by the ongoing declines of some Hawaiian birds. Captive-breeding programs exist for ten species (including the Palila, Hawaiian Crow, and Hawaiian Goose), and provide stock for reintroducing birds back into the wild.

Although Hawaiian water-birds are distributed among wetlands throughout the islands, many of Hawaii's endemic forest birds are only found on single islands. Therefore, a network of conservation sites spanning the major islands, as well as different mountain peaks on individual islands, is required. The most significant tracts of remaining native forest have been protected by TNC and state and federal agencies. The Kamehameha Schools pioneered koa reforestation techniques and demon-

The Akikiki's declining population is estimated at 1,300 birds. ABC has recently petitioned to list the species under the ESA.

PHOTO: JACK JEFFREY

The difficulty of removing feral goats from Kauai is evident from the steep terrain shown in this photograph of Kalalau Valley.

strated benefits for native birds 25 years after planting. Invasive grazing animals have been removed from fenced areas in some parks and reserves. Increased abundance of Elepaio and Akiapolaau was observed following local rat and cat removal at locations on Oahu and Hawaii, providing hope that efforts to control invasive predators can benefit native birds. Some native species have developed partial resistance to avian malaria. In 2009, ABC began its Hawaii program, focusing new political, financial, and conservation attention on the plight of Hawaii's birds.

Actions:

- Increase federal funding for Hawaiian birds.
- Control or eradicate invasive species and prevent the introduction of additional species.
- Eliminate the threat of diseases such as avian malaria and pox.
- Expand captive-breeding and reintroduction programs.
- Protect remaining habitat and restore forests on degraded pastures at highest elevations; fence key areas and remove introduced grazing animals.
- Expand monitoring of bird populations to assess trends, and conduct searches for species feared extinct.
- Reduce future impacts of climate change by reducing greenhouse gas emissions.

Once common in mid-elevation forests, like most other native birds, the Iiwi is now confined to the Hawaiian highlands.

IMPORTANT BIRD AREA
HAWAII UPLANDS

BIRDSCAPE: **HAWAIIAN UPLANDS**
BCR: **67** STATE: **HI**
AREA: **720,000 ACRES**

WATCHLIST SPECIES
RED: **16**
YELLOW: **1**

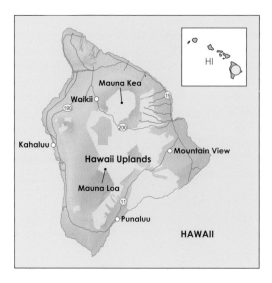

Birds: Endemics include three subspecies of the Elepaio, as well as the WatchListed Omao, Palila, Akiapolaau, Akepa, Hawaiian Hawk, and Hawaii Creeper. The Hawaiian Goose and petrel, Newell's Shearwater, the Hawaiian subspecies of the Short-eared Owl, and Iiwi also occur, as did several species that are now likely extinct, such as the Ou and Nukupuu. The Hawaiian Crow was last recorded in the wild from this island, but is presently only found in captivity. Two less-threatened forest birds, the Hawaii Amakihi and Apapane, also occur.

Threats: Similar to those elsewhere in Hawaii. Molten lava flows here are unique among the islands though, and these have damaged some important areas. The invasive fungus *Armillaria mellea* may be killing mamane trees, and alien wasps parasitize caterpillars that Palila feed to their nestlings.

Active volcanoes, barren lava fields, grasslands, and both dry and wet forests characterize the natural habitats of Hawaii's uplands. Mamane-naio forest grows on the dry slopes of Mauna Kea. On the wetter slopes of both Mauna Kea and Mauna Loa, koa trees dominate above 6,000 feet with ohia at lower elevations.

Conservation: The largest tract of natural forest is dominated by ohia and koa trees and lies on the eastern slope of Mauna Kea. It is partially protected within the Hakalau Forest NWR, and several state reserves. This refuge supports over half the global

The dry slopes of Mauna Kea are home to the Palila, which depends on arid mamane-naio forest for nesting and foraging.

PHOTO: GEORGE WALLACE

PHOTO: MICHAEL WALTHER, OAHU NATURE TOURS

The Palila's habitat is being degraded by introduced grazing animals. A protective fence urgently needs to be completed and repaired.

populations of the Akiapolaau, Hawaii Creeper, and Akepa. A drier forest of mamane and naio grows at high elevations on Mauna Kea. This is the stronghold for the Palila, and is protected within the Mauna Kea Forest Reserve. Protected areas on Mauna Loa, Kilauaea, and Hualalai volcanoes include Hawaii Volcanoes NP, the Kona Forest NWR, the Puu Waawaa State Wildlife Sanctuary, and TNC lands. Habitat restoration underway includes fencing, the removal of grazing animals, and reforestation. The fence protecting the Palila habitat is incomplete and in disrepair, however.

Actions:
- Control and remove invasive species.
- Reforest areas where grazing animals have been removed; erect fences to prevent repeat incursions.
- Increase funding for captive-breeding and release programs for the Hawaiian Crow, Akiapolauu, Hawaii Creeper, and Akepa.
- Develop sustainable koa forestry.

PHOTO: MICHAEL WALTHER, OAHU NATURE TOURS

Three subspecies of the Elepaio occur on the island of Hawaii; this is a male of the "Volcano" race.

IMPORTANT BIRD AREA
HALEAKALA

BIRDSCAPE: **HAWAIIAN UPLANDS**
BCR: **67** STATE: **HI**
AREA: **115,000 ACRES**

WATCHLIST SPECIES
RED: **11**
YELLOW: **1**

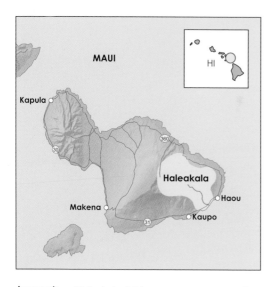

The Haleakala Volcano rises 10,000 feet above sea level in eastern Maui. Forests of ohia and koa trees cover much of its slopes, though its crater is rocky and barren. Its forests are the global stronghold for three unique bird species.

Birds: The WatchListed Maui Alauahio, Maui Parrotbill, and Akohekohe are endemic to the island of Maui, and now restricted to the slopes of Haleakala Volcano. Other WatchListed species include the Hawaiian Goose and petrel, Band-rumped Storm-Petrel, the Hawaiian race of the Short-eared Owl (in open areas), and the colorful Iiwi and Akepa, though the latter is likely now extinct on this island. The Nukupuu also likely became extinct here in 1996, and the Poo-uli in 2004. This area was also once home to the Bishop's Oo, last reported in 1981. The Apapane and Hawaii Amakihi are two additional honeycreepers that occur here, but are less threatened.

Threats: Very similar to those faced on other islands, and detailed under the threats section of the Birdscape account (p. 132). One additional issue specific to Maui is the impact of invasive predatory snails on native snails. This may have affected the Poo-uli which fed on snails.

Conservation: The East Maui Watershed is managed by local, state, and federal agencies, as well as private landowners. It conserves native forests and secures water supplies for Maui's agricultural, commercial, and residential areas. Protected areas

Like many native honeycreepers, the Maui Alauahio (endemic to Maui) is limited by the prevalence of avian malaria at low altitudes.

PHOTO: JACK JEFFREY

The Akohekohe is found only in higher-altitude forests on the island of Maui. Its population has remained stable over recent years.

PHOTO: JACK JEFFERY

include Haleakala NP, Waikamoi Preserve, Hanawi State Natural Area Reserve, and several other state forest reserves and recreation areas. Efforts to control predators, to eradicate pigs, and to fence off areas from grazing animals are underway. Captive-breeding programs were initiated for the Maui Parrotbill and Akohekohe, but the program for the latter has not yet been successful. Hawaiian Geese were introduced to Haleakala Crater in 1985, and now number roughly 250.

Actions:

- Remove remaining grazing animals from forested areas.
- Eradicate invasive predators from critical bird habitat.
- Prevent re-incursions of grazing mammals with fencing.
- Increase funding for captive-breeding of the Maui Parrotbill.
- Release captive-bred or translocated wild birds to bolster populations.

The highlands of Maui present logistical challenges to conservationists; some areas can only be reached by helicopter.

PHOTO: GEORGE WALLACE

IMPORTANT BIRD AREA
KAUAI UPLANDS

BIRDSCAPE: **HAWAIIAN UPLANDS**
BCR: **67**   STATE: **HI**
AREA: **170,000 ACRES**

WATCHLIST SPECIES
RED: **15**
YELLOW: **1**

KAUAI

HI

Hanalei   Kilauea

Kauai Uplands

56

Kapaa

50

Kauai's Alakai Plateau provides the last refuge for some of the rarest bird species on Earth. Despite its relatively small size, the plateau's topography and massive rainfall make field work extremely difficult, meaning that species' trends are hard to track.

Birds: Several native species have not been seen for many years and are likely extinct. These include the Kauai Akialoa (last seen 1967), Kauai Oo (1987), Kamao (1980s), Nukupuu (1975), and Ou (1989). One solitaire thrush, the Puaiohi, and several honeycreepers still remain, including the four WatchListed species endemic to the island: the Akekee, Akikiki, Kauai Amakihi, and Anianiau. The Elepaio, Iiwi, and Apapane are also present. The Hawaiian Petrel, Newell's Shearwater, and Band-rumped Storm-Petrel nest on cliffs within or above forests. Small numbers of Hawaiian Geese have been reintroduced to open areas.

Threats: Malaria is already found throughout Kauai, although its spread is currently inhibited by low temperatures at higher altitudes where there are also fewer mosquitoes. Invasive species, such as bamboo, are also a problem. Between 1979 and 2006, approximately 30,000 Newell's Shearwaters struck power lines and other structures, representing perhaps the largest unauthorized take of an ESA-listed species. Other threats are common to all Hawaiian forests (see p. 132), though Kauai appears to be free of mongooses.

Conservation: Upland forests in Kauai are protected within the Hono O Na Pali Natural Area Reserve, Waimea Canyon SP, and Kaluahonu Preserve, but Kokee SP and the Alakai Wilderness Preserve contain the most important tracts for endemic birds. Since 1999, a captive-breeding program for the Puaiohi has released 113 birds into the Alakai and some have since bred in the wild. The State of Hawaii runs the "Save Our Seabirds" program to rehabilitate "downed" Newell's Shearwaters and Hawaiian Petrels.

Actions:
- Prevent the introduction of mongooses to Kauai.
- Expand reintroduction programs to include the Akekee and Akikiki.
- List the Akekee and Akikiki under the ESA.
- Reduce collisions of Newell's Shearwaters and Hawaiian Petrels with power lines and lighted structures through siting and lighting changes.

PHOTO: PETER LATOURRETTE

The Anianiau currently numbers more than 44,000 and is one of the few native forest birds that appears to be increasing.

The Kauai Amakihi appears to be developing resistance to avian malaria, and benefits from its ability to feed on an introduced vine.

PHOTO: JACK JEFFRREY

Rainfall in the remote Alakai Wilderness often exceeds 20 feet per year. It is the last refuge for five unique bird species.

PHOTO: BILL HUBICK

ARCTIC TUNDRA

AREA: **1,239,034** SQ. MILES
(INC. CANADA)
STATE: **AK**
BCRs: **1, 2,** AND **3**

WATCHLIST SPECIES
RED: **10**
YELLOW: **31**

The tundra is a major nursery for the continent's shorebirds and waterfowl. Within the U.S., it is found only in the extreme north and west of Alaska, where it faces significant threats from oil development and climate change. Permafrost lies below much of the surface, and the landscape is characterized by plants and lichens that tolerate long, dark winters, and grow during brief, productive summers with long hours of daylight. Occasional shrubby willows and spruces are also found, particularly in the transitional zone between tundra and boreal forest. Wetlands and bogs abound, and rich marine habitats occur along the coasts.

Birds: During winter, only the hardiest ptarmigans, ravens, and redpolls can find enough food to survive on land. The summer burst of productivity though attracts migratory shorebirds and waterfowl, as well as some passerines and raptors. Seabirds dominate coastal and marine habitats, forming immense breeding colonies. Only the MacKay's Bunting is entirely restricted to this region, breeding on Hall and St. Matthew Islands and wintering along the coast of the Alaskan mainland. Several Eurasian species extend their breeding ranges from the west, but return to the Old World after nesting. These include the Arctic Warbler, Bluethroat, Northern Wheatear, Yellow Wagtail,

ILLUSTRATION: C. VEST

Birds: 1. Smith's Longspur; 2. Bar-tailed Godwit; 3. Willow Ptarmigan; 4. Snowy Owl; 5. Long-tailed Jaeger; 6. Buff-breasted Sandpiper; 7. Yellow-billed Loons; 8. American Golden-Plover; 9. Spectacled Eider; 10. Northern Pintails. **Mammals:** 11. caribou; 12. wolf. **Vegetation**: 13. cotton grass; 14. cushion plant. **Threats:** Habitat disturbance and pollution from oil and gas development including: 15. pipelines; 16. drill-sites; 17. tankers; 18. over-grazing by overabundant Snow Geese.

ARCTIC TUNDRA HABITATS	CONDITION	THREAT LEVEL	FLAGSHIP BIRD
Coastal tundra	Good	High	Yellow-billed Loon
Upland tundra	Good	Medium	Bristle-thighed Curlew
Glaciers and scree	Fair	High	Kittlitz's Murrelet

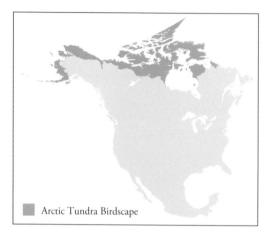

Arctic Tundra Birdscape

White Wagtail, and Red-throated Pipit. The few WatchListed land-birds that breed here include the Short-eared Owl, Smith's Longspur, and McKay's Bunting. Other WatchList species include the Emperor Goose, Trumpeter Swan, Steller's and Spectacled Eiders, Yellow-billed Loon, Red-faced Cormorant, Hudsonian and Bar-tailed Godwits, Wandering Tattler, Black Turnstone, Surfbird; Buff-breasted, Rock, Semipalmated, Western, White-rumped, and Stilt Sandpipers; Red Knot, and Sanderling. The Eskimo Curlew is likely to be extinct, but hopes remain that a small population may continue to breed somewhere on the tundra. Breeding seabirds include the Aleutian Tern, and Marbled, Kittlitz's, and Ancient Murrelets.

Threats: Tundra habitats are generally not threatened by development, but mining and drilling activities impact some areas through road construction, disturbance, toxic spills, and increases in local predator densities due to human activity. Climate change is altering arctic habitats faster than most on Earth; permafrost is thawing, glaciers melting, and summer sea ice rapidly disappearing. Toxins, such as PCBs, DDT, dioxins, and mercury, accumulate in the Arctic via global air currents and contaminate birds. Grazing by overabundant Snow Geese is denuding tundra vegetation in some areas. It is challenging to track changes in bird populations where so few people live, and better monitoring is needed for many species such as the Harris's Sparrow and McKay's Bunting.

Conservation: Large areas of North America's Arctic remain undeveloped, roadless, and retain herds of migratory caribou and musk ox, as well as large predators such as grizzly bears and wolves that are now extirpated from many areas further south. Few people mean few invasive species, though arctic islands are an exception. Fur farmers introduced red and arctic foxes to more than 190 of the Aleutian Islands to the detriment of the native land-birds and seabird colonies. Efforts to eradicate foxes, rats, and other invasive species on these islands are ongoing.

PHOTO: RALPH WRIGHT

95% of all Emperor Geese nest in Alaska's Yukon-Kuskokwim Delta, and most stage or winter at Izembek NWR.

Frozen for much of the year, the tundra is still the nursery for many of North America's shorebirds (Hudson Bay tundra pictured).

Key protected areas include Alaska's Arctic and Yukon Delta NWRs, as well as the Alaska Maritime NWR, which protects both marine and tundra habitats on Alaska's Aleutian and Bering Sea islands. Other important areas include Izembek NWR, and Teshekpuk Lake Special Management Area within the National Petroleum Reserve.

Canada has much larger areas of tundra than the U.S., and in 2008, announced that it would protect over 1.1 million acres of arctic wilderness in the Nunavut territory, including important seabird breeding sites, by establishing new National Wildlife Areas there.

Actions:
- Combat climate change through policies to reduce greenhouse gas emissions.
- Restrict development in important areas for WatchList species, including near Teshekpuk Lake in the National Petroleum Reserve, Izembek NWR, and Arctic NWR.

- Improve monitoring for breeding birds in this remote region.
- Remove invasives from important breeding islands.
- Reduce Snow Goose populations to historic levels.

Recent surveys estimated the McKay's Bunting population at 31,200, more than five times previous estimates.

IMPORTANT BIRD AREA
ARCTIC NWR

BIRDSCAPE: **ARCTIC TUNDRA**
BCR: **3**   STATE: **AK**
AREA: **19.8 MILLION ACRES**

WATCHLIST SPECIES
RED: **5**
YELLOW: **20**

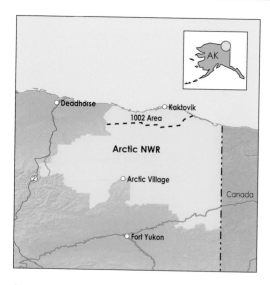

The Arctic NWR (ANWR) is probably the most controversial refuge in the NWR system due to conflicts that pit oil drilling and exploration against wildlife protection.

Birds: ANWR is an important area for breeding and migrating shorebirds and waterfowl.

WatchList species occurring here include the Yellow-billed Loon, Trumpeter Swan, Steller's and Spectacled Eiders, American Golden-Plover, Wandering Tattler, Bar-tailed Godwit, Surfbird, Red Knot, Sanderling; Semipalmated, Western, White-rumped, Stilt, and Buff-breasted Sandpipers; Thayer's Gull (rare), Ross's Gull (fall migrant), Short-eared Owl, and Smith's Longspur. On the south side of the Brooks Range, the WatchListed Olive-sided Flycatcher, Varied Thrush, and Rusty Blackbird breed. Several species that mainly occur in Asia also breed here, including the Gray-headed Chickadee, which is a local resident in the Brooks Range. The Eskimo Curlew may once have bred here.

Threats: Political pressure to allow oil companies to build roads and drilling stations to extract oil from ANWR is a persistent threat to the Coastal Plain ecosystem within the 1.5 million acre portion of the refuge known as the 1002 Area. This is where most of the WatchListed shorebirds and waterfowl breed, as well as being the calving ground for 129,000 caribou. Scientists warn that oil ex-

The Stilt Sandpiper migrates from its tundra breeding grounds to winter as far south as Argentina.

PHOTO: ROBERT ROYSE

Most Snowy Owls remain in the far north all year, but they can occur far to the south, likely due to changes in lemming abundance.

PHOTO: ROBERT ROYSE

traction and development would greatly damage tundra and wetland habitats, risk polluting pristine wilderness areas, and disrupt breeding activities for birds, caribou, and other wildlife. Climate change, associated sea level rise, and the reduction of arctic sea ice also threaten wildlife and habitats in ANWR.

Conservation: ANWR was established in 1980, protecting the largest Wilderness Area in the U.S. It spans much of the Brooks mountain range and Arctic Coastal Plain. Except for a few exploratory test wells, conservationists have so far been successful in blocking oil drilling in the 1002 Area, despite intense political and corporate pressure.

Actions:
- Permanently prevent oil drilling in ANWR by declaring it a NM or Wilderness Area.

The Smith's Longspur breeds at low densities in the Brooks Range of the Arctic NWR, and winters as far south as Texas.

PHOTO: ROBERT ROYSE

IMPORTANT BIRD AREA

YUKON DELTA NWR

BIRDSCAPE: **ARCTIC TUNDRA**
BCR: **2**   STATE: **AK**
AREA: **19.5 MILLION ACRES**

WATCHLIST SPECIES
RED: **4**
YELLOW: **21**

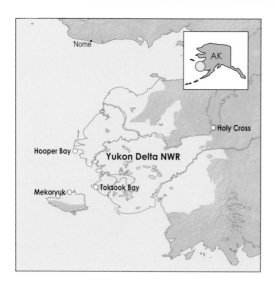

T his refuge encompasses nearly 20 million acres of coastal and mountain tundra, mudflats, wetlands, and boreal forest at the deltas of the Yukon and Kuskokwim Rivers, along with the offshore islands of Nelson and Nunivak. It is the single most important site for breeding waterfowl in the U.S.

Birds: One of the largest aggregations of waterbirds in the world occurs at this refuge. More than 1.6 million ducks, geese, swans, loons, grebes, and cranes breed here. Fully 100,000 WatchListed Spectacled Eiders once bred at the refuge, but this population has declined to 7,250 pairs. As many as 20 million shorebirds breed or stage on the refuge, and up to one-million seabirds nest on Nunivak Island. Roughly half the global population of the WatchListed Bristle-thighed Curlew nests in the Nulato Hills, and virtually all stage on the Delta. The majority of the world's Emperor Geese and Black Turnstones nest here. It is also the primary wintering area for the McKay's Bunting. Other WatchList species include the Red-faced Cormorant, American Golden-Plover, Wandering Tattler, Hudsonian and Bar-tailed Godwits, Surfbird, Red Knot, Sanderling; Western, Semipalmated, and Rock Sandpipers; Aleutian Tern, Olive-sided Flycatcher, Willow Flycatcher, Varied Thrush, and Rusty Blackbird.

Threats: Sea level rise due to climate change could inundate the coastal habitats that are the most pro-

The Delta's population of the Spectacled Eider declined by over 96% from the 1970s to the 1990s; fortunately it has since stabilized.

The Red Phalarope visits coastal tundra to breed, but spends up to 11 months at sea, often foraging near whales.

ductive for shorebirds and waterfowl. Expansion of human settlements near waterfowl breeding areas boosts populations of gulls, ravens, and foxes, increasing predation of eggs and young. Reasons behind population declines in Steller's and Spectacled Eiders are unclear, but may be partly related to residual lead shot contamination in wetlands.

Conservation: The borders of this refuge include roughly seven-million acres of private land, including those belonging to 25,000 Yup'ik people. Subsistence hunting of waterfowl, salmon, and other wildlife by the Yup'ik is permitted. Refuge management activities mostly involve wildlife monitoring rather than habitat manipulation, and this information is used to manage hunting and fishing activities.

Actions:
- Combat climate change with policies to reduce greenhouse gas emissions.

- Identify ways to increase the populations of Steller's and Spectacled Eiders.
- Work with Yup'ik communities to reduce lead contamination and address the problem of trash dumps that boost local predator populations.

The Bristle-thighed Curlew stages in the Delta and winters on Pacific islands, becoming flightless during its molt.

NORTHERN FORESTS

AREA: **2,717,957 SQ. MILES**
(INC. CANADA)
STATES: **AK, CT, MA, ME, MI,**
MN, NH, NY, VT, WI
BCR: **4, 6, 7, 8, 12, AND 14**

WATCHLIST SPECIES
RED: **3**
YELLOW: **9**

The northern forests are a major nursery for North America's neotropical migratory songbirds. Fortunately, they are still among the largest and most intact forest ecosystems on Earth. Habitats are dominated by conifers (spruces, firs, tamarack, and pines) interspersed with northern hardwoods (birch, aspen, poplar, alder, and willow), and wetlands. At the northern edge of the boreal region, stunted conifers give way to tundra, while at the southern edge, aspen parklands grade into grasslands and deciduous forests. Conifer forests also mix with deciduous hardwoods in parts of the Northeast, though conifers still pre-

dominate in the mountains there. Permafrost underlies much of the northern portion of this region.

Birds: More than 300 bird species breed in the Northern Forests Birdscape, including many of our migrant waterfowl, shorebirds, and songbirds. Some birds exhibit nomadic behavior to take advantage of periodic booms in food availability. For instance, the density of many warblers (including the Tennessee, Cape May, and Bay-breasted) skyrockets in areas with outbreaks of spruce budworms, and crossbills search for conifers with good cone crops. Breeding WatchList species include the

ILLUSTRATION: C. VEST

Birds: 1. Spruce Grouse; 2. Cape May Warbler; 3. Sandhill Crane; 4. Black-backed Woodpecker; 5. White-winged Crossbill; 6. Olive-sided Flycatcher; 7. Common Raven; 8. Common Goldeneye; 9. Barrow's Goldeneye 10. Evening Grosbeak; 11. Boreal Chickadee; 12. Bay-breasted Warbler; 13. Canada Warbler; 14. Boreal Owl. **Vegetation**: 15. spruce; 16. sedge marsh. **Threats**: 17. unsustainable old-growth timber harvest; 18. oil extraction from tar sands; 19. climate change; 20. drying wetlands.

Northern Forests Habitats	Condition	Threat Level	Flagship Bird
Mountaintop spruce fir	Good	High	Bicknell's Thrush
Jack pine	Fair	High	Kirtland's Warbler
Boreal forest	Good	Medium	Bay-breasted Warbler
Northern hardwoods	Fair	Low	Black-throated Blue Warbler

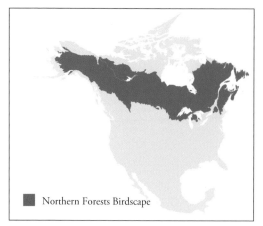

Northern Forests Birdscape

Trumpeter Swan, Yellow Rail, Whooping Crane, Hudsonian Godwit, Olive-sided Flycatcher, Bicknell's Thrush; Kirtland's, Bay-breasted, and Canada Warblers; Nelson's Sparrow, and Rusty Blackbird. Several more common species that breed in the northern forests have experienced largely unexplained population declines of more than 50% over the last 40 years, including the Greater Scaup, Ruffed Grouse, Common Tern, Boreal Chickadee, and Evening Grosbeak.

Threats: Fortunately, the short growing season and distance from most human population centers have protected large areas from being cleared for agriculture and urban development. This distance does not safeguard northern forests from the effects of air pollution, including acid rain, however. Climate change is also affecting these forests. Trees adapted to growing in permafrost soils are dying in areas where the permafrost is melting. Expanding oil extraction in tar sands and oil shales clears habitat, and creates open waste pits that attract and drown waterfowl, contaminate watersheds, and release tremendous amounts of greenhouse gases. One major tar sand development area is just south of Wood Buffalo NP in Canada, home to

the only naturally occurring breeding population of the Whooping Crane. Clear-cutting trees can eliminate mature forest patches (>150 years old), removing dead snags whose cavities are required by some breeding birds. Large areas of forestry lands are sprayed with pesticides to control spruce budworms that damage trees. Unfortunately, the budworm is an important food supply for many birds, and one of the pesticides used (fenitrothion) is toxic to birds, such as the Bay-breasted and Cape May Warblers. Gray Jays store food to consume during lean winters, but their food is now rotting in some places due to warmer winter temperatures.

Conservation: Despite the threats facing northern forests, conservationists have succeeded in improving policies and increasing protected acreage. The U.S. Clean Air Act and similar regulations in Canada have significantly reduced airborne pollutants, resulting in a 35% decrease in acid rain in

The use of insecticides to control budworm can impact species such as the Cape May Warbler which depends on spruce-fir.

PHOTO: ROBERT ROYSE

The U.S. imports $20 billion in timber products annually from Canada, mostly from northern forest areas.

the northeast U.S. over the last 20 years—though its effects still live on in some northern forests and wetlands. Though the U.S. northern forests are extensive, a much larger area occurs in Canada. In 2008, Ontario declared that it would protect 55.6 million acres (43% of its area), and Quebec, 4.4 million acres. Many U.S. state and national forests, NWRs, and parks also protect northern forests, including Yukon Flats NWR (Alaska), Adirondack Park (New York), the Kirtland's Warbler Management Area (Michigan); Chippewa and Superior (Minnesota), Hiawatha and Nicolet (Michigan), Green Mountain (Vermont), and White Mountain (New Hampshire) NFs; and Isle Royale NP (Michigan).

Actions:

- Protect large tracts of remaining intact forest.
- Implement forest management practices that protect old-growth, address pesticide spraying, and maintain habitat for declining species.
- Reduce demand for wood pulp from northern forests through programs that reduce paper use and increase recycling.

The Rusty Blackbird nests in dense forest near wetlands. Why its population has rapidly declined is unknown.

IMPORTANT BIRD AREA
YUKON FLATS NWR

BIRDSCAPE: **NORTHERN FORESTS**
BCR: **4**   STATE: **AK**
AREA: **8.6 MILLION ACRES**

WATCHLIST SPECIES
RED: **1**
YELLOW: **12**

13 bird species spend the winter here, but bird diversity tops 150 species during the spring and summer. The area supports over two-million ducks, including most of Alaska's Canvasbacks, Northern Pintail, and Greater Scaup, whose populations have severely declined over the last 40 years. Forests support the WatchListed Olive-sided Flycatcher and Rusty Blackbird, and Smith's Longspurs breeds in the alpine tundra of the Brooks Range. Other WatchList species that breed or migrate through include the Trumpeter Swan, Swainson's Hawk, American Golden-Plover, Wandering Tattler, Hudsonian Godwit, Surfbird, Sanderling; and Semipalmated, Western, and Stilt Sandpipers.

Threats: The isolation of this refuge provides protection from most forms of development. Interest in developing oil and gas outside the refuge in the White Mountains, along with pipelines that would have run through the refuge, ended efforts to designate 650,000 acres as a Wilderness Area in 1987. The effects of climate change on this refuge are, as yet, unknown.

This enormous refuge protects a vast tract of boreal forest dominated by spruce, birch, and aspen trees, interspersed by lakes, braided rivers, and streams.

Birds: The refuge is one of the most important waterfowl breeding areas in North America. Just

Extensive wetlands in the Yukon Flats support over two-million breeding ducks, such as these Northern Pintail.

PHOTO: BILL HUBICK

The American Golden-Plover undertakes one of the longest migrations of any shorebird, wintering in southern South America.

PHOTO: ROBERT ROYSE

Conservation: Two-and-a-half-million acres of land within the refuge is owned by the state, various indigenous groups, and native regional corporations. A major hydroelectric project proposed during the 1950s would have flooded the refuge, but broad opposition stopped the project. Refuge management includes monitoring wildlife to gauge the impacts of subsistence and sport hunting. Scientists from the University of Alaska Fairbanks lead research on fire management, and on waterfowl and other birds. Refuge managers cooperate with state agencies and local people to maintain a near-natural fire regime through prescribed burns that create a dynamic mosaic of habitats and age classes among trees.

Actions:

- Continue successful research, monitoring, and fire management partnerships with local and state groups.
- Monitor for impacts of emerging threats, such as climate change.

The Olive-sided Flycatcher is associated with burned forests, where dead trees can provide flycatching perches.

PHOTO: ROBERT ROYSE

IMPORTANT BIRD AREA

ADIRONDACK PARK

BIRDSCAPE: **NORTHERN FORESTS**
BCR: **14**   STATE: **NY**
AREA: **6.1 MILLION ACRES**

WATCHLIST SPECIES
RED: **1**
YELLOW: **5**

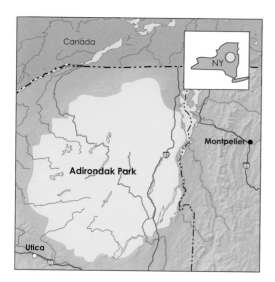

Birds: The Park supports the largest population of the Bicknell's Thrush in the U.S. Other Watch-List species include the Olive-sided Flycatcher, Wood Thrush, Canada Warbler, and Rusty Blackbird. Small numbers of the WatchListed Golden-winged and Bay-breasted Warblers also breed here. Other breeding species include the Ruffed Grouse, Whip-poor-will, Boreal Chickadee, and Evening Grosbeak, which all have continental populations that have declined by more than 50% over the last 40 years. The park is at the southern edge of the range of several boreal birds, including the Spruce Grouse, Three-toed and Black-backed Woodpeckers, Yellow-bellied Flycatcher, Gray Jay, and Palm and Blackpoll Warblers.

Adirondack Park is roughly the size of Vermont, and contains the largest Wilderness Area east of the Mississippi. Approximately 137,000 people live year-round inside the park. They are joined annually by 200,000 seasonal residents, and millions of tourists.

Threats: Air pollution from coal-fired power plants in the Midwest has damaged both wetlands and forests within the park through the deposition of heavy metals and acid rain. These pollutants stunt and kill spruce forests on western slopes, and reduce the abundance of snails and other invertebrates required by thrushes. More than 500 of the

The breeding habitat of the Bicknell's Thrush is restricted to coniferous forests. It winters in the Caribbean, particularly on Hispaniola.

Adirondack Park includes the largest Wilderness Area east of the Mississippi, but suffers from acid rain (Chapel Pond pictured).

Adirondack's 2,800 lakes and ponds are now too acidic to support much aquatic life.

Conservation: The State of New York created the Park in 1892 to protect the Hudson River and Erie Canal from siltation caused by widespread logging. The park includes 2.6 million acres of state forest, and roughly three-million acres of privately owned land (including TNC). The Adirondacks High Peaks Wilderness Area covers 226,435 acres. Strict regulations govern development. Plans to expand skiing at Whiteface Mountain were recently changed to avoid damaging breeding areas for the Bicknell's Thrush.

Actions:
- Further reduce acid rain and other damaging pollutants from Midwestern coal-fired power plants.
- Monitor Bicknell's Thrush and other boreal species for potential impacts of climate change.

The Evening Grosbeak vanished from 50% of traditional bird feeding sites between 1988 and 2006.

IMPORTANT BIRD AREA
KIRTLAND'S WARBLER MA

BIRDSCAPE: **NORTHERN FORESTS**
BCR: **12**   STATE: **MI**
AREA: **150,000 ACRES**

WATCHLIST SPECIES
RED: **1**
YELLOW: **0**

The Kirtland's Warbler Management Area is one of the great success stories for Endangered birds in the U.S. Intensive habitat management efforts, combined with cowbird control, have restored a dynamic jack pine ecosystem and greatly increased the population of the Endangered warbler.

Birds: This is the most important breeding site for the WatchListed Kirtland's Warbler. Habitat management efforts also benefit other species that need dense thickets, open areas, or snags, including the Ruffed Grouse, American Kestrel, Upland Sandpiper, Eastern Bluebird, Eastern Towhee, and Grasshopper Sparrow.

Threats: Breeding habitat for the Kirtland's Warbler was once maintained by natural wildfire that created large stands of young jack pines. Habitat loss and fragmentation due to agriculture and development reduced the natural frequency of fires, and facilitated the overabundance of Brown-headed Cowbirds which parasitize the warbler's nests. Without intensive management to create appropriate habitat and to control cowbirds, the Kirtland's Warbler would have become extinct.

Conservation: The area is administered by Seney NWR in cooperation with the USFS and Michigan Department of Natural Resources. Approximately 4,000 Brown-headed Cowbirds are trapped and removed annually from Kirtland's Warbler nesting areas. This practice has successfully reduced nest parasitism rates from 70% to 3%. Kirtland's Warblers require jack pine stands that are larger than 80 acres and between six and 20 years old. Managers provide roughly 38,000 acres of appropriate habitat for Kirtland's Warblers at all times by logging, burning, seeding, and replanting jack pines on a 50-year rotational basis. These efforts have helped the Kirtland's Warbler population grow from a low of just 200 singing males in the 1970s and 1980s to more than 1,700 males today. The species is now even beginning to expand its range

PHOTO: ROBERT ROYSE

Prescribed burns create open habitats used by Upland Sandpipers, which have declined 32% range-wide since 1966.

Thanks to conservation efforts, the Kirtland's Warbler has increased almost tenfold in recent decades (male pictured).

PHOTO: ROBERT ROYSE

into Wisconsin and Ontario, Canada. Members of the refuge staff lead guided tours to limit disturbance to breeding birds. The annual Kirtland's Warbler Festival celebrates the species' spring arrival, encourages ecotourism in the area, and provides educational outreach opportunities.

Actions:

- Maintain and increase breeding habitat for Kirtland's Warblers on public lands through clearcutting, managed burns, and the planting of jack pines.
- Continue cowbird removal from Kirtland's Warbler nesting areas.

PHOTO: DANIEL LEBBIN

Maintaining young jack pine stands (above) and controlling cowbirds are key Kirtland's Warbler management techniques.

WESTERN FORESTS

AREA: **957,099** SQ. MILES
(INC. CANADA)
STATES: **AK, AZ, CA, CO, ID, MT, NM, NV, OR, UT, WA, WY**
BCRs: **5, 10, 15, 16,** AND **34**

WATCHLIST SPECIES
RED: **4**
YELLOW: **28**

Western forests are naturally more frag-mented than those of other regions, in part due to the West's fire ecology and complicated topography, where open areas, grass-lands, or deserts often separate forests that grow on mountain slopes. They vary from extremely arid and open to dense temperate rainforests, boasting trees that are the oldest (bristlecone pines), tallest (California redwoods), and largest (giant sequoias) in the world. Although conifers tend to dominate, large areas also consist of deciduous aspens, oaks, cottonwoods, and sycamores. The frequency of wildfire varies, and contributes to the forests' het-erogeneity. Some pine forests experience frequent low-intensity burns that do not kill adult trees (e.g., ponderosa pine); some experience intense stand-re-placing burns (e.g., lodgepole pine); and yet others (e.g., Pacific rainforests) burn only rarely, perhaps once every 400 years.

Birds: Western forests support a wide diversity of WatchList bird species, many of which are re-stricted to specific forest types. The Marbled Mur-relet and Sooty Grouse breed only in coastal forests of the Pacific Northwest; the Flammulated Owl and Lewis's Woodpecker nest in mature stands of ponderosa pine; the Pinyon Jay depends on pin-yon pines; the Nutall's Woodpecker, Yellow-billed

ILLUSTRATION: C. VEST

Birds: 1. Black Swift; 2. Hermit Warbler; 3. Sooty Grouse; 4. Lewis's Woodpecker; 5. White-headed Woodpecker; 6. Williamson's Sapsucker; 7. Varied Thrush; 8. Calliope Hummingbird; 9. Flammulated Owl. **Mammals:** 10. elk. **Vegetation:** 11. Douglas fir; 12. dead snag; 13. ponderosa pine. **Threats:** 14. logging; 15. bark beetle outbreaks; 16. fire suppression.

Western Forests Habitats	Condition	Threat Level	Flagship Bird
Douglas fir/redwoods	Fair	Medium	Spotted Owl
Lodgepole pine	Good	Medium	Williamson's Sapsucker
Ponderosa pine	Fair	High	Flammulated Owl
Pinyon-juniper woodland	Fair	Medium	Pinyon Jay
Madrean pine-oak	Good	Low	Red-faced Warbler
West coast oak savanna	Poor	High	Yellow-billed Magpie
Aspen groves	Fair	Medium	Red-naped Sapsucker
Alpine tundra	Good	High	Black Rosy-Finch

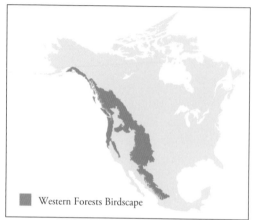

Western Forests Birdscape

Magpie, and Oak Titmouse inhabit oak woodlands; and the Willow Flycatcher, Lucy's Warbler, and Abert's Towhee occur along riparian corridors. Other WatchList species include the Spotted and Elf Owls, Black Swift, Blue-throated Hummingbird, Elegant Trogon, Arizona and White-headed Woodpeckers, Williamson's Sapsucker, Olive-sided Flycatcher, Mexican Chickadee, Varied Thrush; and Virginia's, Hermit, Grace's and Red-faced Warblers. Western Forests also support several more widespread species whose populations have declined by more than 50% in the last 40 years including the Ruffed Grouse, Whip-poor-will (in the Southwest), Rufous Hummingbird, and Evening Grosbeak.

Threats: Western forests have historically been affected by predator control programs that allow grazing animals to become overabundant. They also continue to be degraded by fire suppression, cattle grazing, and logging. Many occur on public or private lands historically managed for unsustainable timber extraction. By the 1980s, the forestry

industry had cleared so much forest in the Pacific Northwest that many feared the extinction of old-growth-dependent species such as the "Northern" Spotted Owl. An introduced fungus (*Phytophthora*) causes sudden oak death syndrome which is affecting large areas of California and Oregon. Climate change is having an impact too. For example, native bark beetles are better able to survive warming winters. Recent outbreaks are occurring further north than they have in the past, for example, killing over half the lodgepole pines in central British Columbia in the last 14 years.

Conservation: Conservation of old-growth forests to benefit birds listed under the ESA such as the "Northern" Spotted Owl and Marbled Murrelet has generated significant conflict with the timber

PHOTO: GREG LAVATY

The Red-faced Warbler nests on the ground but forages in the canopy of southwestern pine-oak forests.

PHOTO: MIKE PARR

California's oak savannas occur in dry areas with frequent low-intensity fires, and harbor species such as the Yellow-billed Magpie.

industry. Although millions of acres have been designated as Critical Habitat and protected from logging, many scientists believe current federal restrictions on logging are still insufficient to fully protect endangered birds.

Despite this controversy, western forests have experienced dramatic restoration in some areas. For example, wolves were reintroduced to the Greater Yellowstone ecosystem in the 1990s, and have helped control elk numbers, allowing riparian woodlands and their associated songbirds to rebound. Raptors and ravens also benefit from carcasses left behind by wolves. Efforts to restore ponderosa and pinyon pine, aspen, oak, and riparian forests are occurring throughout the West. Both the timber industry and USFS have moved away from clear-cuts and even-aged management toward more diverse systems that mimic natural disturbance and restore habitat structure.

Western forests occur on public lands including BLM lands, state forests, NWRs; and NPs such as Yellowstone (Wyoming, Montana, Idaho), Olympic (Washington), and Rocky Mountain (Colorado); as well as Wallowa-Whitman (Oregon) and Coronado (Arizona, New Mexico) NFs, and the parks and forests of the Sierra Nevada.

Actions:
- Implement forestry practices that maintain or enhance habitat for declining birds, including protecting all old-growth forests on public lands and restoring mature, open ponderosa pine forests for cavity-nesting species through thinning and prescribed burns.
- Improve bird monitoring, particularly for nomadic species such as the Pinyon Jay, and for cavity-nesting birds.
- Combat climate change by mandating reductions in greenhouse gas emissions.

IMPORTANT BIRD AREA
OLYMPIC PENINSULA

BIRDSCAPE: **WESTERN FORESTS**
BCR: **5**   STATE: **WA**
AREA: **1.5 MILLION ACRES**

WATCHLIST SPECIES
RED: **3**
YELLOW: **18**

Birds: The area supports breeding populations of several WatchListed species typical of Pacific temperate rainforests, including the Sooty Grouse, Marbled Murrelet, "Northern" Spotted Owl, Black Swift, Olive-sided and Willow Flycatchers, and Varied Thrush. During migration and winter, many other WatchList species use the area's rocky shores and coastal waters, including the American Golden-Plover, Wandering Tattler, Black Turnstone, Surfbird, Red Knot, Sanderling, Semipalmated and Western Sandpipers, Heermann's and Thayer's Gulls, Ancient Murrelet, and an occasional Yellow-billed Loon. The Ruffed Grouse, Rufous Hummingbird, and Evening Grosbeak all breed here; each has declined by over 50% nationally in the last 40 years.

Olympic NP and Olympic NF together protect a vast wilderness on the Olympic Peninsula featuring 73 miles of coastline, and a spectacular temperate rainforest, where Sitka spruce, Douglas-fir, western hemlock, and western red cedar are nourished by over 12 feet of annual rainfall, and tower to heights of up to 300 feet.

Threats: The Northwest Forest Management Plan allowed for the extraction of eight- to-ten-million board feet of timber from roughly 125,000 acres of the NF. This plan has been widely criticized by scientists as offering insufficient protection for endangered birds, however. Barred Owls appear to have benefitted from anthropogenic habitat changes that have enabled them to expand their range and

Once they have been logged, it takes more than one-hundred years for these old-growth forests to re-establish (Hoh rainforest pictured).

PHOTO: ©ISTOCKPHOTO.COM / KEN CANNING IMAGES

The Olympic Peninsula's heavy rainfall nourishes waterfalls behind which Black Swifts nest safe from predators.

displace or hybridize with Spotted Owls. Climate change threatens to melt the park's 60 glaciers.

Conservation: More than two-million acres of forest were originally protected in 1897; 95% (876,669 acres) of the NP and 15% (88,265 acres) of the NF are designated as Wilderness Areas. Management includes regulating logging and suppressing some forest fires. The National Park Service is restoring 65 miles of the Elwha River by removing two dams and replanting vegetation within the drained reservoirs.

Actions:
- Ensure logging does not occur in critical old-growth habitats required by Spotted Owls and Marbled Murrelets.
- Research the impacts of the increasing penetration of Barred Owls into Spotted Owl habitat, and investigate potential measures to mitigate or limit these impacts, including potentially the

experimental removal of Barred Owls in areas of high competition.
- Evaluate the reintroduction of wolves to control elk, and the eradication of introduced mountain goats that damage alpine meadows.

Fewer than 2,300 pairs of "Northern" Spotted Owls remain in the Pacific Northwest due to logging and other threats.

IMPORTANT BIRD AREA
SIERRA NEVADA

BIRDSCAPE: **WESTERN FORESTS**
BCR: **15**   STATE: **CA**
AREA: **8.7 MILLION ACRES**

WATCHLIST SPECIES
RED: **5**
YELLOW: **19**

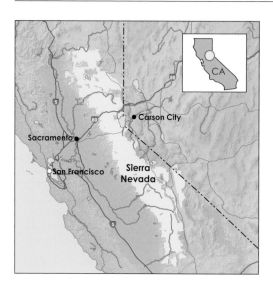

pines grow on eastern slopes, with spruce-fir forests and alpine tundra at high elevations.

Birds: The area supports many WatchList species including the California Condor, Sooty Grouse, Mountain Quail; Flammulated, "California" Spotted, and Short-eared Owls; Black Swift, Calliope Hummingbird; Nutall's, Lewis's, and White-headed Woodpeckers; Williamson's Sapsucker, Olive-sided Flycatcher, Pinyon Jay, Varied Thrush (during migration), Oak Titmouse, Wrentit, Hermit Warbler; and Black-chinned, Brewer's, and Sage Sparrows. High elevations support a variety of boreal species near their southern range limits, including the American Pipit, Gray-crowned Rosy-Finch, Evening Grosbeak, and a small population of the Great Gray Owl. Meadows provide nesting habitat for Willow Flycatchers, which are declining in this region.

A pproximately 8.7 million acres of NP and NF lands protect this area, in a nearly continuous 350 mile stretch from north to south. Oak savannas and ponderosa pines grow at lower elevations on the western slopes, with groves of giant sequoias at mid elevations. Lodgepole

Threats: Decades of fire suppression have favored dense forests of drought-vulnerable trees (e.g., larch, fir) over open stands of hardier pines. As a re-

The Sierra Nevada includes 8.7 million acres of conifer forest and tundra habitat where development is highly restricted.

PHOTO: GEORGE NF IMANTOWICZ

The Williamson's Sapsucker nests in lodgepole pines and aspens up to 7,000 feet in elevation. Eastern populations winter in Mexico.

PHOTO: ROBERT ROYSE

sult, forests are now more vulnerable to catastrophic wildfires that can destroy rather than rejuvenate them, as well as to outbreaks of native bark beetles. Air pollution damages vegetation in the southern Sierra, and invasive plants and cattle also damage native habitats.

Conservation: Conservation actions began with federal protection of Yosemite in 1864. Today, the federal government manages several NFs, including Lassen, Sierra, and Sequoia; and NPs, including Yosemite, Kings Canyon, and Sequoia. Balancing recreation and wildlife management is a challenge, with more than 3.5 million visitors annually to Yosemite alone. Management includes thinning forests to protect against pest outbreaks and catastrophic fires, controlling invasive plants such as Scotch broom and yellow starthistle, and monitoring tree pests and diseases.

Actions:
- Restore natural fire regimes to reduce the risk of catastrophic wildfires and bark beetle outbreaks.
- Increase old-growth habitat.
- Identify and protect key meadows.
- Stem losses of oak woodlands in foothills.
- Control and eradicate invasive plants.

PHOTO: PETER LATOURETTE

The Calliope Hummingbird forages in early-successional shrublands up to 15 years of age.

IMPORTANT BIRD AREA
WALLOWA-WHITMAN NF

BIRDSCAPE: **WESTERN FORESTS**
BCR: **12**   STATE: **OR**
AREA: **2.3 MILLION ACRES**

WATCHLIST SPECIES
RED: **1**
YELLOW: **14**

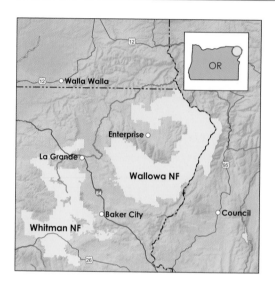

This National Forest is divided into several disjunct units in northwestern Oregon. Elevations range from 875 to 9,845 feet, and vegetation varies accordingly, from arid grasslands at low elevations to alpine vegetation on peaks. In between, high-elevation prairie playas, dry ponderosa pine woodlands, and deciduous riparian hardwoods (aspen, cottonwood, birch, and alder) can be found. The Forest includes Hell's Canyon National Recreation Area, which contains North America's deepest river gorge.

Birds: The forests of ponderosa pine support WatchListed cavity-nesters including the Flammulated Owl, Lewis's and White-headed Woodpeckers, and Williamson's Sapsucker. Other WatchList species occurring here include Mountain Quail, Long-billed Curlew, Swainson's Hawk, Short-eared Owl, Calliope Hummingbird, Olive-sided and Willow Flycatchers, Varied Thrush, Sage and Brewer's Sparrows, and wintering Black Rosy-Finch. An endemic subspecies of the Gray-crowned Rosy-Finch breeds only here. Several more common species with continental populations that have declined by more than 50% since 1967 also occur, including the Northern Bob-

white, Ruffed Grouse, Rufous Hummingbird, Loggerhead Shrike, Horned Lark, and Grasshopper and Lark Sparrows. The forest is also home to various raptors, including the Ferruginous Hawk, Northern Goshawk, and Great Gray Owl.

Threats: This is a multi-use area for wildlife, logging, grazing, and recreation. Fire suppression has created a dense understory of firs in ponderosa habitat that restricts open foraging areas for Flammulated Owls, and reduces suitability for White-headed Woodpeckers. Off-road vehicle use along roads and trails disturbs wildlife. Vehicles venturing off trails can also spread invasive species and damage vegetation, soil, and waters.

Conservation: Prescribed burns are used to manage ponderosa pine woodlands, restoring more open conditions and clearing understory dominated by Douglas-fir. The "Columbia" Sharp-tailed Grouse was reintroduced in 1990.

Actions:
- Restore ponderosa pine woodlands for Watch-Listed cavity-nesters.
- Reduce habitat damage caused by off-road vehicles.

The Lewis's Woodpecker uses ponderosa pines, and has declined by more than 50% since the 1960s.

PHOTO: BILL HUBICK

Great Gray Owls can sometimes be seen perching on roadside poles, watching for rodents. They will also use artificial nest cavities.

IMPORTANT BIRD AREA
YELLOWSTONE

BIRDSCAPE: **WESTERN FORESTS**
BCR: **10**   STATE: **WY, MT, ID**
AREA: **>9.6 MILLION ACRES**

WATCHLIST SPECIES
RED: **1**
YELLOW: **15**

T he dramatic scenery, numerous geysers, and abundant wildlife of the Yellowstone ecosystem attract tourists from all over the world. Much of the landscape is forested with lodgepole pine, but other habitats include alpine tundra, sagebrush, grasslands, riparian thickets, and aspen groves.

Birds: WatchList species include the Clark's Grebe, Trumpeter Swan, Swainson's Hawk, Greater Sage-Grouse, Long-billed Curlew, Short-eared Owl, Calliope Hummingbird, Lewis's Woodpecker, Williamson's Sapsucker, Olive-sided and Willow Flycatchers, Pinyon Jay, Sage and Brewer's Sparrows, Lark Bunting, and Black Rosy-Finch. High concentrations of Peregrine Falcons, Great Gray Owls, and Barrow's Goldeneyes also occur. Black-backed and Three-toed Woodpeckers frequent recently burned forests. More common species with significant continental population declines include the American Bittern, Ruffed Grouse, Rufous Hummingbird, Loggerhead Shrike, Horned Lark, Lark Sparrow, Snow Bunting, and Evening Grosbeak.

Threats: Over-browsing of habitat by elk, and higher than normal predation of birds by coyotes due to the absence of wolves are problems that are improving following wolf reintroductions. Fire suppression has led to massive fires that have destroyed significant areas of habitat (1.4 million acres in 1988), though these policies are now changing. Climate change threatens alpine habitats, and the use of snowmobiles within park boundaries produces pollution and noise that disturbs wildlife.

Conservation: Most of the greater Yellowstone ecosystem is protected on federal lands, anchored at the center by the 2.2 million-acre Yellowstone NP (the world's oldest NP, established in 1872). Adjacent are Grand Teton NP; and Bridger Teton, Shoshone, and Custer NFs. In Shoshone NF, prescribed burns are used to enhance grasslands, sagebrush, and aspen habitats. Wolf reintroduc-

The gregarious Pinyon Jay caches seeds in the ground for later use. Seeds left uneaten help regenerate pinyon pine forests.

PHOTO: ROBERT ROYSE

Yellowstone is famous for its geysers and dramatic scenery, but it is also an important area for some waterfowl and many land-birds.

PHOTO: ©ISTOCKPHOTO.COM

tions began in Yellowstone in 1995. The wolves have since reduced numbers of coyote (which had suppressed bird populations through predation), and significantly thinned elk herds, allowing over-browsed riparian and aspen woodland habitats to recover. Carcasses left behind by wolves also benefit scavengers, such as ravens, eagles, and Black-billed Magpies.

Actions:
- Continue to restore natural fire regimes, large predators, migration patterns of bison and elk, and other natural processes.

PHOTO: ROBERT ROYSE

The Rufous Hummingbird winters primarily in western Mexico; it has declined by 55% range-wide since 1966.

IMPORTANT BIRD AREA
ROCKY MOUNTAIN NP

BIRDSCAPE: **WESTERN FORESTS**
BCR: **16**   STATE: **CO**
AREA: **265,769 ACRES**

WATCHLIST SPECIES
RED: **1**
YELLOW: **19**

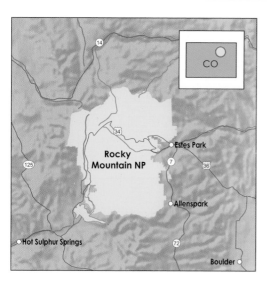

This park ranges in altitude from 7,500 to 14,259 feet, and includes coniferous forests (ponderosa pine, lodgepole pine, Douglas-fir, and spruce); and riparian corridors of willow, aspen, sagebrush, open meadows, and alpine tundra.

Birds: The park is an important breeding area for the WatchListed Black Swift and Brown-capped Rosy-Finch. Other WatchListed breeding species include the Clark's Grebe, Swainson's Hawk, Flammulated Owl, Calliope Hummingbird, Williamson's Sapsucker, Olive-sided Flycatcher, and Virginia's Warbler. During winter or migration, other WatchList species occur, including the Lewis's Woodpecker, Marbled Godwit, Western Sandpiper, Grace's Warbler, and Black Rosy-Finch, as well as the occasional Long-billed Curlew and Lark Bunting. Other species of interest include the Dusky Grouse, White-tailed Ptarmigan, Band-tailed Pigeon, and Evening Grosbeak.

Threats: Elk were hunted out of the park by the 1870s, and wolves by 1900. Elk were reintroduced beginning in 1913, but without wolves to

PHOTO: ROBERT ROYSE

Virginia's Warbler nesting success is reduced during dry years; the species has declined by 30% since 1966.

PHOTO: ©ISTOCKPHOTO.COM / SHERWOOD IMAGERY

Warming temperatures can facilitate bark beetle outbreaks in conifers and also threaten alpine tundra (Hallett Peak pictured).

keep their populations in check, they have become overabundant, placing unnatural grazing pressure on native habitats. Fire suppression has increased the risk of severe forest fires through the buildup of fuel loads. High-altitude glaciers and tundra are vulnerable to visitor over-use and climate change, and warmer temperatures also facilitate bark beetle outbreaks in pine forests. Other threats include air pollution, invasive plants in high-use areas, recreational disturbance, and the expansion of urban areas outside the park's boundaries.

Conservation: Tourists are limited to assigned areas. Approximately 95% of the park is roadless, and either designated or recommended as Wilderness. Much of the boundary (62%) is buffered by the Arapaho NF. Park managers now use fencing and culling to control elk populations, and hence restore vegetation. The National Park Service is attempting to control beetle outbreaks by using insecticides and removing infested trees.

Actions:
• Restore natural vegetation through prescribed burns, thinning, and controlling elk (by culling or excluding with fences).

• Control bark beetle outbreaks through tree management and the careful application of pesticides.

PHOTO: MICHAEL WOODRUFF

The Flammulated Owl nests in ponderosa pine cavities and forages in open areas that are reduced by fire suppression.

IMPORTANT BIRD AREA:
CORONADO NF

BIRDSCAPE: **WESTERN FORESTS**
BCR: **34**   STATE: **AZ, NM**
AREA: **1.8 MILLION ACRES**

WATCHLIST SPECIES
RED: **3**
YELLOW: **30**

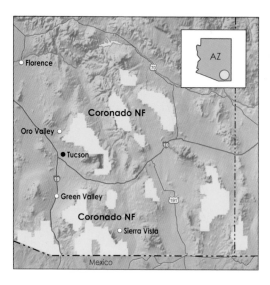

C oronado spans 12 mountain ranges in southeastern Arizona and adjacent southwestern New Mexico. The mountains rise more than 5,000 feet above surrounding grasslands and deserts to form "sky islands" of madrean pine-oak forest, oak savanna, riparian sycamore forest, juniper and chaparral brush-land, ponderosa and Chihuahuan pine forest, and spruce-fir forest.

Birds: Arizona's sky islands support many primarily Mexican species found nowhere else in the U.S. Breeding WatchList species include the Swainson's Hawk, Montezuma and Scaled Quail; Flammulated, Elf, and Spotted Owls; Blue-throated and Costa's Hummingbirds, Elegant Trogon, Arizona Woodpecker, Bendire's Thrasher, Mexican Chickadee; Virginia's, Lucy's, Grace's, and Red-faced Warblers, Varied Bunting, Abert's Towhee; and Black-chinned, Rufous-winged, and Five-striped Sparrows. Other WatchList species pass through during migration or spend the winter, including the Calliope Hummingbird, Williamson's Sapsucker, Olive-sided Flycatcher, Pinyon Jay, Gray Vireo, Sprague's Pipit, Brewer's Sparrow, Lark and Painted Buntings, and Chestnut-collared

Past fire suppression policies have led to catastrophic fires capable of killing adult trees, such as this burn in the Chiricahuas.

PHOTO: DANIEL LEBBIN

PHOTO: PETER LATOURETTE

The Montezuma Quail relies on tubers and bulbs for food. Birds require territory sizes of 120 acres or more when breeding.

Longspur. Other principally Mexican species found here include the White-eared Hummingbird, Buff-breasted Flycatcher, and Olive Warbler. The Thick-billed Parrot also used to occur.

Threats: Proposed copper mining in Coronado threatens to destroy natural habitats and leak contaminated water. Thirty-five-thousand cattle are permitted to graze within the forest, impacting bird habitat. Poor fire management has led to catastrophic forest fires. Much of the NF occurs near to or along the U.S.-Mexico border, and associated law enforcement, fence construction, and illegal immigrant traffic affect land management. Many bird populations in this area depend on source populations or winter habitat in nearby Mexico, where habitat loss and degradation can be severe.

Conservation: Most of these lands were first protected as Forest Reserves beginning in 1902. Other protected areas buffer the forest, including Chiricahua NM and Fort Huachuca. Management now prioritizes biodiversity, fire safety (including prescribed burns), and recreation. Attempts at reintroducing Thick-billed Parrots conducted from 1986-1993 failed due to problems with disease, social behavior, and predators.

Actions:
- Continue fire management, including prescribed burns, to prevent catastrophic fires.
- Ensure that additional border security operations, such as fences and road construction, undergo stringent environmental review prior to approval.
- Reassess possible reintroduction of the Thick-billed Parrot.
- Implement safeguards to reduce the impact of mining.

EASTERN FORESTS

AREA: **548,830** SQ. MILES
STATES: **AL, CT, DE, GA, KY,
IL, IN, IO, MA, MD, ME, MI,
MN, NC, NH, NJ, NY, OH, PA,
RI, SC, TN, VA, VT, WI, WV**
BCRs: **13, 23, 24, 28, 29,** AND **30**

WatchList Species
Red: **1**
Yellow: **10**

Eastern forests are dominated by deciduous trees (oaks, hickories, beeches, maples, etc.), though pine barrens and savannas grow locally on poor soils, or in recently burned areas, and isolated spruce-fir forests grow atop the highest Appalachian Mountains. Storms, beavers, and infrequent forest fires create natural disturbances that "reset" forest succession. Early successional meadows and shrub-lands provide habitat for a different community of birds than do mature forests. Wetlands include streams, rivers, beaver ponds, coastal marshes, and many man-made lakes and ponds, but few natural lakes.

Birds: Eastern forests support a high diversity of breeding songbirds, most of which migrate to the tropics for the winter, including the spectacular wood-warblers. Eastern forests are also important for other migratory birds that pass through on the way to and from their breeding grounds to the north. The Red-headed Woodpecker, Willow Flycatcher, Wood Thrush; Blue-winged, Golden-winged, Prairie, Cerulean, Prothonotary, Swainson's, Kentucky, and Canada Warblers are all WatchList species that breed in eastern forests. The Northern Bobwhite, Ruffed Grouse, American Woodcock, Whip-poor-will, and Field

ILLUSTRATION: C. VEST

Birds: 1. American Woodcock; 2. Yellow-breasted Chat; 3. Golden-winged Warbler; 4. Blue-winged Warbler; 5. Broad-winged Hawks; 6. Scarlet Tanager; 7. Cerulean Warbler; 8. Worm-eating Warbler; 9. Wood Thrush. **Vegetation** (from early succession on left to mature forest on right): 10. sumac; 11. yellow poplar; 12. white oak; 13. dogwood. **Threats**: 14. habitat loss, fragmentation, and degradation due to development; 15. mountaintop removal mining; 16. acid rain; 17. birds suffer collisions with towers and other structures; 18. overbrowsing by overabundant white-tailed deer; 19. cowbird nest parasitism.

Eastern Forests Habitats	Condition	Threat Level	Flagship Bird
Mature upland hardwoods	Good	Low	Cerulean Warbler
Glades and barrens	Fair	Medium	Prairie Warbler
Appalachian mixed hardwoods	Fair	Medium	Canada Warbler
Early succession	Poor	High	Golden-winged Warbler

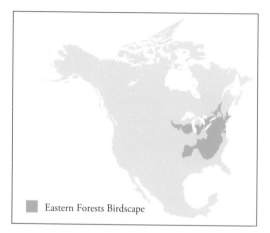

Eastern Forests Birdscape

Sparrow are more common species of early successional habitats that have suffered >50% population declines over the last 40 years. The WatchListed Henslow's Sparrow breeds in grasslands within this Birdscape.

Threats: Large portions of the eastern forests were historically cleared for agriculture and more recently, for urban development. Development permanently reduces forest, and fragments remaining areas; large sections of the coastal plain from Boston south to Richmond, and areas around Atlanta are now urbanized. Surface coal mining (including mountaintop removal mining) also contributes to habitat loss in some portions of the Appalachian Mountains. Poor forest management is currently a major issue, and most Appalachian forests are fairly homogenous in age and structure, requiring prescribed fires or thinning to stimulate understory regeneration and stand diversity. Eastern deciduous forests once burned naturally, though with a longer cycle than most western forests. This burning was encouraged by native Americans, and fire suppression has not benefitted the forest. Since European colonization, introduced diseases have ravaged trees, wiping out American chestnuts

and elms, and more recently, attacking hemlocks, dogwoods, ashes, and beeches. Outbreaks of introduced gypsy moths continue to defoliate large areas. Balsam wooly adelgids have killed most mature Fraser firs in their narrow range, and the loss of hemlocks to the hemlock wooly adelgid will drastically change our eastern forests. Browsing by overabundant white-tailed deer impedes forest regeneration and reduces understory for birds that depend on this vegetation for nesting. Acid rain damages northeastern forests and those in the Appalachians. Birds in forest fragments, or near urban and agricultural areas may suffer higher rates of cowbird nest-parasitism, as well as higher predation rates. As farms were abandoned over the last hundred years, land began to revert back into forest. At first, this benefitted early successional birds such as the Brown Thrasher and Prairie Warbler (and even the Henslow's Sparrow whose habitat further west had been decimated by agriculture),

Despite the recovery of eastern forests, the Cerulean Warbler population is still declining by 4.3% annually.

PHOTO: ROBERT ROYSE

Oak-hickory forests such as these in Monongahela NF, West Virginia, now dominate in areas once teeming with American chestnuts.

PHOTO: MIKE PARR

but later, as forests matured, many grass- and shrub-land birds declined.

Conservation: Forest cover in the eastern U.S. has increased over the past century, with much of the forest regeneration occurring on abandoned farmlands. Meanwhile, some abandoned strip mines have been restored as grasslands and shrub-lands. Concern over declining neotropical migrants has helped motivate a large increase in bird monitoring (and research into the effects of forest fragmentation) since the 1980s. Clean Air Act regulations have reduced acid rain by up to 35%, though acid rain's effects are still impacting forests and wetlands. Beavers have returned to much of the region and are creating additional wetlands and successional habitats. There is evidence of native dogwood resistance to anthracnose, the disease that decimated them in recent years.

Most of the largest tracts of forest remain within NFs in the Ozark and Appalachian Mountains, including Ozark (Arkansas), Allegheny (Pennsylvania), Daniel Boone (Kentucky), Monongahela (West Virginia), George Washington and Jefferson (Virginia, West Virginia, Kentucky), Nantahala (North Carolina), Chattahoochee (Georgia), Francis Marion and Sumter (South Carolina), Mark Twain (Missouri), and William B. Bankhead (Alabama) NFs. Other important sites include Shenandoah (Virginia) and Great Smoky Mountain (North Carolina) NPs, and Hawk Mountain (Pennsylvania) on the Kittatinny Ridge. Fewer large tracts of forest remain in the northern reaches of this Birdscape, and on the Mid-Atlantic coastal plain.

Actions:
- Reduce fragmentation from urban development.
- Prevent the introduction of more invasive species.
- Maintain early successional shrub-lands, including along power line cuts.
- Increase thinning and prescribed burns to restore forest structure and diversity.
- Halt mountaintop removal coal mining.
- Continue to restore mine sites and reforest reclaimed mine-lands.

IMPORTANT BIRD AREA
GREAT SMOKY MOUNTAINS

BIRDSCAPE: EASTERN FORESTS
BCR: 28   STATES: NC, TN
AREA: 521,490 ACRES

WATCHLIST SPECIES
RED: 1
YELLOW: 11

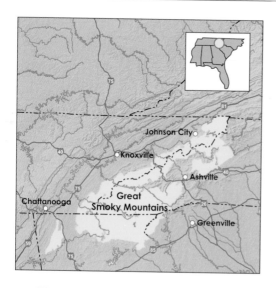

G reat Smoky Mountains NP is one of the largest protected areas in the eastern U.S., and is the most visited NP in the country. Along with the surrounding NFs, it provides important habitat for many bird species that require

undisturbed forest interior to breed. Most of the area is dominated by diverse deciduous forests, but small areas of spruce-fir forest occur at high elevations.

Birds: Breeding forest interior and early successional WatchList species include the Red-headed Woodpecker, Wood Thrush; and Golden-winged, Blue-winged, Prairie, Cerulean, Prothonotary, Swainson's, Kentucky, and Canada Warblers. During migration, additional WatchList species pass through, including the Olive-sided Flycatcher and Bay-breasted Warbler. Several typically northern species extend their ranges south here, including the Northern Saw-whet Owl, Black-capped Chickadee, Red-breasted Nuthatch, Winter Wren, Golden-crowned Kinglet, and Red Crossbill.

Threats: Originally named for the haze created by natural mists, the smoky appearance has more recently been rooted in a different cause: air pollu-

Large tracts of forest in the Great Smoky Mountains are critical for the Worm-eating Warbler and other forest-interior birds.

PHOTO: ROBERT ROYSE

Air pollution from power plants outside the park causes acid rain, which pollutes streams and reduces breeding success for birds.

PHOTO: ©ISTOCKPHOTO.COM / CHERYL E DAVIS

tion from coal-fired power plants and cars outside the park. Consequently the Park has high sulfur concentrations, which acidify rainwater, soil, and streams. Air pollution may also weaken the immune systems of trees and make them more vulnerable to disease and infestation. The park has lost all of its mature American chestnuts to blight, and 95% of its mature Fraser firs at higher elevations to wooly adelgids. Feral pigs are also a growing problem.

Conservation: The Park is contiguous with several large NFs, including Cherokee, Pisgah, Nantahala, and Chattahoochee. Together they protect over 3.3 million acres. Native fish, river otters, and Peregrine Falcons were reintroduced during the 1980s, and elk were reintroduced in 2001, but efforts to reintroduce red wolves in the 1990s were unsuccessful. Clean Air Act regulations have reduced pollutants, but better enforcement is needed. Plans to build a 30-mile-long road through the park were blocked in 2007.

Actions:
- Enforce Clean Air Act regulations to further reduce acid rain.
- Increase funding to increase park staffing.
- Control invasive species in the park.
- Restore populations of native animals (elk, wolves, fishers) and plants to help re-establish natural ecosystem function.

Mountaintop conifer forests are home to the most southerly Red Crossbills in the East.

PHOTO: ALAN WILSON

IMPORTANT BIRD AREA

OZARK NF

BIRDSCAPE: **EASTERN FORESTS**
BCR: **24**   STATE: **AR**
AREA: **1.1 MILLION ACRES**

WATCHLIST SPECIES
RED: **2**
YELLOW: **12**

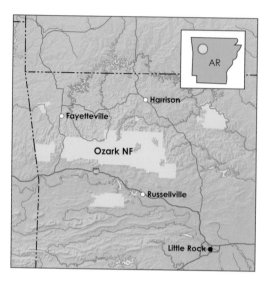

Ozark NF is one of the largest and least fragmented tracts of forest in the Central Hardwoods, and serves as a population source for many forest interior breeding birds. Hardwood forests dominated by oaks and hickories cover 70% of Ozark NF, with pine forest and glades occurring in some of the remaining areas.

Birds: Breeding WatchList species include the Red-headed Woodpecker, Wood Thrush; Blue-winged, Prairie, Cerulean, Prothonotary, Swainson's, and Kentucky Warblers; Painted Bunting, and Bachman's Sparrow. The WatchListed Olive-sided Flycatcher; and Golden-winged, Bay-breasted, and Canada Warblers also pass through during migration. Ozark NF also supports a large proportion of the eastern subspecies of the Bewick's Wren, which has disappeared from most of its range. Isolated populations of the Black-throated Green Warbler and Chestnut-sided Warbler breed here, far from their normal range to the north and east. Species with declining continental populations, such as Northern Bobwhite, American Woodcock, and Whip-poor-will also occur.

Threats: This NF is managed for multiple purposes, including recreation, timber production, and grazing, all of which can conflict with the requirements of birds. Additionally, much of the land is open for oil and gas development leases. Little of the original short-leaf pine forests persist today, and Red-cockaded Woodpeckers have disappeared from the area, though a small population of the Brown-headed Nuthatch remains in the Sylamore District. Housing development has also impacted habitat in the Ozark region.

Conservation: The forest was first protected in 1908. Currently, 66,728 acres are designated as Wilderness Areas. Low-intensity prescribed burns are used to maintain fire-dependent ecosystems and restore natural fire regimes.

Actions:

- Promote more widespread use of thinning and prescribed burns to restore forest structure and diversity, including both hardwood forests and short-leaf pine woodlands, for WatchList species and the Bewick's Wren.

Overabundant deer have dramatically thinned forest understory, impacting Kentucky Warbler breeding habitat.

PHOTO: ROBERT ROYSE

The Wood Thrush and other forest-interior species are able to nest most successfully in large undisturbed tracts.

The 1.1 million-acre Ozark NF was created in 1908 by a proclamation from President Theodore Roosevelt (Alley Spring pictured)

IMPORTANT BIRD AREA
KITTATINNY RIDGE

BIRDSCAPE: **EASTERN FORESTS**
BCR: **28**   STATE: **PA**
AREA: **2,400 ACRES**

WATCHLIST SPECIES
RED: **1**
YELLOW: **11**

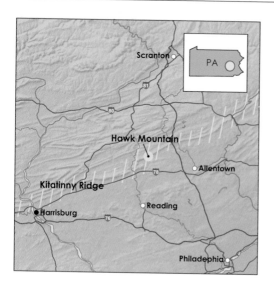

Pennsylvania's Kittatinny Ridge is a major migration corridor for birds, bats, butterflies, and dragonflies. Several important hawk watches occur along the ridge (such as Waggoner's Gap), with Hawk Mountain being the most famous.

Birds: The ridge is well-known among birdwatchers who come to witness the spectacle of the roughly 20,000 raptors of 16 species that pass overhead during each fall migration. In addition, another 140 or so other species are recorded regularly, including the WatchListed Red-headed Woodpecker, Olive-sided Flycatcher, Wood Thrush, and several migrant warbler species.

Threats: Government bounties once encouraged people to shoot raptors by the thousand, and DDT and other organochlorine pesticides took a heavy toll on many birds of prey. Now, communication towers and wind turbines along the Kittatinny Ridge and other nearby mountains are potential threats to migrating raptors, as well as songbirds, owls, and bats that migrate at night.

Conservation: Hawk Mountain Sanctuary was first established in 1934 following an early grassroots effort to stop the excessive killing of raptors during migration. Supporters of Hawk Mountain successfully advocated for changes in laws from those that offered bounties on hawks, to those that protected them. This is also the oldest operating hawk watch site in the world, providing long-term data on raptor trends. Rachel Carson cited declines in Bald Eagle numbers recorded at Hawk Mountain in her book *Silent Spring* as evidence of DDT's harm to birds. Today, the Sanctuary coordinates raptor monitoring projects across the hemisphere and helps train raptor biologists from around the world, including from Veracruz, Mexico, where more than one-million North American raptors "bottleneck" during fall migration.

Actions:
• Continue public education and outreach efforts at Hawk Mountain Sanctuary, as well as intern training programs for raptor biologists internationally.

PHOTO: GREG LAVATY

The Red-tailed Hawk is one of the most familiar eastern raptors and can be seen passing through Hawk Mountain in the fall.

PHOTO: GREG LAVATY

Early fall is the best time to watch large numbers of Broad-winged Hawks pass by on their way to South America.

PHOTO: DANIEL LEBBIN

Visitors to Hawk Mountain Sanctuary counted more than 15,000 raptors flying overhead in 2009.

SOUTHERN FORESTS

AREA: **354,535** SQ. MILES
STATES: **AL, AR, FL, GA, IL, IN, KY, MO, MS, NC, OK, LA, SC, TN, TX, VA**
BCRs: **25, 26, 27, AND 31**

WATCHLIST SPECIES
RED: **4**
YELLOW: **28**

Upland pine forests, evergreen oak scrub, and deciduous bottomland swamps characterize this Birdscape. Frequent but low-intensity fires maintain open pine savannas on uplands by burning away most deciduous trees and favoring the growth of fire-resistant ground cover (e.g., wire grass, palmettos), over-topped by a canopy of fire-resistant pine trees (e.g., longleaf pine). A unique evergreen oak scrub grows in Peninsular Florida, where sandy soils combined with frequent tree-destroying fires are inhospitable to even the hardy pines. Just as few trees can withstand regular fires, floods along major rivers such as the Missis-

sippi and its tributaries create some of the world's greatest bottomland swamps that favor tree species tolerant of inundated soils, such as tupelos, black gums, and coniferous cypresses. These floods also influence the understory, and in places, facilitate the growth of native canebreaks.

Birds: WatchList species that breed in southern forests include the Swallow-tailed Kite, Red-headed and Red-cockaded Woodpeckers, Willow Flycatcher, Florida Scrub-Jay, Wood Thrush; Prairie, Cerulean, Prothonotary, Swainson's, and Kentucky Warblers; Bachman's Sparrow, and

Birds: 1. Prothonotary Warbler; 2. Wood Duck; 3. Great Egret; 4. Swainson's Warbler; 5. Northern Bobwhite; 6. Red-headed Woodpecker; 7. Brown-headed Nuthatch; 8. Red-cockaded Woodpecker; 9. Bachman's Sparrow. **Reptiles**: 10. American alligator. **Vegetation**: 11. Spanish moss; 12. sycamore; 13. cypress; 14. water tupelo; 15. giant cane; 16. longleaf pine. **Threats**: 17. invasive kudzu; 18. nutria; 19. flood control; 20. loblolly pine plantations; 21. fire suppression.

SOUTHERN FORESTS HABITATS	CONDITION	THREAT LEVEL	FLAGSHIP BIRD
Bottomland hardwoods	Fair	Medium	Prothonotary Warbler
Cypress-tupelo-gum swamps	Good	Medium	Swallow-tailed Kite
Subtropical forest/hammocks	Poor	Medium	White-crowned Pigeon
Florida scrub	Poor	Critical	Florida Scrub-Jay
Longleaf pine	Poor	High	Red-cockaded Woodpecker

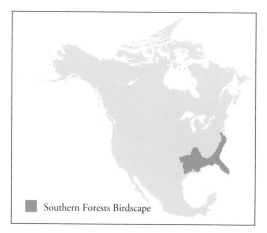

Southern Forests Birdscape

longleaf pine savanna, and 10-15% of the Florida scrub remain today, compared to the original extent of these forests 500 years ago. Fire suppression and flood control have tamed the two mighty disturbance processes which drove the evolution of southern forests, degrading much of the habitat that remains.

Conservation: Despite severe loss of forest and ongoing threats, conservation efforts are yielding success in stabilizing the populations of some endangered birds, and in restoring habitat in key areas of the southern forests. Thanks to management

Painted Bunting. Large numbers of birds that breed far to the north spend the winter in the bottomland swamps and in the grass of pine savannas, including the WatchListed Rusty Blackbird and Henslow's Sparrow. The southern forests were also once home to the extinct Carolina Parakeet, as well as the probably extinct Ivory-billed Woodpecker and Bachman's Warbler.

Threats: Historically, large areas of both the bottomland hardwoods and pine savannas were cleared for agriculture, or logged and replaced by plantation forests that were cut on short rotations to maximize wood production. These plantation forests are typically of lower value to specialized birds. For example, they are often harvested before trees can grow large enough to support cavity-nesting species such as the Red-cockaded Woodpecker. The vast canebreaks that once grew on rich alluvial soils were almost completely converted to agriculture. Large portions of Florida's oak scrub have been replaced by urban development, citrus orchards, and cattle ranches. Roughly 10% of the bottomland hardwood forest, less than 5% of

PHOTO: ROBERT ROYSE

The Red-headed Woodpecker can benefit from floods that create snags with plentiful nesting cavities.

Spanish moss hanging from trees conceals the nests of Northern Parulas and other birds of the southern swamps.

of old-growth longleaf pine savannas within NFs and military bases, and on commercial forestry lands through Safe Harbor Agreements, populations of the Red-cockaded Woodpecker have been increasing since the 1990s. Bottomland forests in the Mississippi Alluvial Valley are also recovering and increasing in acreage thanks to restoration projects. Populations of some WatchList birds of the bottomland swamps (such as the Swainson's Warbler) are also increasing as a result. Hopes for the survival of the Ivory-billed Woodpecker have been largely dashed, however. Since 2004, intensive searches have been conducted for this species throughout bottomland forests of the Southeast, yielding no confirmed sightings. Many parks, military bases, and public lands protect southern forests, including Apalachicola (Florida), Ouachita (Arkansas), Croatan (North Carolina), Francis Marion (South Carolina), Okefenokee (Florida),

Davy Crockett (Texas), Sam Houston (Texas), and Ocala (Florida) NFs; Avon Park Air Force Range (Florida), Eglin Air Force Base (Florida), Fort Benning and Fort Stewart (Georgia), Fort Bragg (North Carolina); Big Thicket National Preserve (Texas), Congaree Swamp NM (South Carolina), as well as the Big Woods of Arkansas, and Florida's Lake Wales Ridge.

Actions:

- Restore and prevent further loss of mature bottomland forest in the Lower Mississippi Alluvial Valley and coastal plains.
- Restore Florida scrub and longleaf pine savannas through prescribed burns, land protection, and incentive programs for private landowners.
- Develop and implement management plans for public lands that maintain or enhance habitat for WatchList species.

Important Bird Area
Big Woods

Birdscape: **Southern Forests**
BCR: **26**   State: **AR**
Area: **>500,000 acres**

WatchList Species
Red: **2**
Yellow: **21**

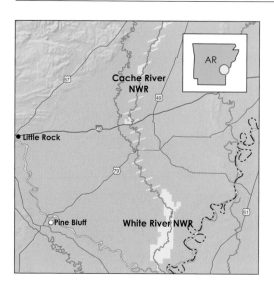

The Big Woods region of Arkansas contains two large NWRs and some of the best remaining bottomland hardwood forests in the Mississippi Alluvial Valley.

Birds: The Big Woods supports breeding populations of many WatchList species, including the King Rail, Red-headed Woodpecker, Wood Thrush; and Cerulean, Prothonotary, and Swainson's Warblers. During winter, large numbers of Rusty Blackbirds occur in the flooded forests, and Sprague's Pipits, Le Conte's Sparrows and Short-eared Owls use grassy areas. During migration, the WatchListed "Interior" Least Tern, American Golden-Plover, Western and Stilt Sandpipers, Painted Bunting; and Blue-winged, Golden-winged, Prairie, Bay-breasted, and Canada Warblers pass through. This area is also home to several species with declining continental populations, including the American Bittern, American Kestrel, Northern Pintail, American Woodcock, Whip-poor-will, Bewick's Wren (eastern subspecies), and Loggerhead Shrike. Over one-million waterfowl winter here, including 10% of the Mississippi flyway population of Mallard. The Ivory-billed Woodpecker and Bachman's Warbler once lived here, but have not been confirmed in many decades.

Threats: Dams, levees, and irrigation projects have drastically altered natural flood regimes along the Mississippi River and its tributaries, causing river

Wood Ducks have benefited from nest box programs that make up for a shortfall in natural tree cavities.

PHOTO: BILL HUBICK

Flooding plays a key role in the establishment of cypress swamps.Cypress is tolerant of deep water whereas other trees are not.

PHOTO: BILL HUBICK

channelization. Fortunately, the Cache and White Rivers retain much of their natural flooding characteristics, but future water management projects remain serious potential threats.

Conservation: The White River NWR was established in 1935 and protects 90 miles of forest along the river. Contiguous with its northern boundary is the 56,000-acre Cache River NWR. TNC had replanted 50,000 acres of bottomland hardwood forest there by 2005, and has plans to restore another 200,000 acres by 2015, while working with the U.S. Army Corps of Engineers and local farmers to restore water flows. Recent unconfirmed sightings of the Ivory-billed Woodpecker within the Cache River NWR galvanized the formation of the Big Woods Conservation Partnership, a coalition of government agencies and conservation groups, to restore bottomland forests in the area.

Actions:

- Restore 200,000 acres of additional bottomland hardwoods by 2015, through land acquisition, reforestation, and restoring water flows to degraded agricultural lands.

PHOTO: ROBERT ROYSE

Healthy stands of native giant cane occur within the Big Woods, providing nesting habitat for the Swainson's Warbler.

IMPORTANT BIRD AREA
APALACHICOLA NF

BIRDSCAPE: **SOUTHERN FORESTS**
BCR: **27**   STATE: **FL**
AREA: **836,961 ACRES**

WATCHLIST SPECIES
RED: **8**
YELLOW: **29**

L ocated in the Florida panhandle, this site protects one of the best remaining examples of southern pine forest (longleaf, slash, and loblolly) and pine savanna, along with swamp and bog habitats. The forest has almost 500,000 visitors each year.

Birds: This area supports 12% of the global population of the WatchListed Red-cockaded Woodpecker. Several other WatchList species breed in the area's forests and savannas including the Swallow-tailed Kite, Red-headed Woodpecker, Wood Thrush, Painted Bunting; and Prothonotary, Swainson's, and Kentucky Warblers. Wetlands and coastal habitats support the WatchListed Mottled Duck; Black, Clapper, and King Rails; Snowy, Wilson's, and Piping Plovers; Gull-billed Tern, and Black Skimmer. During winter and migration, additional WatchList species use the area, including the Yellow Rail, American Golden-Plover, Long-billed Curlew, Marbled Godwit, Red Knot, Sanderling, Western Sandpiper, Short-eared Owl, Rusty Blackbird; Henslow's, Le Conte's, and Nelson's Sparrows; and Golden-winged, Blue-winged, Prairie, Bay-breasted, Cerulean, and Canada Warblers. This area is also home to several species with declining continental populations, including the American Bittern, American Kestrel, Northern Bobwhite, American Woodcock, Whip-poor-will, Brown-headed Nuthatch, Loggerhead Shrike,

Forest succession has contributed to the decline of the Bachman's Sparrow, but prescribed burns can aid in its recovery.

PHOTO: ROBERT ROYSE

These healthy longleaf pines have blackened bark from periodic fires that clean out competing hardwoods.

PHOTO: MIKE PARR

Worm-eating Warbler, and Field and Grasshopper Sparrows.

Threats: Residential development on private land within and adjacent to public lands restrict the Forest Service's efforts to conduct controlled burns. Roads and transmission lines for electricity and natural gas pass through the area and fragment habitat. Other threats include surface water pollution, political pressure to increase timber sales, and additional road construction.

Conservation: Apalachicola NF was established in 1936. The forest management plan emphasizes conserving mature forest for Red-cockaded Woodpeckers, and requires prescribed burns on 30,000-50,000 acres annually. Red-cockaded Woodpeckers have been translocated from this healthy population to boost existing populations, and to establish new ones elsewhere in the Southeast.

Actions:
- Manage public land using frequent prescribed burns to maintain healthy pine savannas.
- Acquire private in-holdings to facilitate prescribed burns over larger areas.
- Restrict road construction or other activities that would disturb wildlife and fragment habitat.

PHOTO: ROBERT ROYSE

Red-cockaded Woodpeckers nest in mature pines, and use tree sap to defend their cavities from snakes.

IMPORTANT BIRD AREA
LAKE WALES RIDGE

BIRDSCAPE: **SOUTHERN FORESTS**
BCR: **31**　STATE: **FL**
AREA: **517,300 ACRES**

WATCHLIST SPECIES
RED: **6**
YELLOW: **12**

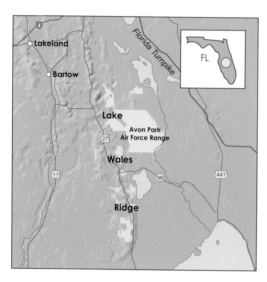

Lake Wales Ridge is an ancient sand dune in central Florida that supports a unique, arid, oak scrub and many endemic threatened plants and animals. Flatwood pine savannas, bayheads and other swamps, marshes, and seasonal ponds are additional habitats.

Birds: The ridge supports roughly 250 groups of the WatchListed Florida Scrub-Jay and small numbers of the Red-cockaded Woodpecker, Swallow-tailed Kite, and Bachman's Sparrow. King Rails are rare, year-round wetland inhabitants, and the Prairie Warbler winters here. In addition to WatchList species, Florida specialties include the "Florida" Sandhill Crane and Short-tailed Hawk. Birds with declining continental populations include the Northern Bobwhite, American Kestrel, American Woodcock, Burrowing Owl, Whippoor-will, Loggerhead Shrike, Field and Grasshopper Sparrows, and Eastern Meadowlark.

Threats: Much of the ridge and its surrounding areas have been cleared for agriculture, cattle ranches, and residential and commercial development. By 1990, 78% of the arid ridge habitat was lost, surpassing 85% by 2008. Habitat conversion further contributes to reduced fire frequency that degrades the remaining habitat. Invasive plants are also a threat, with climbing fern rapidly spreading through central Florida. Scrub-jays are also threatened by free-roaming cats.

Conservation: Approximately 11% of the Lake Wales Ridge is protected. The Archbold Biological Station anchors the southern end, and is the epicenter of research on the Florida Scrub-Jay. TNC is actively acquiring land, including the Tiger Creek Preserve. The state's Lake Placid Scrub Wildlife Management and Environmental Area was joined to the Archbold Station through acquisition in 2002. Lake Wales Ridge State Forest covers 26,563 acres. In 1991 the Lake Wales Ridge Ecosystem Working Group was formed to coordinate management of the area. Climbing fern monitoring and eradication work is underway. Despite all these valuable efforts, however, the overall scrub-jay population is still declining.

Actions:

- Expand protected areas within Lake Wales Ridge through land acquisition.
- Restore degraded lands on recently acquired properties, including the eradication of invasive plants.
- Use prescribed burns to maintain healthy oak scrub and pine habitat.
- Protect remaining parcels of natural scrub on private lands using Safe Harbor Agreements and Farm Bill incentives (see p. 433).
- Upgrade the Florida Scrub-Jay's status from Threatened to Endangered under the ESA.
- Require that cats to be kept indoors; halt any Trap Neuter and Release programs in the Lake Wales region.

The Florida Scrub-Jay is a cooperative breeder; the number of family groups continues to decline as central Florida is developed.

PHOTO: ROBERT ROYSE

Oak scrub (foreground) is maintained by periodic prescribed burns at Archbold Biological station.

PHOTO: DANIEL LEBBIN

GRASSLANDS AND PRAIRIES

AREA: **1,091,029** SQ. MILES
STATES: **CO, IL, IN, IO, KS, LA, MI, MN, MO, MT, ND, NE, NM, OH, OK, SD, TX, WY**
BCRs: **11, 17, 18, 19, 21, 22,** AND **37**

WATCHLIST SPECIES
RED: **14**
YELLOW: **25**

The prairies are dominated by grasses and forbs with scattered or little woody vegetation. Some, such as those in the "prairie potholes" region, also include important wetlands. Subtle changes in moisture and disturbance by wild fires and native grazing animals, such as bison, pronghorn, prairie dogs, and locusts, create an ever-shifting mosaic of different prairie habitats. Certain bird species (e.g., the Mountain Plover, McCown's Longspur) prefer recently burned or heavily grazed patches, while others (e.g., the Henslow's Sparrow) occupy grasslands left undisturbed for many years. Some birds such as prairie-chickens require com-

binations of both tall, dense prairies and shorter, open grasslands. To cope with changing habitats, many grassland birds have become nomadic to find the best breeding sites each season.

Birds: Grasslands support many WatchList species, including the Greater and Lesser Prairie-Chickens, Scaled Quail, Swainson's Hawk, Mountain Plover, Long-billed Curlew, Marbled Godwit, Short-eared Owl, Sprague's Pipit, Lark Bunting; Brewer's (winter), Baird's, Henslow's, Le Conte's, and Nelson's Sparrows; and Smith's (winter) and Chestnut-collared Longspurs. WatchListed breed-

ILLUSTRATION: C. VEST

Birds: 1. McCown's Longspur; 2. Ferruginous Hawk; 3. Burrowing Owl; 4. Greater Prairie-Chicken; 5. Mountain Plover; 6. Baird's Sparrow; 7. Lark Bunting; 8. Sprague's Pipit. **Mammals:** 9. black-tailed prairie dog; 10. bison. Threats: 11. overabundant Brown-headed Cowbirds; 12. over-grazing 13. agriculture; 14. fire suppression; 15. collisions with barbed-wire fences.

GRASSLANDS AND PRAIRIES HABITATS	CONDITION	THREAT LEVEL	FLAGSHIP BIRD
Prairie-oak savanna	Poor	High	Greater Prairie-Chicken
Tall-grass prairie	Poor	Critical	Henslow's Sparrow
Mixed-grass prairie	Fair	High	Long-billed Curlew
Short-grass prairie	Poor	Medium	Mountain Plover
Gulf coast prairie	Poor	Critical	"Attwater's" Greater Prairie-Chicken

Grasslands and Prairies Birdscape

been most extensive in the tall-grass prairies, with less than 4% of the original 148 million acres remaining; roughly 25% of the original >400 million acres of short-grass and mixed-grass prairies remain somewhat intact spatially if not ecologically. Many remaining fragments, often small and isolated, are unsuitable for most grassland bird species, however. Other threats include habitat degradation from over-grazing, mining, oil and gas exploration and extraction, and wind turbines. Invasive plants such as purple loosestrife are a problem throughout the Great Plains, clogging wetlands and creating habitat unsuitable for virtually all birds. The intensity and rotational period of grazing can make a huge difference to the value of habitat for grass-

ing birds closely tied to prairie wetlands include the Piping and Snowy Plovers, Clark's Grebe, Mottled Duck; and King, Yellow, and Black Rails. The WatchListed Whooping Crane, and shorebirds such as the American Golden-Plover, Hudsonian Godwit; and Semipalmated, Western, and Stilt Sandpipers also occur during migration. Aside from WatchList species, there are many more widespread grassland birds whose populations are steeply declining, including the Upland Sandpiper, Ferruginous Hawk, Eastern Meadowlark, Horned Lark, Dickcissel, McCown's Longspur; and Field, Grasshopper, Lark, and Harris's Sparrows.

Threats: Habitat loss and the disruption of natural disturbance regimes are the greatest threats to prairie birds. During the 1800s, most of the native grazing animals were eradicated as settlers converted western prairies to cattle grazing or agriculture. During the late 1900s, many of the grasslands east of the 100th meridian that could still support specialized birds were converted to row crops—monocultures with reduced habitat value. Other fields were planted with hay or alfalfa, the early harvesting of which destroys nests. Habitat loss has

PHOTO: GREG LAVATY

The subtle but stunning colors of the Le Conte's Sparrow allow it to blend into the wet grasslands it inhabits.

PHOTO: ©ISTOCKPHOTO.COM / KEN CANNING IMAGES

Grazing herds of nomadic bison once created a mosaic of short- and tall-grass prairie that provided diverse bird habitats.

land birds, and regimes that mimic natural systems can support good populations of grassland species. Climate change may alter precipitation patterns, drying some prairie wetlands out.

Conservation: Healthy grasslands need to be large enough to support the dynamic disturbance processes that leave patches in different successional stages. Otherwise, management is required to mimic these natural processes. Some of our healthiest prairies remain in protected areas, such as the Flint Hills-Konza Prairie (Kansas), Comanche National Grassland (Colorado), TNC's Tallgrass Prairie National Preserve (Oklahoma), and Lostwood NWR (North Dakota), where managers are restoring the natural disturbance regimes through prescribed burns, mowing, selective rotational grazing of cattle (to mimic the foraging pattern of bison), and recovering prairie dog and bison populations. In addition to grasslands protected on public lands, roughly 30 million grassland acres on private lands in the U.S. are enrolled in the CRP. Under this voluntary program, the federal government pays farmers to take marginal farmlands out of production under ten-to-15 year contracts. Many grassland birds do better on CRP lands than on croplands, particularly where native seed mixtures are used. However, recent mandates for increased biofuel production, combined with rising prices for corn and other agricultural products, are weakening the incentives for CRP enrollment, and the acreage is declining as contracts expire and are not renewed.

Actions:
- Increase the area of grassland managed to mimic natural disturbance regimes.
- Restore native grazing animals where possible
- Reduce over-grazing on public lands.
- Strengthen incentives for grassland conservation under the CRP and other Farm Bill programs.
- Maintain ranching on private lands to prevent their conversion to more intensive uses that are less compatible with wildlife.
- Control invasives; minimize the impact of energy extraction and wind farms.

IMPORTANT BIRD AREA

FLINT HILLS

BIRDSCAPE: **GRASSLANDS AND PRAIRIES**
BCR: **22** STATES: **KS, OK**
AREA: **8 MILLION ACRES**

WATCHLIST SPECIES
RED: **6**
YELLOW: **13**

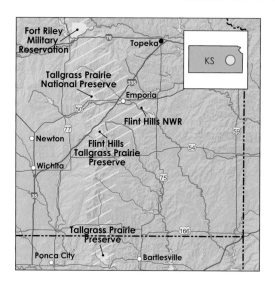

Birds: The area supports important populations of WatchListed grassland birds such as the Greater Prairie-Chicken, Swainson's Hawk, Short-eared Owl, Red-headed Woodpecker, Bell's Vireo, Painted and Lark Buntings, and Henslow's Sparrow. Many other WatchListed species occur during migration or winter including the King Rail; American Golden-, Snowy, and Piping Plovers; Long-billed Curlew, Hudsonian and Marbled Godwits; Semipalmated, White-rumped, Stilt, and Buff-breasted Sandpipers; Sprague's Pipit, Baird's and Le Conte's Sparrows, Smith's and Chestnut-collared Longspurs, and Rusty Blackbird. Other species with declining North American populations include the American Kestrel, Northern Bobwhite, American Woodcock, Whippoor-will, Loggerhead Shrike; Field, Grasshopper, and Harris's Sparrows; Dickcissel, and Eastern Meadowlark.

The Flint Hills protect one of the best remnant tall-grass prairies, spanning roughly eight-million acres of land under varying public and private ownerships, including the Fort Riley Military Reservation, Konza Prairie Research Natural Area, Flint Hills NWR, and the Flint Hills Tallgrass Prairie Preserve.

Threats: Historically, the tall-grass prairies were largely converted to agriculture, or degraded by fire suppression and the loss of native grazers such as bison. Restoring these disturbance processes, and

The Konza Prairie has never been plowed, and research conducted here informs grassland management well beyond Kansas.

PHOTO: DANIEL LEBBIN

Burning and grazing can create a mix of different grass heights used by Greater Prairie-Chickens for nesting and lekking.

controlling invasive plants remains a challenge in surviving prairie fragments.

Conservation: More than half of Fort Riley is native or restored prairie, where habitat is managed under a hay-lease program with prescribed burns. TNC purchased the Konza Prairie Research Natural Area, the Flint Hills Tallgrass Prairie Reserve, and the Tallgrass Prairie Preserve, which are carefully managed for native grasslands, and are also used for studies on prairie restoration and management. The 18,500-acre Flint Hills NWR was established in 1966, and adjacent farmlands are managed on a share basis with private owners. The Tallgrass Legacy Alliance, a coalition of government agencies, conservation groups, and ranchers, facilitates tall-grass prairie conservation on private lands through education, easements, and other voluntary incentive programs.

Actions:
- Continue prescribed burns and rotational grazing of cattle or bison.
- Maintain bird-compatible ranching on private lands to prevent their conversion to uses that are less beneficial to wildlife.
- Strengthen Farm Bill or other incentives for habitat conservation on private lands.

The Swainson's Hawk nests in the Flint Hills and winters as far south as Argentina.

IMPORTANT BIRD AREA

COMANCHE NATIONAL GRASSLAND

BIRDSCAPE: **GRASSLANDS AND PRAIRIES**
BCR: **18**   STATE: **CO**
AREA: **435,707 ACRES**

WATCHLIST SPECIES
RED: **7**
YELLOW: **18**

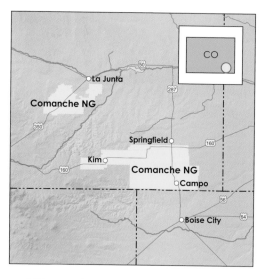

omanche NG is a fine example of short-grass prairie, and an important site for the Lesser Prairie-Chicken. It is managed for wildlife, livestock grazing, recreation, and mining.

Birds: This site supports approximately 500 Lesser Prairie-Chickens with ten leks, as well as breeding WatchListed Scaled Quail, Swainson's Hawk, Mountain and Snowy Plovers, Long-billed Curlew, "Interior" Least Tern, Red-headed Woodpecker, Willow Flycatcher, Pinyon Jay, Bell's Vireo, and Lark Bunting. Mountain Plovers also use the area during migration. Other migrants or wintering WatchList species include the Whoop-

ing Crane, Piping Plover; Semipalmated, Western, White-rumped, and Stilt Sandpipers; Short-eared Owl, Lewis's Woodpecker, Olive-sided Flycatcher, Virginia's Warbler, and Brewer's Sparrow. This site also supports declining grassland species such as the Ferruginous Hawk, American Kestrel, Loggerhead Shrike, and Grasshopper Sparrow.

Threats: Energy production threatens both habitat and birds. Comanche's Campo Field has six oil wells nearing the end of their productive lifespan. In 2005, a court ruled that a revised management plan for Comanche violated several environmental laws and needed to be reworked. The development of wind energy could affect this area in the future.

Conservation: This NG was established to restore degraded lands following the "Dust Bowl" of the 1930s. Rotational grazing is used to enhance habitat, and prescribed burns can, for example, double the number of breeding Mountain Plovers in a given area. The eradication of invasive salt cedar (in riparian areas) and cheatgrass (and replacement with cottonwoods, bluestem grass, and switchgrass) is ongoing. Blinds allow birders to visit and observe prairie-chicken leks without disturbing the birds.

Actions:

- Ensure that the management plan complies with existing environmental laws.
- Enhance habitat for the Lesser Prairie-Chicken and assure that traditional leks are not harmed by energy extraction.
- Expand the size of the breeding Mountain Plover population through prescribed burns in prime areas.
- Ensure grazing is managed to enhance habitat.
- Continue efforts to eradicate invasive salt cedar and cheatgrass, and restore natural vegetation and fire cycles.

PHOTO: ROBERT ROYSE

Rotational grazing and prescribed burns can create short-grass conditions preferred by the McCown's Longspur.

Expanding energy development threatens the Lesser Prairie-Chicken. Its range has declined by 90% from its historical distribution.

PHOTO: NOPPADOL PAOTHONG

The Mountain Plover prefers bare ground in recently burned areas, or grasslands clipped short by prairie dogs.

PHOTO: ROBERT ROYSE

IMPORTANT BIRD AREA
LOSTWOOD NWR

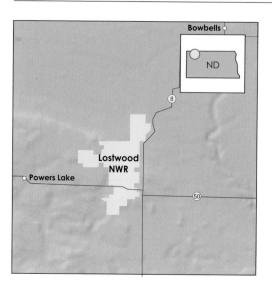

Lostwood protects short- and mixed-grass prairie, interspersed with "prairie pothole" wetlands. An estimated 70% of the refuge is virgin prairie. Most of the wetlands have never been drained.

Birds: Lostwood supports a variety of Watch-Listed breeding species, including the Piping Plover (30 pairs), Marbled Godwit, Short-eared Owl, Willow Flycatcher, Sprague's Pipit, Lark Bunting; Baird's, Le Conte's, and Nelson's Sparrows; and the Chestnut-collared Longspur. Other breeding species include the Sharp-tailed Grouse (one of the largest U.S. populations), Ferrugi-

nous Hawk, Northern Pintail, Canvasback, Willet, Upland Sandpiper, Wilson's Phalarope, Black Tern, Grasshopper and Clay-colored Sparrows, and Bobolink. Other WatchList species occurring during winter or migration include the Whooping Crane, Yellow Rail, American Golden-Plover; Semipalmated, White-rumped, and Stilt Sandpipers; Red-headed Woodpecker, Smith's Longspur, and Rusty Blackbird. The Greater Prairie-Chicken once occurred on the refuge, but has not been reported since 1952.

Threats: Fire suppression and loss of bison and other native grazers have altered natural disturbance cycles, and management is required to mimic these processes. In the surrounding landscape, wetland drainage, fire suppression, and invasive plants are also problematic. Coal-burning power plants pollute the air and water, contaminating wetlands with mercury and other toxins. According to the EPA, the average temperature in North Dakota has increased by more than a degree over the last century, and could increase three to four over the next. This could exacerbate droughts and significantly reduce pothole ponds. Although periodic drying is necessary to maintain healthy potholes, climate change threatens to cut wetlands and the number of breeding ducks by half.

Conservation: Refuge managers use a combination of prescribed burns and selective grazing to combat the spread of exotic plants, inhibit woody vegetation, and create favorable breeding conditions for Baird's Sparrows and other birds.

Actions:
- Use cattle grazing and prescribed burns to mimic natural disturbance cycles.
- Reduce emissions from coal-fired power plants in North Dakota; favor energy production from emission-free sources.
- Evaluate the feasibility of reintroducing the Greater Prairie-Chicken to the refuge.

Much of the Baird's Sparrow's habitat has been degraded by invasive weeds, fire-suppression, or conversion to crops.

PHOTO: ROBERT ROYSE

The prairie potholes are the nation's "duck factory", and also provide habitat for shorebirds such as the Wilson's Phalarope.

PHOTO: ROBERT ROYSE

Bobolinks find refuge at Lostwood, but have declined range-wide as hayfields are converted to crops or other land uses.

PHOTO: ROBERT ROYSE

SOUTHERN ARID LANDS

AREA: **308,676** SQ. MILES
(INC. MEXICO)
STATES: **AZ, NM, TX**
BCRs: **20, 35,** AND **36**

WATCHLIST SPECIES
RED: **8**
YELLOW: **19**

The arid lands of southern and western North America are bisected by the Rocky Mountains and Sierra Madre Occidental. The Southern Arid Lands Birdscape lies to the east of the Rockies and contains three distinct habitat types. To the west, the dry Chihuahuan Desert vegetation includes grasses, shrubs such as creosote and acacia; along with yuccas, agaves, prickly pear cacti, ocotillo, Mormon tea, tarbush, and mesquite (along washes). The Tamaulipan Brushlands are grassy shrub-lands with thorny brush and trees, including mesquite and acacia. Lush forests of sabal palm and Montezuma bald cypress once grew along the meandering Rio Grande, a riparian corridor that flows through the Tamaulipan Brushlands, but

only remnants now remain. Both habitats extend significantly into Mexico. In the third habitat type, the Edward's Plateau, mesquite and acacia savannas transition into oak- and ashe-juniper and shinnery oak woodlands across limestone karst.

Birds: This Birdscape supports many breeding WatchList species, including the Scaled and Montezuma Quail, Mountain Plover, Green Parakeet, Red-crowned Parrot, Elf Owl, Blue-throated Hummingbird, Willow Flycatcher; Black-capped, Gray, and Bell's Vireos; Virginia's, Colima, Lucy's, and Golden-cheeked Warblers; Black-chinned Sparrow, Varied and Painted Buntings, and Audubon's Oriole. Several WatchList species that breed

ILLUSTRATION: C. VEST

Birds: 1. Varied Bunting; 2. Colima Warbler; 3. Scaled Quail; 4. Zone-tailed Hawk; 5. Greater Road-runner; 6. Black-capped Vireo; 7. Golden-cheeked Warbler. **Vegetation**: 8. mesquite; 9. prickly pear cactus; 10. oak scrub; 11. ashe-juniper. **Threats**: 12. invasive salt cedar; 13. red fire ants; 14. development; 15. water drawdown.

Southern Arid Lands Habitats	Condition	Threat Level	Flagship Bird
Ashe-juniper	Poor	High	Golden-cheeked Warbler
Oak scrub	Poor	High	Black-capped Vireo
Cliffs and caves	Good	Medium	Cave Swallow
Lower Rio Grande woodlands	Poor	Critical	Altamira Oriole
Tamaulipan brush-land	Good	Low	Olive Sparrow
Chihuahuan desert	Fair	Medium	Scaled Quail

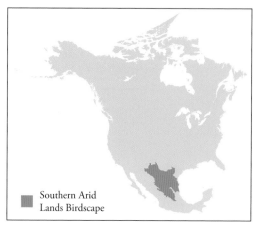

Southern Arid
Lands Birdscape

in prairies to the north winter here, including the Sprague's Pipit; Brewer's, Sage, Baird's, and Le Conte's Sparrows; Lark Bunting, and Chestnut-collared Longspur. Several more widespread species that have experienced >50% population declines over the last 40 years also occur in this region, including the Eastern Meadowlark, Loggerhead Shrike; and Field, Grasshopper, Black-throated, and Lark Sparrows.

Threats: Much of the habitat has been converted to agriculture, cleared for urban development, or degraded by over-grazing. For example, 90% of the natural habitat in the Edwards Plateau has been developed, and much of the remaining area is fragmented and suffers from fire suppression. Over-grazing, and the clearance of riparian habitats has had the largest impact on the Tamaulipan region. Groves of sabal palm and bald cypress were once found throughout much of the lower Rio Grande Valley, but are now rare. Grass was the dominant land cover across much of the Chihuahuan Desert during the 1850s, but well-drilling provided greater access to water for cattle, leading to over-

grazing which has since favored creosote and mesquite shrubs over desert grasses. Other threats to habitats in this Birdscape include road building, mining, invasive species, and water diversion and withdrawal activities.

Conservation: Protected areas in the U.S. portion of the Chihuahuan Desert include Bosque del Apache NWR, White Sands NM, Carlsbad Caverns NP, and Lincoln NF (New Mexico), and Guadalupe Mountains NP and Kickapoo Cavern SP (Texas). Lands owned by the government and TNC protect much of the Davis Mountains in southwest Texas. Protected areas in the Edward's Plateau and Tamaulipan Brushlands tend to be smaller, including Balcones Canyonlands, Lower Rio Grande Valley, and Santa Ana NWRs, and Bentsen-Rio Grande Valley SP (all Texas).

Actions:

- Increase acreage of oak-juniper habitat in the Edwards Plateau through the expansion of Balcones Canyonlands NWR, and through fire management and incentives for private landowners.
- Reduce grazing pressure to restore desert grasslands and riparian woodlands.

Habitat degradation due to over-grazing has contributed to the Scaled Quail's 66% population decline since 1966.

PHOTO: TOM GREY

Water diversion for agriculture and urban areas can impact natural ecosystems in arid areas such as along the Rio Grande Valley.

Greater Roadrunners build their nest platforms of sticks low down in cacti or bushes. They can run at speeds of up to 20 mph.

IMPORTANT BIRD AREA:
BIG BEND NP

BIRDSCAPE: **SOUTHERN ARID LANDS**
BCR: **35**   STATE: **TX**
AREA: **801,163 ACRES**

WATCHLIST SPECIES
RED: **0**
YELLOW: **24**

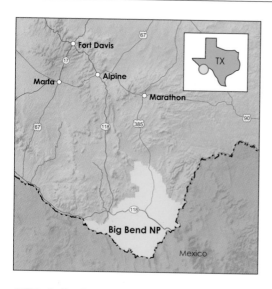

Big Bend spans 118 miles of the Rio Grande River in southwest Texas, and encompasses a variety of habitats from river floodplains, to shrub desert, grasslands, pinyon-oak-juniper woodlands, and moist conifer forests in the high Chisos Mountains. It is the only place to reliably find the Colima Warbler in the U.S.

Birds: Breeding WatchList species include the Scaled Quail, Flammulated and Elf Owls, Blue-throated Hummingbird, "Southwestern" Willow Flycatcher, Black-capped Vireo (rare), Colima and Lucy's Warblers, Varied and Painted Buntings, and Black-chinned Sparrow. Other WatchList species occur here during winter and migration, including the Swainson's Hawk, Long-billed Curlew, Short-eared Owl, Calliope Hummingbird, Olive-sided Flycatcher, Sprague's Pipit, Virginia's Warbler, Brewer's Sparrow, Lark Bunting, and Chestnut-collared Longspur. Many of the species on the park's list occur as vagrants, including species more commonly found in the Rio Grande Valley or the sky islands of southeastern Arizona.

Threats: Air and water pollution degrade visibility and water quality, and damage vegetation. Water

diversion to urban areas from the Rio Grande and its tributaries reduces water flow and increases the concentration of pollutants in the remaining water, stressing both vegetation and wildlife. Reduced flows and drought have reduced breeding success of some birds, including the Black-capped Vireo.

Conservation: Big Bend was first protected as a State Park in 1933, and later established as a NP in 1944. The National Park Service has recently proposed adding an adjacent 9,263-acre tract encompassing the Christmas Mountains. TNC has also protected 64,000 acres here. The Park is part of the largest trans-boundary park in North America, connecting with protected areas in Mexico, and covering two-million acres in total.

Actions:
- Reduce air pollution by cleaning up coal-fired power plants in Texas and Mexico.
- Ensure adequate water flows through the Rio Grande by restricting water diversions upstream.
- Add the Christmas Mountains to Big Bend NP.

The first nest of the Colima Warbler was discovered in Big Bend; this remains the only known breeding locality in the U.S.

PHOTO: GREG LAVATY

The stark beauty of Juniper Canyon belies the richness and complexity of Big Bend's ecosystems; this is a must-go for American birders.

PHOTO: ©ISTOCKPHOTO.COM / DAVID HUGHES PHOTOGRAPHY

PHOTO: ROBERT ROYSE

Although the Varied Bunting may be expanding its range northward in Arizona, it appears to be declining in Texas.

PHOTO: GREG LAVATY

Habitat loss, cowbird parasitism, and trapping for the Mexican bird trade have reduced Painted Bunting numbers.

IMPORTANT BIRD AREA
FORT HOOD

BIRDSCAPE: **SOUTHERN ARID LANDS**
BCR: **20**   STATE: **TX**
AREA: **217,000 ACRES**

WATCHLIST SPECIES
RED: **2**
YELLOW: **4**

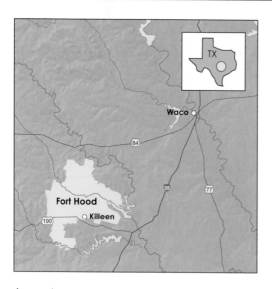

The fort is located on the Edwards Plateau of central Texas. Approximately 65% of the property is grassland and 30% is forest, dominated by oaks and junipers in varying stages of succession. The largest breeding populations of the Black-capped Vireo and Golden-cheeked Warbler under a single management authority occur here.

The Golden-cheeked Warbler breeds only in Texas, where it prefers mature woodlands of oak and ashe-juniper.

PHOTO: GREG LAVATY

Birds: Breeding WatchList species include Black-capped Vireo (8,000 singing males) and Golden-cheeked Warbler (5,400 singing males), as well as the Swainson's Hawk, Prothonotary Warbler, and Painted Bunting. The WatchListed Le Conte's Sparrow occurs during migration and winter. Species with declining continental populations also occur on the fort, including the Northern Bobwhite, Loggerhead Shrike, and Grasshopper Sparrow.

Threats: Nest parasitism by Brown-headed Cowbirds threatens both the Black-capped Vireo and Golden-cheeked Warbler. Invasive fire ants also cause nest failure among the vireos.

Conservation: Fort Hood is managed for both military training and for wildlife. Early successional shrub-lands are maintained by prescribed burns and fires caused by exploding ordinance during training. TNC has a cooperative agreement with the Fort, helping to remove cowbirds and to conduct prescribed burns. The Fort's cowbird control program reduced Black-capped Vireo nest parasitism rates from 90% to <10% from 1988-1998. Outside the Fort, cowbirds have been observed to parasitize 68% of Golden-cheeked Warbler nests in Kendall County, but no parasitism has been observed at the Fort since the cowbird control program began, and the warbler population has increased. Both the Black-capped Vireo and Golden-cheeked Warbler are threatened range-wide by habitat loss. The vireo prefers early successional habitat, two to 15 years after fire, whereas the warbler moves in as the oak-juniper scrub matures, creating a management challenge that requires an ongoing effort to provide for both species.

Actions:
- Manage burns and control cowbirds to benefit both Black-capped Vireos and Golden-cheeked Warblers.
- Control fire ants and other invasive species.

Wildfires at Fort Hood help maintain habitat for Black-capped Vireos; fire suppression prevents this habitat from regenerating.

PHOTO: JERRY R. OLDENETTEL

Thanks to intensive conservation efforts, more than 8,000 Black-capped Vireos have been counted at Fort Hood (female pictured).

PHOTO: GREG LAVATY

IMPORTANT BIRD AREA
LOWER RIO GRANDE VALLEY

BIRDSCAPE: **SOUTHERN ARID LANDS**
BCR: **36**   STATE: **TX**
AREA: **1.2 MILLION ACRES**

WATCHLIST SPECIES
RED: **14**
YELLOW: **32**

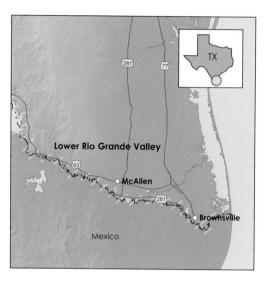

The Lower Rio Grande Valley stretches 275 miles and spans 1.2 million acres along the border between Texas and Mexico. It is the only place in the U.S. to find some primarily Mexican bird species, and is thus a major draw for birders. Habitats include sabal palm groves, riparian woodlands, oxbow lakes, salt flats, deserts, coastal grasslands, marshes, mangroves, beaches, and mudflats.

Birds: This region supports breeding populations of the WatchListed Mottled Duck, Scaled Quail, Swainson's Hawk, Reddish Egret, Clapper Rail, Gull-billed and Least Terns, Green Parakeet, Red-crowned Parrot, Elf Owl, Bell's Vireo, Painted and Varied Buntings, Seaside Sparrow, and Audubon's Oriole. During winter and migration, many other WatchList species occur, including the Swallow-tailed Kite; Yellow, Black, and King Rails; American Golden-, Mountain, Piping, Snowy, and Wilson's Plovers; Long-billed Curlew, Hudsonian and Marbled Godwits; Semipalmated, Western, White-rumped, Stilt, and Buff-breasted Sandpipers; Olive-sided and Willow Flycatchers, Wood Thrush, Sprague's Pipit; Blue-winged, Golden-winged, Cerulean, Prothonotary, Kentucky, and

Canada Warblers; Lark Bunting, and Le Conte's Sparrow. Aplomado Falcons have been reintroduced.

Threats: More than 95% of the valley has been converted to agriculture or residential and commercial development, and only remnants of riparian forest are left. The remaining habitat is now highly fragmented, which makes birds vulnerable to nest predation, and to parasitism by Bronzed Cowbirds. Agriculture depends on water diverted from the Rio Grande, and this has drastically reduced natural flooding. As a result, wetlands have dried up and riparian forests have been replaced by scrub. Additional threats include pesticides and invasive species (e.g., exotic grasses and feral pigs). Border fence construction is also fragmenting remaining patches of habitat.

Conservation: Protected areas include Falcon SP, Bentsen-Rio Grande Valley SP, Santa Ana and Lower Rio Grande Valley NWRs, and the Sabal Palm Grove Audubon Center and Sanctuary (which protects the best remaining 172-acre palm grove). Management typically includes removal of invasive species and habitat restoration, including water pumping to mimic natural floods at Santa Ana. Birders contribute over $100 million to the local economy, and are attracted year round to birding festivals in McAllen, Brownsville, and Harlingen.

Actions:
- Acquire and protect habitat, including expanding the Lower Rio Grande Valley NWR to its full acquisition goal.
- Control invasive vegetation and manage water to mimic natural floods.
- Ensure that the environmental impact of border security measures is minimized.

This palm-bordered lake at Sabal Palm Grove is the last best example of this habitat on the lower Rio Grande.

PHOTO: DANIEL LEBBIN

In the U.S., the Audubon's Oriole occurs only in the Rio Grande Valley; it is a victim of habitat loss and Bronzed Cowbirds parasitism.

PHOTO: GREG LAVATY

WESTERN ARID LANDS

AREA: **511,732 SQ. MILES**
STATES: **AZ, CA, ID, NV, OR, UT, WA**
BCRs: **9, 32, AND 33**

WATCHLIST SPECIES
RED: **10**
YELLOW: **24**

The Western Arid Lands Birdscape is found west of the Rocky Mountains. Habitats are united by arid climates, but differ in seasonal temperatures, rainfall, and vegetation. Sagebrush and other dry shrub-lands dominate most of the Great Basin. To the south, these give way to deserts characterized by yuccas and cacti, including the Joshua tree (the signature yucca of the Mojave), and the saguaro cactus (emblematic of the Sonoran desert). In coastal California, a Mediterranean climate of hot, dry summers and cool, wet winters encourages low-lying chaparral and coastal sage scrub. Grasslands are also found along the coast as well as further inland.

Birds: Sagebrush supports a variety of specialist breeders, including the Sage Thrasher, the Watch-Listed Greater and Gunnison Sage-Grouse, and Sage and Brewer's Sparrows. The deserts and their remaining riparian corridors support many Watch-List breeders, including the Gilded Flicker, Costa's Hummingbird, "Southwestern" Willow Flycatcher, Bendire's and Le Conte's Thrashers, Lucy's Warbler, Abert's Towhee, and Rufous-winged Sparrow. The WatchListed Allen's Hummingbird, "Least" Bell's Vireo, California Gnatcatcher, California Thrasher, Wrentit, Black-chinned and Sage Sparrows, and Lawrence's Goldfinch breed in California's chaparral. Several more widespread birds

ILLUSTRATION: C. VEST

Birds: 1. Brewer's Sparrow; 2. Sage Thrasher; 3. Sage Sparrow; 4. Greater Sage-Grouse; 5. Le Conte's Thrasher; 6. Costa's Hummingbird; 7. Elf Owl; 8. Gilded Flicker. **Vegetation**: 9. sagebrush; 10. rabbit-brush; 11. Joshua tree; 12. yucca; 13. saguaro; 14. ocotillo. **Threats**: 15. over-grazing; 16. energy development; 17. water diversion; 18. off-road vehicles; 19. buffelgrass; 20. cheatgrass; 21. Russian thistle.

WESTERN ARID LANDS HABITATS	CONDITION	THREAT LEVEL	FLAGSHIP BIRD
Chaparral	Fair	Medium	Wrentit
Coastal sagebrush	Fair	Critical	California Gnatcatcher
Interior sagebrush	Poor	Critical	Greater Sage-Grouse
Mojave Desert	Good	Low	Le Conte's Thrasher
Sonoran Desert	Fair	Medium	Elf Owl
Rocky canyons	Good	Low	Canyon Wren
Riparian cottonwood and sycamore	Poor	Critical	Elegant Trogon
Riparian willow and alder	Poor	Critical	Willow Flycatcher
Mesquite washes	Fair	Medium	Lucy's Warbler
Desert grassland	Poor	Medium	Bendire's Thrasher

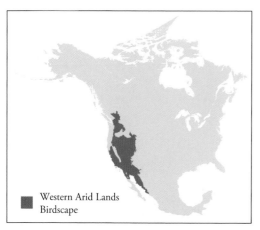

Western Arid Lands Birdscape

found in the region have experienced population declines of greater than 50% over the last 40 years, including the Whip-poor-will, Loggerhead Shrike, and Grasshopper, Black-throated, and Lark Sparrows.

Threats: Habitat has been converted to ranching or cleared for urban development, with many remaining areas fragmented, degraded by overgrazing, or damaged by invasive species and altered fire regimes. Fire management is perhaps the most complex issue, as frequent fires in sagebrush can facilitate the spread of non-native cheatgrass, whereas fire suppression can allow pinyon-juniper woodland to colonize. In coastal chaparral, frequent fire encourages invasive buffelgrass, but fire suppression can lead to catastrophic fires that completely destroy vegetation. Rapid expansion of natural gas drilling in the Great Basin threatens to disturb and fragment habitat, and interfere with Greater Sage-Grouse mating displays. Other threats include

wind farm development, West Nile virus, mining, off-road vehicle use, and barbed wire fences that pose a collision risk for grouse.

Conservation: Important protected areas in this Birdscape include Mojave National Preserve (California), Joshua Tree (California) and Great Basin (Nevada) NPs, Gunnison NF (Colorado), Camp Pendleton and Vandenberg Air Force Base (California), Los Padres NF (California), Cabeza Prieta NWR (Arizona), and Sonoran Desert and Organ Pipe Cactus NMs (Arizona). More than 50% of the remaining sagebrush habitat in the U.S. is on BLM land. The potential for listing the Greater Sage-Grouse under the ESA is currently driving significant conservation effort aimed at

PHOTO: ROBERT ROYSE

The Wrentit, characteristic of West Coast chaparral, is the sole member of the babbler family in North America.

Desert birds survive in extreme environments, but they are also sensitive to human disturbance (Organ Pipe Cactus NM pictured).

PHOTO: ©ISTOCKPHOTO.COM / ERIC FOLTZ PHOTOGRAPHY

precluding this possibility. The "Coastal" California Gnatcatcher is already protected by the ESA. Although Critical Habitat has been designated to protect this threatened subspecies, there is significant competition for real estate in coastal California. One potential solution is the implementation of Habitat Conservation Plans that allow for the development of gnatcatcher habitat while requiring mitigation projects to offset losses. The net effect of these plans needs careful evaluation, however.

Actions:
- Incorporate the needs of birds such as sage-grouse into BLM management plans and leases.
- Determine the net effect of Habitat Conservation Plans for the "Coastal" California Gnatcatcher.
- Increase land protection using conservation easements, improve grazing practices, and use prescribed burns to enhance habitat on private lands.

- Install fence markers that are visible to grouse.
- Limit the impact of wind development on sensitive bird species.

PHOTO: ROBERT ROYSE

The Lucy's Warbler is confined to the southwestern U.S. and parts of northern Mexico.

Important Bird Area:
Great Basin NP

BIRDSCAPE: **Western Arid Lands** WATCHLIST SPECIES
BCR: **9**   STATE: **NV**   RED: **1**
AREA: **77,180 ACRES**   YELLOW: **16**

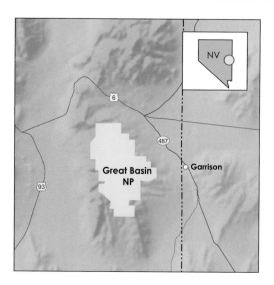

Great Basin NP rises 8,000 feet from the surrounding "sagebrush sea" to the 13,063-foot summit of Wheeler Peak. Habitats range from sagebrush to alpine tundra, through pinyon-juniper woodlands, forests of spruce and aspen; and ponderosa, limber, and bristlecone pines (some more than 4,000 years old).

Birds: Several WatchListed species breed in the park's sagebrush, including the Greater Sage-Grouse, and Brewer's and Sage Sparrows, while the park's alpine tundra supports breeding Black Rosy-Finches. Other breeding WatchList species include the Clark's Grebe, Swainson's Hawk, Long-billed Curlew, Flammulated Owl, Calliope Hummingbird, Williamson's Sapsucker, Gray Vireo, Pinyon Jay, and Virginia's Warbler. In winter or during migration, other WatchList species occur here, including the Marbled Godwit, Lewis' Woodpecker, Olive-sided Flycatcher, and Hermit Warbler. The park also supports several species experiencing continental-level population declines, including the Ferruginous Hawk, American Kestrel, Ruffed and Dusky Grouse, Rufous Hummingbird, Loggerhead Shrike, Sage Thrasher, and Evening Grosbeak.

Threats: Fire suppression facilitates the invasion of pinyon-juniper woodlands into sagebrush, reducing habitat quality. This can then lead to fuel build-up that precipitates catastrophic fires that destroy vegetation, damage soil, and benefit invasive plants such as cheatgrass, spotted knapweed, crested wheatgrass, and bull thistle. Other threats include groundwater pumping in nearby areas that threatens to dry up park springs, proposed coal-fired power plants that threaten to pollute the park's air, and a changing climate that threatens to alter plant communities—especially alpine communities at high elevations. Gas and wind development also threaten to impact surrounding areas.

Conservation: The 77,180-acre park is surrounded by public lands managed by BLM and Humboldt-Toiyabe NF. Cattle-grazing permits were bought out to compensate ranchers and end

The Sage Sparrow breeds in sagebrush habitat within the park, but is absent from areas invaded by cheatgrass.

PHOTO: GREG LAVATY

The removal of invasive spotted knapweed in the park will help restore native vegetation for the Greater Sage-Grouse.

PHOTO: NOPPADOL PAOTHONG

grazing within the park in 1999, though some sheep grazing continues. Efforts to remove invasive plants are focused on spotted knapweed and bull thistle. In 2005, the park combined seeding native shrubs and grasses with selectively cutting pinyon pines to restore sagebrush around Grey Cliffs Campground.

Actions:

- Control invasive plants.
- Use prescribed burns to restore natural fire regimes and sagebrush communities.

PHOTO: ROBERT ROYSE

The Clark's Grebe breeds on western lakes, where its floating nests are easily disturbed by recreational boaters.

IMPORTANT BIRD AREA
GUNNISON BASIN

BIRDSCAPE: **WESTERN ARID LANDS** WATCHLIST SPECIES
BCR: **16**   STATE: **CO**   RED: **3**
AREA: **>600,000 ACRES**   YELLOW: **16**

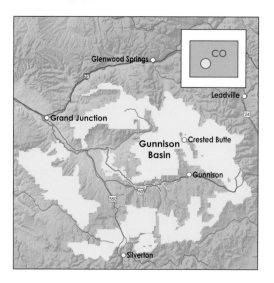

Gunnison Basin contains important tracts of upland sagebrush mixed with grasslands, meadows, and riparian areas. The basin supports c. 2,000 Gunnison Sage-Grouse, or just over 60% of the global population.

Birds: This area is critically important to the WatchListed Gunnison Sage-Grouse. Other WatchList species also occur, including the Swainson's Hawk, Long-billed Curlew, Semipalmated and Western Sandpipers (during migration), Marbled Godwit, Flammulated and Short-eared Owls (the latter in winter), Lewis's Woodpecker, Williamson's Sapsucker, Willow and Olive-sided Flycatchers, Pinyon Jay (rare), Virginia's Warbler; Lark, Sage, and Brewer's Sparrows; and Black Rosy-finch (in winter). Several species with continental population declines exceeding 50% since 1966 are also found, including the Rufous Hummingbird (during migration), Loggerhead Shrike, Horned Lark, Snow Bunting (in winter), and Evening Grosbeak.

Threats: Habitat loss, degradation, and fragmentation from housing developments, roads, grazing, agriculture, and off-road vehicle use are the most severe threats to the Gunnison Sage-Grouse, which has declined by more than 60% within the Basin since the 1950s. Fire suppression and invasive plants (e.g., cheatgrass) also degrade sagebrush habitat.

Conservation: Land ownership in Gunnison Basin is a mixture of private, state, and federal. Federal agency landowners include the USFS (Gunnison NF), National Park Service (Curecanti National Recreation Area), and BLM. Private land owners have granted conservation easements on more than 30,000 acres to protect the Gunnison Sage-Grouse, covering at least 18% of lek sites. Agencies are also cooperating to conserve the grouse, and to map its habitat and lek sites; conservation plans are also being developed. The Waunita lek is the only one available for public viewing. The hunting of Gunnison Sage-Grouse is not permitted.

Actions:
- Conserve habitat for the Gunnison Sage-Grouse, especially by protecting its lek sites.
- Manage fire regimes and control invasive species to maintain natural habitat conditions.

PHOTO: NOPPADOL PAOTHONG

Allowing easy access to the Waunita Gunnison Sage-Grouse lek minimizes disturbance at other display sites.

During the Gunnison Sage-Grouse's elaborate display, the male inflates air sacs on its neck to make weird popping noises.

Around 2,000 Gunnison Sage-Grouse remain in their small area of western sagebrush habitat.

IMPORTANT BIRD AREA

CABEZA PRIETA NWR AND ORGAN PIPE CACTUS NM

BIRDSCAPE: **WESTERN ARID LANDS**   WATCHLIST SPECIES
BCR: **33**   STATE: **AZ**   RED: **3**
AREA: **1.19 MILLION ACRES**   YELLOW: **16**

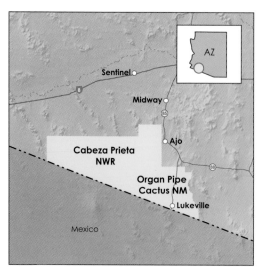

Cabeza Prieta NWR and the adjacent Organ Pipe Cactus NM protect the largest contiguous tract of Sonoran Desert in the U.S. Habitats include saguaro and organ pipe cactus desert scrub, creosote brush-lands, mesquite forests along washes important for migrants, and juniper-oak woodlands in the rugged Ajo Mountains.

Birds: Several WatchList species breed here, including the Gilded Flicker, Elf Owl, Costa's Hummingbird, Bendire's and Le Conte's Thrashers, Lucy's Warbler, and Varied Bunting. Other WatchList species occur during winter or migration, including the Swainson's Hawk, Allen's and Calliope Hummingbirds, Sprague's Pipit, Willow Flycatcher, Gray Vireo, Virginia's Warbler, Abert's Towhee; Rufous-winged, Brewer's, and Sage Sparrows; Lark Bunting, Chestnut-collared Longspur, and Lawrence's Goldfinch.

Threats: This area is a hotspot for illegal immigration. Immigrants and the border patrol agents who chase them degrade the fragile desert ecosystem. Border fence construction also threatens habitat, and will block the movements of native grazing animals. Fire prone invasive buffelgrass poses an additional threat.

Conservation: Cabeza Prieta NWR was established in 1939, and in 1990, 803,418 acres of the refuge were designated as Wilderness and closed to vehicle traffic. Organ Pipe NM, established in 1937, has the majority of its 330,000 acres designated as Wilderness. In 2006, the National Park Service finished building a steel barrier fence to prevent vehicle access along the 30-mile southern boundary of Organ Pipe. A similar barrier along Cabeza-Prieta NWR's 56-mile border with Mexico was completed in 2008. The area lies across the border from the El Pinacate and Gran Desierto de Altar Biosphere Reserve in Mexico.

Actions:
- Increase resources to help reduce impacts of vehicular traffic within Wilderness areas.
- Monitor, and if needed control, the spread of buffelgrass.

The reclusive Crissal Thrasher inhabits riparian habitat and desert washes. It is a reluctant flier, preferring to run to cover.

PHOTO: ROBERT ROYSE

Cavities excavated by Gilded Flickers (above) and Gila Woodpeckers in large cacti may eventually house tiny Elf Owls.

Important Bird Area

San Pedro Riparian NCA

BIRDSCAPE: **WESTERN ARID LANDS**   WatchList Species
BCR: **34**   State: **AZ**   Red: **5**
Area: **>56,000 acres**   Yellow: **26**

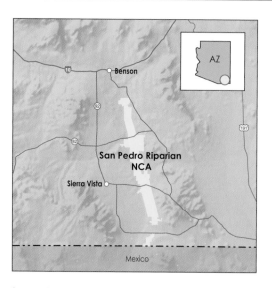

The San Pedro Riparian NCA protects a narrow corridor of riparian forest along 40 miles of the San Pedro River. It is a critical corridor for neotropical migrants.

Birds: Several WatchList species breed here, including the Scaled Quail, Swainson's Hawk, Elf Owl, Gilded Flicker, "Southwestern" Willow Flycatcher, Bell's Vireo, Bendire's Thrasher, Lucy's Warbler, Varied Bunting, and Abert's Towhee. During winter or migration, additional WatchList species use the area, including: Costa's, Calliope, and Allen's Hummingbirds; Williamson's Sapsucker, Olive-sided Flycatcher, Gray Vireo; Virginia's, Hermit, Grace's, and Red-faced Warblers; Painted Bunting; Rufous-winged, Black-chinned, Brewer's, Sage, and Baird's Sparrows; Lark Bunting, and Chestnut-collared Longspur. This river corridor provides critical stopover habitat for roughly four-million songbirds each spring during their migration north through inhospitable desert. The area is also home to many regionally distinct populations, including the "Mexican" Mallard, "Western" Yellow-billed Cuckoo, and "Lillian's" Eastern Meadowlark.

Threats: Ninety percent of Arizona's riparian ecosystems have been destroyed through the eradication of beavers by fur traders, and through timber extraction and cattle grazing. Invasive salt cedar has also replaced native cottonwood in many areas. Groundwater pumping, and pollution from urban areas and mines upstream also threaten the area. Finally, plans to build security walls and other infrastructure along the U.S.-Mexico border threaten to impact the area, especially if environmental regulations are waived.

Conservation: The NCA was established in 1988 and is administered by BLM. All farming and cattle grazing has been halted within its boundaries, allowing much of the riparian vegetation to recover. Beavers were reintroduced in 1999 to restore pools. Since 2002, TNC has worked to restrict groundwater pumping on more than 5,000 acres along the river.

Actions:
- Ensure adequate water flow along the river.
- Ensure that border security infrastructure abides by existing environmental laws.
- Protect and restore lands along the San Pedro River through acquisitions and easements.

Riparian forests provide vital travel corridors and stopover sites for migrating birds, such as the Wilson's Warbler.

PHOTO: ROBERT ROYSE

PHOTO: ©ANNIE GRIFFITHS BELT / NATIONAL GEOGRAPHIC STOCK

Aerial view of San Pedro Riparian NCA; these woodlands provide an important wildlife corridor amidst an otherwise arid landscape.

PHOTO: BILL HUBICK

The "Lillian's" Eastern Meadowlark, a paler subspecies found in southwestern grasslands, may be split into a separate species.

IMPORTANT BIRD AREA
CAMP PENDLETON

BIRDSCAPE: **WESTERN ARID LANDS**   WATCHLIST SPECIES
BCR: **32**   STATE: **CA**   RED: **8**
AREA: **>126,000 ACRES**   YELLOW: **25**

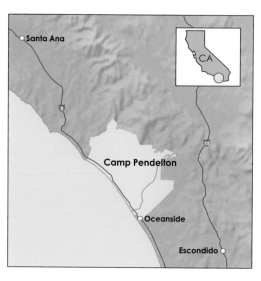

Threats: As on other military bases, there is a potential for conflict between military and wildlife needs, for example where beaches and dunes used by endangered shorebirds might be needed for combat training. Invasive plants also degrade habitat, and cowbirds parasitize the nests of "Least" Bell's Vireos. Urban development is eventually expected to completely surround the base (except for adjacent portions of Cleveland NF) and thereby reduce and fragment habitat. Upstream development threatens the local hydrology and could increase flash flood risk and sedimentation in the Santa Margarita River estuary. Other threats include feral cats, proposals to construct power lines and to channelize Murrieta Creek upstream of the Santa Margarita River, and outbreaks of oak disease.

Camp Pendleton protects habitat in four major watersheds, including that of the Santa Margarita River, as well as the largest stretch of undeveloped coastline in southern California. Habitats include saltmarsh, coastal dunes, sage-scrub, chaparral, grasslands, and oak and riparian woodland.

Birds: Large populations of the "California" Least Tern and "Least" Bell's Vireo breed on the base, as well as a significant population of the "Coastal" California Gnatcatcher. Other WatchList species include the Mountain Quail, "Light-footed" Clapper Rail, "Western" Snowy Plover, Costa's Hummingbird, Nuttall's Woodpecker, "Southwestern" Willow Flycatcher, Oak Titmouse, Wrentit, California Thrasher, Black-chinned Sparrow, Lawrence's Goldfinch, and Tricolored Blackbird. During winter and migration, the WatchListed Clark's Grebe, American Golden-Plover, Wandering Tattler, Marbled Godwit, Black Turnstone, Surfbird, Red Knot, Rock and Western Sandpipers, Sanderling, Swainson's Hawk, Short-eared Owl, Allen's Hummingbird, Olive-sided Flycatcher, Hermit Warbler, and Sage Sparrow occur, while the Black-vented Shearwater and Black Storm-Petrel can be seen offshore.

Conservation: Management for birds includes the fencing of dunes for Least Terns, trapping cowbirds to increase breeding success for Bell's Vireos, and the control of invasive plants. A portion of Camp Pendleton's northern boundary is shared with Cleveland NF, partly protecting this area from the effects of urbanization. The public is allowed access with permission. In 2007, the Trust for Public Land purchased the 1,206 acre Santa Margarita Peak parcel along the base's northern boundary to buffer habitat from encroaching urbanization.

Actions:
• Expand protected areas surrounding Camp Pendleton to buffer it from urbanization; improve connectivity with Cleveland NF.
• Control invasive species including feral cats; control cowbirds; protect habitat on the base.

So much of the California Gnatcatcher's coastal sage-scrub habitat has been lost that the species was listed under the ESA in 1993.

The California Thrasher has declined more than 50% since 1966 due to habitat loss and fragmentation.

WETLANDS

AREA: **169,776 SQ. MILES**	WATCHLIST SPECIES
STATES: **ALL**	RED: **15**
BCRs: **ALL BCRs**	YELLOW: **17**

Wetlands vary from springs and desert streams that may not flow in most years, to ephemeral pools present seasonally, to permanent lakes, rivers, and marshes. Some inland lakes are saline due to the concentration of salts caused by evaporation. Large numbers of birds commonly gather in wetlands due to abundant food availability.

Birds: More than 30 WatchListed bird species inhabit wetlands while breeding, during migration, or in winter, including the Emperor Goose, Trumpeter Swan; Mottled, Hawaiian, and Laysan Ducks; Steller's and Spectacled Eiders, Yellow-billed Loon, Reddish Egret; Yellow, Black, and King Rails; Hawaiian Coot, Whooping Crane, Snowy and Piping Plovers, Long-billed Curlew; Hudsonian, Bar-tailed, and Marbled Godwits; Semipalmated, Western, and Stilt Sandpipers; Least Tern, Nelson's Sparrow, and Tricolored Blackbird. Several more common wetland species have experienced >50% population declines over the last 40 years, including the Northern Pintail, Greater Scaup, Little Blue Heron, and American Bittern.

Threats: Around half of the original wetlands in the "Lower 48" states have been destroyed or al-

ILLUSTRATION: C. VEST

Birds: 1. American Bittern; 2. King Rail; 3. Northern Harrier; 4. Mottled Duck; 5. Northern Rough-winged Swallow; 6. Red-winged Blackbird; 7. White Ibis; 8. Sedge Wren; 9. Yellow Rail. **Vegetation**: 10. cattails; 11. pickerel weed. **Threats**: 12. *Phragmites*; 13. purple loosestrife; 14. garbage; 15. pollution in runoff; 16. development.

WETLANDS HABITATS	CONDITION	THREAT LEVEL	FLAGSHIP BIRD
Lakes and ponds	Fair	Medium	Clarke's Grebe
Salt and playa lakes	Poor	High	Snowy Plover
Sawgrass marshes	Poor	High	"Cape Sable" Seaside Sparrow
Freshwater marshes	Poor	Medium	King Rail
Prairie potholes	Fair	High	Marbled Godwit
Boreal wetlands and bogs	Good	Medium	Lesser Yellowlegs
Streams	Good	Medium	American Dipper
Rivers	Fair	Medium	Least Tern
Waterfalls	Fair	High	Black Swift

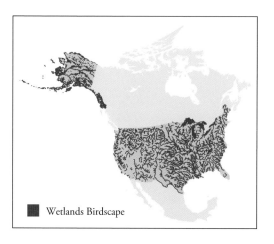

Wetlands Birdscape

tered. Marshes and ponds have been drained for agriculture and urban development; 2.5 million dams obstruct 20% of U.S. river miles; aquifers are drawn down to supply drinking water to cities; and farms, industries, and urban centers pollute streams. The total acreage of wetlands in the "Lower 48" states declined by more than 806,600 acres from 1954-1997 (followed by a small recent net increase). Natural flooding processes have been reduced by 50-90% in the Mississippi Alluvial Valley, and greatly reduced in other dammed river systems. Wetland loss can be locally extreme. For example, 91% of California's Central Valley wetlands have been destroyed or converted to agriculture. Beavers create small wetlands and though now recovering from historical over-hunting, they have still not fully returned to some regions. Warming climate threatens wetlands in areas such as the boreal forest and prairie potholes, and changes to snowmelt and receding glaciers are altering flows of mountain streams. Wetlands are also often impacted by invasive plants (e.g., *Phragmites* and purple loosestrife) and herbivores (e.g., Mute Swan and nutria). Finally, some wetland birds, such as Caspian Terns and Double-crested Cormorants, are locally persecuted as pests due to their perceived impacts on fisheries.

Conservation: Many wetlands are managed or protected by NWRs and other federal protected areas, as well as SPs and water management areas. Waterfowl hunting licenses raise funds for the acquisition, restoration, and management of wetlands and adjacent habitats through the NWR sys-

PHOTO: ROBERT ROYSE

The American Avocet and other wetland birds benefit from programs that encourage landowners to protect their habitat.

PHOTO: STEVE COLLINS

Wetlands occupy just 4.8% of U.S. territory but are vital for birds. Here Sandhill Cranes gather on the Platte River during migration.

tem. The Land and Water Conservation Fund (see p. 432) is a major source of support for wetland conservation. Grants under the North American Wetlands Conservation Act (see p. 433) implemented through the Joint Venture system (see p. 411) have also conserved millions of wetland acres, and this is largely credited with restoring mid-continent waterfowl numbers since the 1980s. Many wetlands on private lands have been conserved under the USDA Wetland Reserve Program (e.g., 1.9 million acres in 2002). Wetlands are also protected by a variety of federal, state, and local laws restricting water pollution, runoff, and development (e.g., Section 404 of the Clean Water Act). A wetland mitigation banking system administered by EPA also requires that developers compensate for projects that impact wetlands by creating new ones in other locations towards a goal of "no net loss" set by President Bush in 1989. The results of the program are hard to quantify however. Some wetland

species have greatly benefited from management actions. For example, the Wood Duck was severely depleted by over-hunting in the early 1900s, but a ban on hunting from 1918-1941, conservative bag limits since, an artificial nest box program begun in the 1930s, and increasing numbers of beaver ponds and other habitat improvements have helped the population to recover in many areas.

Actions:

- Protect remaining wetland systems from pollution and development.
- Restore wetlands in areas vulnerable to flooding to benefit both people and birds.
- Strengthen regulations limiting the development, draining, filling or diversion of wetlands.
- Strengthen the Wetland Reserve and Conservation Reserve Programs.
- Increase support for wetland conservation through the Joint Ventures.

IMPORTANT BIRD AREA
THE EVERGLADES

BIRDSCAPE: **WETLANDS**
BCR: **31**   STATE: **FL**
AREA: **4 MILLION ACRES**

WATCHLIST SPECIES
RED: **10**
YELLOW: **29**

The Everglades is a "river of grass" slowly flowing from Lake Okeechobee to the tip of Florida. Wetland habitats include a variety of marshes, beaches, mangroves, and cypress swamps that each provide important bird habitats. The huge colonies of water-birds that once flourished here have been reduced to a fraction of their former numbers by water diversion projects.

Birds: The area is most famous for its colonial water-birds such as egrets, herons, and spoonbills. It is home to many breeding WatchList species, including the Mottled Duck, Reddish Egret, Swallow-tailed Kite, Clapper and King Rails, Wilson's Plover, Gull-billed Tern, Black Skimmer, White-crowned Pigeon, Mangrove Cuckoo, Prairie Warbler, and the endemic "Cape Sable" subspecies of the Seaside Sparrow. During winter and migration, many additional WatchList species use the area including Black and Yellow Rails; Semipalmated, Western, White-rumped, and Stilt Sandpipers; Short-eared Owl, Prothonotary and Swainson's Warblers; and Saltmarsh and Nelson's Sparrows. Other birds of interest include the "Great White Heron" (a distinctive, local population of the Great Blue Heron sometimes considered a separate spe-

cies) as well as the Snail Kite, and Black-whiskered Vireo.

Threats: The flow of water through the Everglades has been disrupted over the last century by canal construction to control floods, drain areas for agriculture, control mosquitoes, and to provide drinking water. Reduced water flows have dried wetlands, increased wildfire frequency, and facilitated saltwater intrusion. The Everglades is also infested with invasive plant and animal species. Agricultural runoff causes pollution leading to algal blooms and affecting natural vegetation. Numbers of colonial water-birds have fallen from 250,000 to fewer than 100,000.

Conservation: Everglades NP and Big Cypress National Preserve protect much of the southern portion of the Everglades; Loxahatchee and Florida Panther NWRs, Fakahatchee Strand Preserve SP, and other SPs and water management areas occupy much of the northern portion. Management includes prescribed burns, invasive plant removal, and the improvement of water flows. The Comprehensive Everglades Restoration Plan was signed in 2000, but federal budgets have not fully funded it. In June 2008, the state of Florida planned to purchase 187,000 acres owned by U.S. Sugar to restore water flow from Lake Okeechobee, but the proposed acreage has since been cut by more than half, allowing sugar production to continue.

Actions:
- Restore or mimic natural water flows, especially to benefit the "Cape Sable" Seaside Sparrow.
- Control and eradicate invasive species.

Roseate Spoonbills have declined in the Everglades due to drainage for development and mosquito control.

IMPORTANT BIRD AREA
CHEYENNE BOTTOMS
AND QUIVIRA NWR

BIRDSCAPE: **WETLANDS**
BCR: **19**   STATE: **KS**
AREA: **63,000 ACRES**

WATCHLIST SPECIES
RED: **6**
YELLOW: **25**

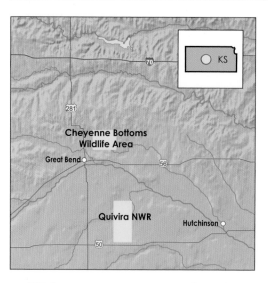

WatchList species include the Greater Prairie-Chicken, Swainson's Hawk, Black and King Rails, Snowy Plover, "Interior" Least Tern, Red-headed Woodpecker, Prothonotary Warbler, and Painted and Lark Buntings. Many additional WatchList species use the area during migration and winter, including the Whooping Crane; American Golden-, Mountain, and Piping Plovers; Long-billed Curlew, Hudsonian and Marbled Godwits; Red Knot, Sanderling; Semipalmated, White-rumped, Stilt, and Buff-breasted Sandpipers; Short-eared Owl, Sprague's Pipit, Henslow's and Le Conte's Sparrows, Chestnut-collared Longspur, and Rusty Blackbird.

Threats: Increased water withdrawals for irrigated agriculture upstream of both Cheyenne Bottoms and Quivira have decreased water flows and this can threaten wetlands during dry years or in late summer. The invasion of mudflats by cattails reduces shorebird habitat. Musk thistle, *Phragmites*, salt cedar, Russian olive, and other invasive plants are problematic and require control. Oil and gas wells also pose a threat to habitat in the region.

Cheyenne Bottoms is one of the largest marshes in the central U.S. Quivira NWR lies just 30 miles to the south, and these wetlands, prairies, croplands, and salt flats serve as critical stopover habitat for many migratory birds.

Birds: Approximately 45% of North America's shorebirds, 500,000 waterfowl, 180,000 Sandhill Cranes, and 5,000 American White Pelicans pass through, particularly during the spring. Breeding

Conservation: Cheyenne Bottoms is a 41,000-acre complex including the state's Cheyenne Bottoms Wildlife Area and the adjacent Cheyenne Bottoms Preserve owned by TNC. Quivira NWR is located 30 miles to the south. The Kansas Department of Wildlife has conducted a multi-million-dollar renovation of Cheyenne Bottoms including subdividing pools to improve water management to benefit shorebirds and waterfowl. Cattails are largely controlled by manipulating water levels. TNC uses grazing and prescribed burns to manage grasslands.

Actions:
- Ensure adequate water flows to Cheyenne Bottoms and Quivira.
- Manage habitat and manipulate water levels to benefit migratory shorebirds and waterfowl.

PHOTO: MIKE PARR

Introduced Phragmites (above) and other invasive plants degrade wetlands by crowding out native species.

American White Pelicans and other water-birds find Cheyenne Bottoms to be an oasis in the vast Great Plains.

PHOTO: ROBERT ROYSE

The wet marshes of Quivira NWR are home to an inland breeding population of the elusive Black Rail.

PHOTO: DAVID SEIBEL

IMPORTANT BIRD AREA

PLATTE RIVER AND RAINWATER BASIN

BIRDSCAPE: **WETLANDS**
BCR: **19**   STATE: **NE**
AREA: **2.7 MILLION ACRES**

WATCHLIST SPECIES
RED: **12**
YELLOW: **30**

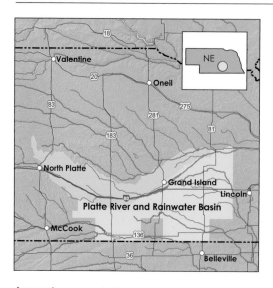

This 80-mile-long stretch of river is a critical staging area during spring migration for millions of birds including waterfowl, shorebirds, and the largest concentration of cranes in the world. Habitats include agricultural fields, marshes, cottonwoods, and braided channels.

Birds: Spectacular concentrations of birds occur here during spring migration, including ten-million waterfowl, 500,000 Sandhill Cranes, and more than 200,000 shorebirds. This includes 90% of the mid-continent population of the Greater White-fronted Goose, 50% of the Mallards, one third of the Northern Pintails, and 500,000 Canada Geese. This is also the primary spring stopover for the WatchListed Buff-breasted Sandpiper in the Great Plains. Other WatchListed species occur during migration or winter, including the Clark's Grebe, Trumpeter Swan, Yellow Rail, Whooping Crane, American Golden- and Snowy Plovers, Long-billed Curlew, Hudsonian and Marbled Godwits, Red Knot, Sanderling; Semipalmated, White-rumped, Western, and Stilt Sandpipers; Short-eared Owl, Sprague's Pipit; Brewer's, Baird's, Le Conte's, and Nelson's Sparrows; Smith's and Chestnut-collared

Longspurs, and Rusty Blackbird. Breeding Watch-List species include Greater Prairie-Chicken, King Rail, "Interior" Least Tern, Piping Plover, and Lark Bunting.

Threats: More than 90% of the original 4,000 major wetlands once present in this region have been converted to agriculture. Now, most wetlands are less than 40 acres in size with few over 1,000 acres. Seventy-five percent of the Platte River's water is removed upstream for agricultural and municipal uses. Without floods and sediment, sand bars disappear, or are overgrown by willows, reducing stopover habitat for migrating cranes. Shallow water also allows predators to gain access to areas previously safe for birds. Avian cholera outbreaks killed over 200,000 ducks and geese from 1974-1994. Snow Geese numbers increased from 15,000 to 353,000 between 1974 and 1985, increasing the risk of cholera outbreaks, as well as competing with other waterfowl and cranes for limited habitat and food.

Conservation: Most of the land is privately owned, but FWS manages a total of 21,742 acres, and the Nebraska Game and Parks Commission manages 6,900 acres. In 1992, a coalition of government agencies, private landowners, and conservation groups formed the Rainwater Basin Joint Venture to restore and permanently protect 37,000 acres of wetlands and 25,000 associated upland acres for birds.

Actions:
• Protect and restore habitat to meet Rainwater Basin Joint Venture goals.
• Restore in-stream flows.

Approximately 500,000 Sandhill Cranes—about 80% of the global population—pass through the Platte River Valley on migration.

Nebraska's Eastern Rainwater Basin is a primary spring stopover site for the dainty Buff-breasted Sandpiper.

IMPORTANT BIRD AREA
GREAT SALT LAKE

BIRDSCAPE: **WESTERN ARID LANDS** WATCHLIST SPECIES
BCR: **9** STATE: **UT** RED: **0**
AREA: **1.1 MILLION ACRES** YELLOW: **18**

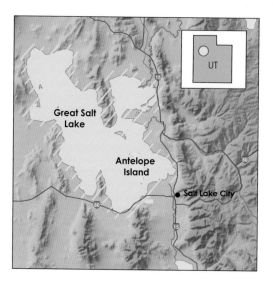

Birds: This is one of the most important sites in the U.S. for shorebirds, ducks, and other waterbirds, with several million using the lake annually. Five- to ten-thousand Snowy Plovers breed along the shore, and the largest breeding colonies of the White-faced Ibis and California Gull occur here. Seventeen-thousand American White Pelicans breed, travelling up to 100 miles to Utah Lake to forage (since the Great Salt Lake is too saline to support fish). The feast of brine shrimp supports huge numbers of Black-necked Stilts, American Avocets, Eared Grebes, Marbled Godwits, and Wilson's and Red-necked Phalaropes. Breeding WatchList species include the Clark's Grebe, Long-billed Curlew, Short-eared Owl, and Brewer's and Sage Sparrows. During winter and migration, the WatchListed Trumpeter Swan, American Golden-Plover, Marbled Godwit, Red Knot, Sanderling; Semipalmated, Western, and Stilt Sandpipers; Virginia's Warbler, and Lark Bunting can be found.

T his is the largest saline lake in the Americas, averaging almost 1.1 million acres, but varying greatly in size between years. It is fed by three major rivers but loses water through evaporation, causing its water to be significantly more salty than the ocean.

Threats: Upstream water withdrawals and drought reduce water flows, lowering lake levels, reducing

PHOTO: ROBERT ROYSE

Eared Grebes gather by the thousand here each fall to feast on brine shrimp, doubling their weight in preparation for migration.

Salt deposits around Antelope Island; the high salinity nourishes billions of brine shrimp and flies, a food source for migrating birds.

PHOTO: ©ISTOCKPHOTO.COM / NOEL COMMUNICATIONS

wetlands, and increasing contaminant concentrations from sewage and industrial effluents. These cause eutrophication, as well as depositing selenium and mercury (levels are so high in some ducks that hunters are advised against eating them). Frequent outbreaks of avian cholera and botulism can kill large numbers of birds. Causeways restrict water flows between portions of the lake, altering its chemistry. Other threats include road construction, development, upstream dams, dumping of soil from a Superfund site, invasive *Phragmites*, and disturbance by off-road vehicles in Snowy Plover nesting areas.

Conservation: The lake is largely state-owned, and several parks and nature preserves protect portions of its shore and islands. These include the Bear River Migratory Bird Refuge, Antelope Island SP, and TNC's Great Salt Lake Shorelands Preserve. Water management at Bear River has reduced the frequency of severe botulism outbreaks.

Actions:
- Ensure adequate fresh water is supplied to the lake.
- Reduce water pollution.
- Reduce avian botulism by manipulating water levels.
- Halt off-road vehicle disturbance near Snowy Plover nesting areas.

PHOTO: GREG LAVATY

The salt pans along the lake's shores are inhabited by 5,000-10,000 breeding Snowy Plovers.

IMPORTANT BIRD AREA

THE GRASSLANDS ECOLOGICAL AREA

BIRDSCAPE: **WETLANDS**
BCR: **32**   STATE: **CA**
AREA: **193,768 ACRES**

WATCHLIST SPECIES
RED: **3**
YELLOW: **16**

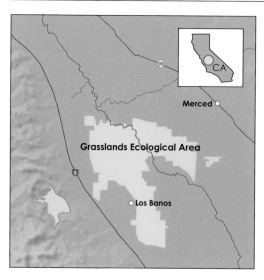

The Grasslands Ecological Area's habitats include seasonally flooded marshes, vernal pools, riparian cottonwoods and willows, oak woodlands, grasslands, and agricultural lands.

Birds: During winter and migration, the site is used by more than one-million shorebirds, cranes,

ducks, and geese, including half the world's Ross's Goose population. Counts of the WatchListed Tricolored Blackbird vary between seasons and years, but 105,000 were estimated to breed in one colony in the mid-1990s. Other breeding Watch-List species include the Clark's Grebe, Swainson's Hawk, Short-eared Owl, Nuttall's Woodpecker, Yellow-billed Magpie, Oak Titmouse, and California Thrasher. Additional WatchList species occur during winter and migration, including American Golden-, Mountain, and Snowy Plovers; Marbled Godwit, Western Sandpiper, Lewis' Woodpecker, Varied Thrush, Hermit Warbler, Sage Sparrow, and Lawrence's Goldfinch.

Threats: The hydrology of California's Central Valley has been altered by intensive agriculture, flood control, and irrigation projects. Agricultural runoff laced with selenium and other contaminants results in bird die-offs. Urban development is occurring rapidly in Merced County.

Conservation: Over 30,000 acres are protected by the San Luis and Merced NWRs, the Volta and

Half the world's Ross's Geese and many other waterfowl winter at this site.

PHOTO: ROBERT ROYSE

The Yellow-headed Blackbird is a conspicuous and numerous resident of prairie wetlands throughout the western U.S.

Los Banos WMAs, and Great Valley Grasslands SP. FWS established the Grasslands WMA in 1979 to protect habitat on more than 65,000 acres of private land through conservation easements and habitat management assistance. The California Department of Fish and Game and several NGO partners actively protect and restore habitat in the area. The Central Valley Improvement Act of 1992 is helping to maintain water supplies for wetlands and other natural habitats.

Actions:

- Protect and restore agricultural lands through easements and acquisition, and by providing habitat management services to private land-owners.
- Reduce water pollution from agricultural runoff and other sources.

Marbled Godwits are among the more than one-million birds that depend on this site.

IMPORTANT BIRD AREA
THE SALTON SEA

BIRDSCAPE: **WETLANDS**
BCR: **33**   STATE: **CA**
AREA: **43,000 ACRES**

WATCHLIST SPECIES
RED: **5**
YELLOW: **20**

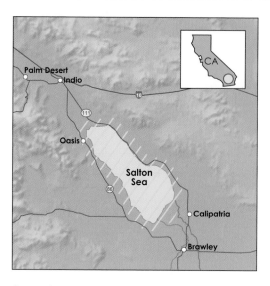

Threats: The lake's salinity is increasing by approximately 1% per year due to evaporation. Inflow comes from three rivers, one of which, the New River, flows north from Mexico and is especially heavily polluted. Agricultural drainage from alkaline soils brings nutrient-rich runoff, causing eutrophication and algal blooms that can lead to outbreaks of avian diseases such as botulism and cholera. The lake also carries an especially high selenium load resulting from high natural concentrations in surrounding soils. Water diversion to San Diego is further concentrating contaminants. Introduced tilapia may soon be the only fish that can survive here; they provide an important food source for pelicans and other birds.

Conservation: Several restoration plans have been recommended, including the possibility of connecting the lake to the ocean. The State of California has recently proposed a $9 billion, 25-year plan that will reduce the surface area by 60%, and calls for diking, sectioning, and evaporating sections of the lake. The 1,800-acre Sonny Bono Salton Sea NWR is found in the southern part of the lake. Other protected areas include the Wister Unit of the Imperial Wildlife Area and the Dos Palmas Nature Preserve. Surrounding lands are managed by BLM and the state of California.

This is the third largest salt lake in North America after the Great Salt Lake and Mono Lake (which is just slightly larger) and is marginally more saline than the ocean. It is a key breeding, wintering, and stopover site for hundreds of thousands of waterfowl and shorebirds. Habitats include desert scrub, agricultural areas, open water, and salt flats. The current lake was formed when the Colorado River breached a dike in 1905, though the area has been periodically flooded then desiccated over thousands of years.

Birds: More than 100,000 waterfowl and close to one-million Eared Grebes winter, along with 1,000 WatchListed Mountain Plovers. This is also the only regular U.S. site for the WatchListed Yellow-footed Gull (non-breeding). Approximately 100,000 shorebirds occur on passage including the WatchListed Western Sandpiper, Long-billed Curlew, and Marbled Godwit. Breeding Watch-List species include the Snowy Plover, Gull-billed Tern, Black Skimmer, the "Yuma" subspecies of the Clapper Rail, and Abert's Towhee. Several other WatchList species have been recorded occasionally or as vagrants.

Actions:
- Reduce contaminants and maintain sufficient fresh water inflow into the Salton Sea to preserve it as a functioning ecosystem.

North America's third-largest saline lake, California's Salton Sea, faces significant threats from runoff and evaporation.

PHOTO: ©ISTOCKPHOTO.COM / BOB REYNOLDS

The Black Skimmer risks picking up contaminants, such as selenium, along with fish from the Salton Sea's waters.

PHOTO: ROBERT ROYSE

IMPORTANT BIRD AREA

JAMES CAMPBELL NWR

BIRDSCAPE: **WETLANDS**
BCR: **67**   STATE: **HI**
AREA: **145 ACRES**

WATCHLIST SPECIES
RED: **3**
YELLOW: **3**

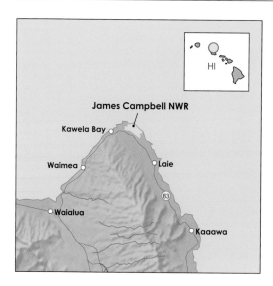

James Campbell NWR

Kawela Bay

Waimea   Laie

83

Waialua

Kaaawa

HI

This is one of the most important of Hawaii's few natural remaining wetlands, and also one of the most accessible to birders. It has small populations of all Hawaii's native water-birds.

Birds: The refuge has breeding populations of the WatchListed Hawaiian Coot and Duck, although introduced Mallards are a serious threat to the purity of the duck's gene pool. The site also hosts the Hawaiian races of the Black-necked Stilt and Common Moorhen. The WatchListed Bristle-thighed Curlew occurs in open areas during the winter, and a few non-breeding individuals often stay throughout the summer. The WatchListed Short-eared Owl (Hawaiian subspecies), Sanderling, and the occasional Laysan Albatross (flying over) also occur.

Threats: Invasive Mallards hybridize with Hawaiian Ducks that were reintroduced to Oahu in the 1960s, and there may no longer be any pure-bred Hawaiian Ducks left on the island. This will be a recurring problem while Mallards exist in Hawaii. Invasive predators such as cats and mongooses threaten adult water-birds, their nests, and chicks;

rats predate eggs, and bullfrogs prey on chicks. Invasive plants such as marsh fleabane and cattail choke off open water unless they are continually controlled. Botulism outbreaks are also a concern.

Conservation: Along with the adjacent Amorient Aquafarm, this is one of the key wetland complexes in Hawaii. Much of the original wetland was drained to form settling ponds for the Kahuku Sugar Mill, and the James Campbell NWR was established in 1977 after the mill closed. The refuge has two units, one of which (Punamano) is permanently closed; the other (Kii) is open from August to mid-February. Efforts are currently underway to expand the refuge to 1,100 acres, which would make it the largest managed wetland in Hawaii. Management activities include predator control and the burning and mechanical removal of invasive plants.

Actions:

- Maintain open water areas by controlling invasive vegetation.
- Control invasive predators.
- Remove non-native Mallards.
- Expand the refuge to 1,100 acres.

PHOTO: BILL HUBICK

Shorebirds such as this Wandering Tattler make the long journey across the Pacific to winter in the Hawaiian Islands.

PHOTO: BILL HUBICK

Resident Hawaiian water-birds have evolved into distinct species or subspecies (such as this "Hawaiian" Black-necked Stilt).

MARINE

AREA: **3,400,000 SQ. MILES (EEZ)**	WATCHLIST SPECIES
STATES: **ALL COASTAL STATES**	RED: **36**
BCRs: **All coastal BCRs**	YELLOW: **51**

The Marine Birdscape includes both oceans, and shoreline habitats that are directly affected by marine waters, including saltmarshes, sandy and rocky beaches, sea cliffs, and islands. Marine habitats vary depending on water depth, temperature, and currents. Different seabird species use cold and warm water, and others use areas where currents mix. Areas where upwelling currents bring nutrients to the surface are particularly productive, and can support huge numbers of seabirds. Beaches and dunes of oceanic and barrier islands, as well as sea cliffs, often provide nesting sites that are safe from predators. Tidal mudflats provide important foraging areas for shorebirds. North America boasts the most extensive saltmarshes on Earth and these provide nesting sites for rails, sparrows, and shorebirds. In tropical estuaries, mangrove forests grow in shallow marine waters, providing nesting and resting sites for a variety of water-birds.

Birds: WatchListed seabirds that only come to land to breed, include the Laysan, Black-footed, and Short-tailed Albatrosses; Bermuda, Black-capped, and Hawaiian Petrels; Cory's, Pink-footed, Greater, Buller's, Sooty, Manx, Newell's, Black-vented, and Audubon's Shearwaters; Ashy, Band-rumped, Tristram's, and Least Storm-petrels; Masked Booby,

ILLUSTRATION: C. VEST

Birds: 1. Black Skimmer; 2. Least Tern; 3. Short-billed Dowitcher; 4. Seaside Sparrow; 5. Saltmarsh Sparrow; 6. Piping Plover; 7. Red Knot; 8. Ruddy Turnstone; 9. Dunlin; 10. Semipalmated Sandpiper; 11. Black Turnstone; 12. Thick-billed Murre; 13. Common Murre; 14. Tufted Puffin; 15. Black-legged Kittiwake; 16. Laysan Albatross; 17. Mottled Petrel. **Mammals:** 18. humpback whale. **Threats**: 19. coastal development; 20. oil spills; 21. bycatch in fisheries; 22. overharvested horseshoe crabs; 23. Herring Gull; 24. Great Black-backed Gull; 25. invasive rats; 26. rising sea level; 27. melting sea ice.

Marine Habitats	Condition	Threat Level	Flagship Bird
Mangroves	Fair	Medium	Mangrove Cuckoo
Saltmarshes	Poor	Critical	Clapper Rail
Mudflats	Poor	High	Western Sandpiper
Sandy beaches and dunes	Poor	High	Piping Plover
Rocky intertidal	Good	Medium	Surfbird
Sea cliffs	Good	Medium	Red-legged Kittiwake
Low sandy atolls	Fair	High	Laysan Albatross
Marine islands	Poor	Critical	Xantus's Murrelet
Sea ice	Fair	High	Ivory Gull
Coastal waters	Fair	High	Roseate Tern
Pelagic waters	Fair	High	Black-capped Petrel

Marine Birdscape

Red-faced Cormorant, Magnificent Frigatebird, Red-legged Kittiwake, Ross's and Ivory Gulls, Bridled and Aleutian Terns, Razorbill; Marbled, Kittlitz's, Xantus's, Craveri's, and Ancient Murrelets; and the Whiskered Auklet. WatchListed water- and shorebirds that breed, migrate or winter along coasts include the American Golden-, Snowy, Wilson's, and Piping Plovers; Bristle-thighed and Long-billed Curlews; Hudsonian, Bar-tailed, and Marbled Godwits; Black Turnstone, Surfbird, Red Knot, Sanderling; Semipalmated, Western, White-rumped, Rock, and Stilt Sandpipers; Heermann's, Thayer's, and Iceland Gulls; Least, Gull-billed, Roseate, and Elegant Terns; and Black Skimmer.

WatchListed birds that breed or winter in tidal marshes include the Reddish Egret; Yellow, Black, and Clapper Rails; Whooping Crane, Short-eared Owl; and Le Conte's, Nelson's, Saltmarsh, and Seaside Sparrows.

Threats: Marine habitats face a barrage of threats. Most of our largest cities occur along the coast, and rivers wash contaminants into the sea. Excess nutrients from agriculture flow into estuaries, feeding algal blooms that rob water of oxygen (eutrophication). As seabirds are often top predators, many experience high levels of contamination, even though they live far from pollution sources. Oil spills kill marine birds and other wildlife. Carbon dioxide pollution threatens marine habitats through ocean acidification. Climate change is predicted to alter or shift ocean currents, increase the frequency of storms, and raise sea levels to inundate coastal habitats. Boat traffic helps spread invasive species from one coast to another and wakes erode shorelines. Disturbance by beachgoers and their pets negatively affects plovers, terns, and other beach-nesting birds. Although the unsustainable exploitation of terrestrial wildlife largely ceased one-hundred years ago in the U.S., it continues today for many marine species. Not only does the unsustainable harvest of fish harm marine ecosystems and reduce important food resources for birds, but fishing equipment such as longlines and gillnets can hook or net seabirds by accident.

Conservation: There are many protected areas, including National Marine and Estuarine Sanctuaries, National Seashores, and NWRs that safeguard marine habitats, including important staging areas for shorebirds and remote seabird breeding colonies. Often, policy solutions are as important as protected areas for marine conservation, how-

Two percent of the global population of the Endangered Short-tailed Albatross gathers around a single fishing vessel (also N. Fulmars).

PHOTO: JOSH HAWTHORNE / ELSEVIER.COM/LOCATE/DSR2

ever. This is particularly true for wide-ranging seabirds, which may spend their lives largely in international waters. Past changes in policies and laws have stopped the overexploitation of shore-birds and seabirds by restricting or ending hunting seasons, preventing damage by off-road vehicles on beaches, and by banning drift nets and reducing seabird bycatch in our oceans. Most recently, bird-scaring devices now in use in U.S. Pacific longline fisheries, promoted by ABC and its partners, have significantly reduced albatross bycatch there.

Actions:

- Promote sustainably managed fisheries, including labeling seafood for consumers.
- Halt unsustainable harvesting with closures until stocks recover (e.g., horseshoe crabs).
- Increase funding to clean up plastic garbage, lead paint (Midway Island), and to eradicate invasive species from seabird nesting islands.
- The U.S. should ratify the Agreement on the Conservation of Albatrosses and Petrels, and take measures to reduce bycatch of seabirds.
- Anticipate and manage the needs of salt marsh, mudflats, beach and other seashore habitats in the face of rising sea levels.
- Balance the needs of energy development, commerce, recreation, and wildlife in marine and coastal areas.
- Reduce terrestrial runoff and pesticide use.
- Enforce the Clean Water Act; review oil spill clean-up plans for key sites to ensure that bird priorities are effectively included; ensure that tankers are double-hulled worldwide.

IMPORTANT BIRD AREA

CAPE COD AND NEARBY ISLANDS

BIRDSCAPE: **MARINE**
BCR: **30**   STATE: **MA**
AREA: **200,000 ACRES**

WATCHLIST SPECIES
RED: **4**
YELLOW: **22**

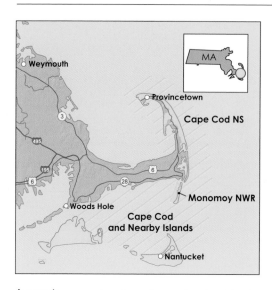

The seas, beaches, dunes, barrier islands, and marshes around Cape Cod provide important breeding and foraging habitat for birds. This is also a critical staging area during fall migration.

Birds: Several breeding WatchList species occur, including important colonies of the Piping Plover and Roseate Tern. The Least Tern, Black Skimmer, Short-eared Owl, Prairie Warbler, Saltmarsh Sparrow, and Seaside Sparrow also breed. Many other WatchList species occur during migration or winter, including Cory's, Greater, and Sooty Shearwaters; Clapper and King Rails, American Golden-Plover, Hudsonian and Marbled Godwits, Red Knot, Sanderling; Semipalmated, White-rumped and Stilt Sandpipers; Razorbill, Nelson's Sparrow, and Rusty Blackbird. In addition to WatchList species, Cape Cod is important for breeding American Black Duck, Common Tern, and American Oystercatcher. During winter, the "Ipswich" Savannah Sparrow winters amongst the dune grass, and half-a-million sea ducks forage offshore, including 250,000 Long-tailed Ducks, as well as all three scoters, eiders, and Harlequin Duck.

Threats: Beach nesting birds are threatened by overabundant predators (large gulls and feral cats), and disturbance from all-terrain vehicles, visitors, and dogs. Oil spills are also a constant threat. An offshore wind farm development, recently approved by the Department of the Interior, could

The sand dunes and beaches of Cape Cod provide important nesting habitat for terns and plovers (Wood End Light pictured).

PHOTO: ©ISTOCKPHOTO.COM / DENIS TANGNEY JR.

Greater Shearwaters wait for an easy fish or squid meal stirred up by a humpback whale.

affect seabirds, sea ducks, and migrating passerines. The invasive Mute Swan is also present. Finally, sea level rise threatens to erode coastal marshes and other habitats.

Conservation: Monomoy NWR protects a series of barrier islands that are sometimes split and re-shaped by severe storms. Since its establishment in 1944, and designation as a Wilderness Area in 1960, thousands of terns have re-colonized. Since 1996, the refuge's staff have protected breeding terns and plovers by controlling large gulls, and eradicating predatory mammals from the refuge. Cape Cod National Seashore protects 43,557 acres, restricts off-road vehicle use, and ropes off tern and plover nesting areas. The Massachusetts Audubon Society manages beaches, such as South Chatham, to protect seabirds, and monitors ducks and terns to collect data on potential off-shore wind development impacts.

Actions:

- Maintain restrictions for off-road vehicles, pedestrians, and dogs to minimize disturbance to beach-nesting birds.
- Control gulls and mammalian predators near tern and plover nesting areas.
- Remove feral cats.
- Determine likely effects of wind farm development on birds.

Closing off dunes and beaches in Piping Plover nesting areas helps the species breed successfully on Cape Cod.

IMPORTANT BIRD AREA
DELAWARE BAY

BIRDSCAPE: **MARINE**
BCR: **30** STATES: **DE AND NJ**
AREA: **>126,000 ACRES**

WATCHLIST SPECIES
RED: **7**
YELLOW: **25**

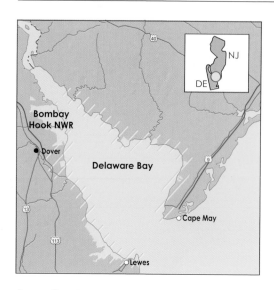

Birds: Several WatchList species breed, including Black, Clapper, and King Rails; Piping Plover, Least Tern, Black Skimmer, and Saltmarsh and Seaside Sparrows. During migration and winter, WatchList species include the Sanderling; Semipalmated, Western, White-rumped, and Stilt Sandpipers; Short-eared Owl, Nelson's Sparrow, and almost the entire population of the *rufa* subspecies of the Red Knot. The bay also provides nesting habitat for the "Coastal Plain" Swamp Sparrow, and wintering habitat for the "Ipswich" Savannah Sparrow, as well as for declining species such as the Northern Bobwhite. Up to 200,000 Snow Geese winter here.

The highest concentration of spawning horseshoe crabs on the Atlantic Coast occurs here, and over one-million shorebirds time their spring migration to feast on the crab's eggs, increasing their weight by as much as 50% before leaving for their arctic nesting grounds.

Threats: Overharvest of horseshoe crabs for use as bait in conch and eel pots is the most serious threat to shorebirds. Between 1990 and 1998, crab egg densities declined by up to 90%, This is believed to have caused a >50% decline in Red Knots; some knots may no longer be able to gain sufficient weight to breed successfully. Other shorebirds such as the Sanderling and Semipalmated Sandpiper

PHOTO: MIKE PARR

Hundreds of thousands of shorebirds (such as the Semipalmated Sandpiper) feast on horseshoe crab eggs to fuel their spring migration.

Hundreds of thousands of primeval-looking horseshoe crabs visit Delaware Bay's beaches each spring to lay their eggs.

PHOTO: MIKE PARR

have also been affected. Coastal development, pets, and off-road vehicles disturb beach nesting birds. Delaware Bay hosts 1,000 oil tankers annually, and spills are a constant threat. Invasive *Phragmites* and overabundant Snow Geese also damage habitat.

Conservation: Approximately 50,000 acres of habitat is protected, including Cape Henlopen SP; Cape May, Bombay Hook, and Prime Hook NWRs; Ted Harvey, Little Creek, Mad Horse Creek, Dix, Egg Island, and Higbee Beach WMAs; and Slaughter Beach, and Mispillion Harbor. Since 1997, ABC and other conservation groups have lobbied to reduce horseshoe crab harvests, and helped to establish a 1,500-square-mile crab sanctuary in the mouth of Delaware Bay. In 2006, both New Jersey and Delaware declared moratoriums on crab harvests.

Actions:

- Maintain restrictions on horseshoe crab harvests.
- List the *rufa* Red Knot under the ESA.
- Control *Phragmites* and other invasive or overabundant species.
- Improve saltmarsh management to benefit sparrows and other marsh birds.

PHOTO: GEORGE JETT

The secretive Saltmarsh Sparrow nests in coastal saltmarshes around Delaware Bay.

IMPORTANT BIRD AREA
GULF STREAM
OFF CAPE HATTERAS

BIRDSCAPE: **MARINE**
BCR: **N/A**   STATE: **NC**

WATCHLIST SPECIES
RED: **3**
YELLOW: **6**

This is an approximately 200-square-mile area that lies 45 miles offshore. Here the warm waters of the Gulf Stream meet the cold waters of the Labrador Current, and mats of sargassum seaweed collect along the surface of frontal boundaries.

Birds: This is an important foraging area for many pelagic seabirds, including the WatchListed Black-capped Petrel; Cory's, Greater, Sooty, Manx, and Audubon's Shearwaters; Band-rumped Storm-Petrel, and Bridled Tern. Other species that occur include the Northern Fulmar, Wilson's and Leach's Storm-Petrels, White-tailed and Red-billed Tropicbirds, South Polar Skua; Pomarine, Parasitic, and Long-tailed Jaegers, and Sooty Tern. Rare species recorded include the Bermuda and Fea's Petrels, Masked Booby, Roseate Tern, and Razorbill.

Threats: Threats are currently few or poorly documented in this area. Potential threats include offshore oil drilling, wind farm development, over-fishing, entanglement in fishing gear, collisions with lighted fishing vessels at night, ocean acidification, contaminants such as oil spills, and

potentially the overharvest of sargassum. Many of the seabirds here likely face more serious threats at nesting islands, such as invasive predators at breeding sites for Bermuda and Black-capped Petrels.

Conservation: Little conservation effort has focused on this remote area, which is not formally protected. Many of this region's species are dependent on conservation efforts at distant breeding colonies, such as artificial burrows and reintroduction efforts for the Bermuda Petrel, and forest conservation and nest protection for the Black-capped Petrel on Hispaniola.

Actions:
- Assess trends in populations of WatchList seabird species.
- Study potential impacts of wind farms, gillnets, sargassum harvest, and other threats to pelagic birds in this region.
- Create a marine reserve in this area.

Endangered Black-capped Petrels travel from colonies in the Caribbean to forage in the rich waters of the Gulf Stream.

PHOTO: GLEN TEPKE

More needs to be known about the population status and distribution of some pelagic birds such as this Audubon's Shearwater.

A Wilson's Storm-Petrel, which comes to land only to nest, elegantly patters across the water surface.

IMPORTANT BIRD AREA
FLORIDA KEYS
AND DRY TORTUGAS

BIRDSCAPE: **MARINE**
BCR: **31**   STATE: **FL**
AREA: **>2.8 MILLION ACRES OF
LAND AND WATER**

WATCHLIST SPECIES
RED: **7**
YELLOW: **33**

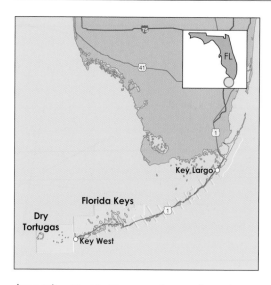

Birds: Several WatchList species breed, including the Magnificent Frigatebird, Masked Booby, Reddish Egret, Wilson's Plover; Roseate, Least, and Bridled Terns; White-crowned Pigeon, Mangrove Cuckoo, Antillean Nighthawk, and local mangrove subspecies of both the Prairie Warbler and Clapper Rail. During winter and migration, additional WatchList species include Greater, Sooty, and Audubon's Shearwaters; Band-rumped Storm-Petrel, Black Rail; American Golden-, Snowy, and Piping Plovers; Red Knot, Sanderling; Semipalmated, Western, White-rumped, and Stilt Sandpipers; and Black Skimmer. Many neotropical migratory songbirds pass through during spring and fall. Large colonies of Sooty Terns and Brown Noddies nest at Bush Key in the Dry Tortugas. The localized "Great White Heron" is also found here.

The Florida Keys stretch west from the tip of Florida 128 miles by road to Key West, and an additional 70 miles further west by boat to the Dry Tortugas. Habitats include pine forests, hardwood hammocks, scrub, mangroves, beaches, mudflats, sea-grass beds, coral reefs, and warm pelagic waters.

Threats: Much of the mangrove forests and hardwood hammocks in the keys have been replaced by development. Garbage has increased the raccoon population, increasing the risk of nest predation for breeding birds. Inadequate waste and storm water infrastructure degrades near-shore water

Red mangroves provide nesting and feeding habitat for birds, while fish breed among the submerged roots and nearby eelgrass beds.

PHOTO: ©ISTOCKPHOTO.COM / ALBERTO POMARES PHOTOGRAPHY

The Reddish Egret actively stalks its prey, running, side-stepping and arching its wings over shallow water in a fascinating "dance".

PHOTO: ROBERT ROYSE

quality. Feral cats and collisions with vehicles kill migrant songbirds. Invasive plants are also a problem (especially the Australian pine). The Dry Tortugas are threatened by growth in visitation, introduced rats, and rising numbers of predatory gulls. Severe weather events can damage or destroy small ephemeral islands.

Conservation: Protected areas include Crocodile Lake, Great White Heron, Key Deer, and Key West NWRs; John Pennekamp Coral Reef, Long Key, Curry Hammock, and Bahía Honda SPs; the large Florida Keys National Marine Sanctuary, and Dry Tortugas NP. The National Park Service works to replace invasive plants with native species. TNC acquires and protects habitat, as well as conducting prescribed burns in pine forests.

Actions:
- Protect, and restore mangrove forest and hardwood hammocks.
- Control invasive species; remove feral cats.
- Restrict visitation to the Dry Tortugas by sport-fishing, commercial, and private boats to sustainable levels.
- Reintroduce the Key West Quail-Dove.

PHOTO: ISTOCKPHOTO.COM / NNEHRING

The male Magnificent Frigatebird inflates its red throat like a balloon to attract a mate.

IMPORTANT BIRD AREA
ARANSAS NWR

BIRDSCAPE: **MARINE**
BCR: **37**   STATE: **TX**
AREA: **>115,000 ACRES**

WATCHLIST SPECIES
RED: **10**
YELLOW: **31**

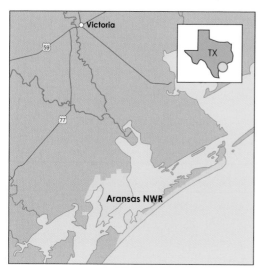

Aransas is most famous as the wintering site for the largest flock of the Endangered Whooping Crane. The refuge is dominated by saltmarsh, but oak woodlands, ponds, mudflats, and open bays also provide habitat for resident and migratory birds. We included this site in the Marine Birdscape, but it is really part marine, and part wetland, like many similar brackish marsh sites around the coasts of the U.S. These tidal marshes are one of the rarest habitats in the country, occupying a total area of less than six-million acres.

Birds: Breeding WatchList species here include the Mottled Duck, Reddish Egret; Black, Clapper, and King Rails; Snowy and Wilson's Plovers, Least Tern, Black Skimmer, and Seaside Sparrow. During migration and winter, many WatchList species use the refuge, including the Yellow Rail, Whooping Crane, American Golden- and Piping Plovers, Long-billed Curlew, Hudsonian Godwit, Red Knot, Sanderling; Semipalmated, Western, White-rumped, and Stilt Sandpipers; Short-eared Owl, and Le Conte's Sparrow.

Threats: Heavy boat traffic along the intercoastal waterway is a constant threat because boat wakes erode saltmarshes and there is the potential for a massive chemical or oil spill that could damage habitats and kill Whooping Cranes and other wildlife. Climate change is a long-term threat, as both sea level rise and increased severe weather events threaten to erode and submerge habitat or ruin nesting attempts by breeding birds. Other threats include human disturbance to birds, feral hogs, and invasive salt cedar on Matagorda Island. Water diversion from the Guadalupe River appears to be raising salinity and reducing food availability for the cranes.

Conservation: Aransas NWR was established in 1937; an additional 11,000 acre tract was purchased by TNC and added during the 1980s. Combined with the adjacent Matagorda Island WMA and SP, Aransas protects the entire 38-mile-long Matagorda Island. The Whooping Crane flock wintering at Aransas and breeding in Canada's Wood Buffalo NP has grown from a low of 16 in 1941 to more than 260 birds today. Much of the refuge is closed to limit disturbance to the cranes, but a refuge watchtower and boat tours offer visitors excellent viewing. The refuge's staff monitors birds, and manages habitat using prescribed burns. The Peregrine Fund reintroduced Aplomado Falcons to Matagorda Island, starting in 1996.

Actions:
- Continue prescribed burns to enhance habitat.
- Minimize risk of oil spills and other pollution from boat traffic through enforcement of bilge-pumping rules.
- Ensure adequate water flow in the Guadalupe River.
- Combat climate change with policies to reduce greenhouse gas emissions.

The only natural, self-sustaining flock of the Whooping Crane breeds in Canada and winters at Aransas NWR.

IMPORTANT BIRD AREA
CALIFORNIA CURRENT

BIRDSCAPE: **MARINE**
BCR: **32**   STATE: **CA**
AREA: **>5.8 MILLION ACRES OF**
LAND AND WATER

WATCHLIST SPECIES
RED: **12**
YELLOW: **27**

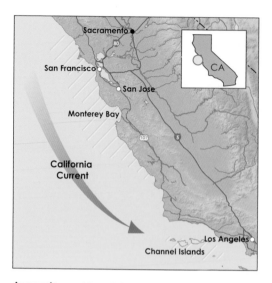

Threats: Introduced cats, rats, and mice; and over-abundant native gulls and foxes predate the eggs and nestlings of vulnerable seabirds. Vegetation has been degraded by introduced grazers, feral pigs, and invasive plants. Seabirds are at risk from contaminants (oil, mercury, plastic garbage) and bycatch in fisheries (gill nets, longlines). Climate change threatens to shift ocean currents which can lead to reproductive failure among seabirds. The "Island" Loggerhead Shrike is on the verge of extinction with just 12 individuals on Santa Cruz and 15 on Santa Rosa.

The cold California Current flows south through deep coastal waters and past groups of islands that support colonies of seabirds. Habitats include open water, kelp beds, rocky intertidal zones, sea cliffs and caves, and beaches.

Birds: The Farallon and Channel Islands host colonies of seabirds, including the WatchListed Ashy and Black Storm-Petrels, and Xantus's Murrelet. Other WatchListed seabirds occur outside the breeding season, including Black-footed and Laysan Albatrosses; Buller's, Pink-footed, Black-vented, and Sooty Shearwaters; Least Storm-Petrel, Elegant Tern; and Marbled, Ancient, and Craveri's Murrelets. Shorebirds present during winter and migration include the American Golden- and Snowy Plovers (also breeds), Marbled Godwit, Long-billed Curlew, Wandering Tattler, Black Turnstone, Surfbird, and Rock Sandpiper. California Condors scavenge along the coasts of Big Sur. Islands are home to the Island Scrub-Jay (Santa Cruz) and unique subspecies of birds such as the Allen's Hummingbird, Pacific-slope Flycatcher, Loggerhead Shrike, and Sage Sparrow.

Pink-footed Shearwaters visit the U.S. West Coast from breeding islands off the coast of Chile.

PHOTO: GLEN TEPKE

Near-shore kelp beds and upwelling currents provide an abundance of food for the seabirds that frequent Monterey Bay.

PHOTO: ©ISTOCKPHOTO.COM / WIDE SIDE PHOTOGRAPHY

Conservation: Protected areas include the Cordell Banks, Gulf of Farallones, Channel Islands, and Monterey Bay National Marine Sanctuaries; TNC's Santa Cruz Island Preserve, Channel Islands NP, and some state marine reserves. San Clemente Island is managed as a Navy bombing range, and following ABC's intervention, also now for endemic subspecies of the Loggerhead Shrike and Sage Sparrow. Island Conservation worked with ABC in 2001 to eradicate rats from Anacapa to successfully restore breeding Xantus's Murrelets.

Actions:

- List the "Island" Loggerhead Shrike under the ESA.
- Eradicate rats and mice; restore vegetation; use artificial nest boxes for burrow nesting seabirds.
- Regulate boat traffic and tourism to minimize disturbance.
- Maintain nest exclosures to reduce gull predation at key sites.
- Restrict fishing near seabird colonies and foraging hotspots.

- Create "wide-berth" zones for ships around nesting islands during the breeding season; enforce bilge-pumping regulations.

PHOTO: ROBERT ROYSE

The Black Oystercatcher favors rocky shores along the Pacific Coast, and is at risk from oil spills and human disturbance.

IMPORTANT BIRD AREA
SAN FRANCISCO BAY

BIRDSCAPE: **MARINE**
BCR: **32**   STATE: **CA**
AREA: **1 MILLION ACRES**

WATCHLIST SPECIES
RED: **4**
YELLOW: **22**

T he bay includes more than one-million acres of open water, mudflats, salt and freshwater marshes, beaches, and coastal scrub, Millions of water-birds occur during migration with 70% of the birds that migrate along the Pacific Flyway stopping here.

Birds: San Francisco Bay provides the most important breeding habitat for "California" subspecies of both the WatchListed Black Rail and Clapper Rail. Other breeding WatchList species include the Clark's Grebe, Snowy Plover, Heermann's Gull, "California" Least Tern, Short-eared Owl, and Tricolored Blackbird. Non-breeding Watchlist species include the American Golden-Plover, Long-billed Curlew, Marbled Godwit, Black Turnstone, Surfbird, Red Knot, Sanderling, Semipalmated and Western Sandpipers, Thayer's Gull, and Elegant Tern. The largest Pacific Coast population of the Canvasback (17,000) winters here.

Threats: Approximately 90% of San Francisco Bay's marshes have been lost to development. Water withdrawals remove roughly half the fresh water from the bay's tributaries facilitating saltwater intrusion and damaging some wetlands. Remaining fresh water entering the bay has high concentrations of pollutants from agricultural, industrial, and urban runoff. Oil spills are a constant threat. Sea level rise threatens to drown the Bay's marshes. Invasive species are also a problem, with smooth cordgrass (*Spartina alterniflora*) colonizing mudflats, pepperweed invading marshes; and introduced foxes, feral cats, and expanding Common Raven populations increasing predation pressure on wetland birds.

Conservation: San Pablo Bay NWR, the state-owned Petaluma Marsh Wildlife Area, and Napa-Sonoma Marshes protect 80% of the "California" Black Rail population. Don Edwards San Francisco Bay NWR is the largest urban wildlife refuge

The Endangered "California" subspecies of Clapper Rail is largely restricted to San Francisco Bay's marshes.

PHOTO: ROBERT ROYSE

San Francisco Bay offers a critical rest-stop for water-birds migrating along the Pacific Flyway.

in the U.S. and spans more than 30,000 acres of the southern bay, including salt ponds purchased in 2002. This refuge protects 60% of the "California" Clapper Rail population and ten percent of all breeding Snowy Plovers in the U.S.

Actions:

- Protect and restore tidal flats and marshes, as well as adjacent upland habitats; restrict further development of these habitats for commercial purposes.
- Eradicate invasive cordgrass from tidal flats, and control other invasive and overabundant species, including feral cats and red foxes.
- Reduce disturbance of shorebirds by pedestrians and off-road vehicles.

The Heermann's is one of North America's most distinctive gulls; 95% nest on a single Mexican island.

IMPORTANT BIRD AREA
COPPER RIVER DELTA AND MONTAGUE ISLAND

BIRDSCAPE: **MARINE**
BCR: **5**   STATE: **AK**
AREA: **569,200 ACRES**

WATCHLIST SPECIES
RED: **2**
YELLOW: **15**

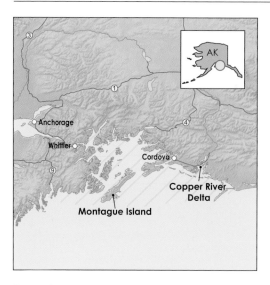

The mudflats, marshes, and rocky intertidal areas of the Copper River Delta attract up to five-million migrant shorebirds each year, with more than 500,000 at Montague Island during peak movement.

Birds: Several WatchList species breed in the wetlands, tundra, and nearby forests including the Trumpeter Swan (8% of its entire population), Wandering Tattler, Surfbird, Kittlitz's and Marbled Murrelets, Short-eared Owl, Olive-sided Flycatcher, and Varied Thrush. During migration and winter, additional WatchList species occur here including the Yellow-billed Loon, American Golden-Plover, Hudsonian Godwit, Surfbird, Black Turnstone (Montague is a major staging area for the previous two species); Semipalmated, Rock, and Western Sandpipers (c. 70% of the latter's population); Red Knot, and Sanderling. These species are joined by most of the Pacific population of Dunlin, as well as breeding "Dusky" Canada Geese.

Threats: Prince William Sound is a busy port for oil tanker traffic and the threat of a catastrophic oil spill is constant. In 1989, the *Exxon Valdez* oil spill narrowly missed the Copper River Delta. Other threats include disturbance to shorebirds by local air traffic, off-road vehicles, and foot traffic. The development of corporately owned in-holdings for oil, coal, and timber extraction is a potential threat that could also lead to road building through the area.

Conservation: The Copper River Delta and Montague Island occur largely within the >5.4 million acre Chugach NF. The USFS has researched and monitored shorebirds here since 1990. The Copper River Delta Shorebird Festival occurs each spring in early May and promotes tourism and appreciation of the region's birdlife.

Actions:

- Ensure that Montague Island is effectively included in oil clean-up plans.
- Research the effects of introduced deer and mink on Montague Island.
- Reduce activities that disturb migrating shorebirds.
- Ensure in-holdings are managed in ways that do not compromise the integrity of the Delta.

The largest shorebird gathering in the Western Hemisphere occurs at the Copper River Delta (Western Sandpiper pictured).

Rocky intertidal zones offer resting and foraging opportunities for both Black Turnstones (left) and Surfbirds (right).

PHOTO: ©MICHAEL WOODRUFF

Tidal mudflats provide forage for shorebirds, while threatened Kittlitz's Murrelets dive for fish near glaciers (at rear of picture).

PHOTO: ©ISTOCKPHOTO / RACLRO

IMPORTANT BIRD AREA
ALASKA MARITIME NWR

BIRDSCAPE: **MARINE**
BCR: **1**   STATE: **AK**
AREA: **>2 MILLION ACRES**

WATCHLIST SPECIES
RED: **8**
YELLOW: **22**

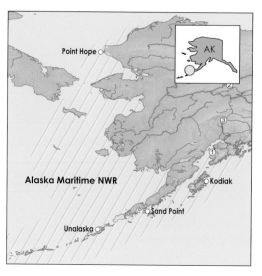

The Aleutian and Bering Sea Islands are protected within the Alaska Maritime NWR, which spans over two-million acres and supports huge colonies of nesting seabirds.

Birds: Over ten-million seabirds nest. Larger colonies include Chagulak (>1 million Northern Fulmars, 100,000 Cassin's Auklets), Kaligagan (100,000 Tufted Puffins), Kiska (1.1 million Least Auklets, 300,000 Crested Auklets), Segula (475,000 Least Auklets), Gareloi (2.3 million Crested and Least Auklets), St. Lawrence Island (3.5 million total seabirds), and the Pribilofs. Breeding Watchlist species include the Red-faced Cormorant, Black Turnstone, Rock Sandpiper, Red-legged Kittiwake, Aleutian Tern, Marbled and Kittlitz's Murrelets, Whiskered Auklet, Short-eared Owl, and McKay's Bunting. Those that occur during winter and migration include the Emperor Goose, Spectacled and Steller's Eiders, Yellow-billed Loon; Short-tailed, Black-footed, and Laysan Albatrosses; Sooty and Pink-footed Shearwaters, American Golden-Plover, Bristle-thighed Curlew, Wandering Tattler, Hudsonian and Bar-tailed Godwits, Sanderling, and Western Sandpiper. These islands are also home to a variety of endemic bird subspecies.

Threats: Foxes were introduced to the Aleutian Islands by fur trappers from 1750-1936. Norway rats on Kiska kill thousands of auklets, and threaten to wipe out seabird colonies if not removed. Introduced caribou on Atka and Adak are degrading native vegetation. Climate change is warming sea temperatures, reducing sea ice, and causing shifts

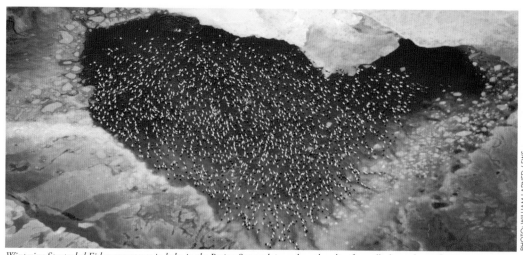

Wintering Spectacled Eiders congregate in holes in the Bering Sea pack ice, where they dive for mollusks on the sea floor.

PHOTO: WILLIAM LARNED / FWS

The Tufted Puffin is one of fourteen species of breeding alcids found here, making the Aleutian Islands the "alcid capital" of the world.

in fish populations. On average, one boat wrecks in the Pribilofs each year, and a 1996 oil spill killed thousands of King Eiders. Other threats include over-fishing, entanglement in fishing gear, and contaminants including oil spills (mostly fuel oil).

Conservation: FWS manages more than two-million acres of land and sea within the Aleutian Islands Unit of the Alaska Maritime NWR, as well as 170,000 acres in the Bering Sea. After foxes nearly wiped out the "Aleutian" Cackling Goose they were removed from most islands. Having survived only on Buldir, the goose population now exceeds 70,000, and it was removed from protection under the ESA in 2001. FWS, TNC, and Island Conservation eradicated rats from Rat Island in 2008; Kiska and other islands are the next targets. Kittlitz's Murrelets have returned to Agattu following fox removal. FWS maintains rat traps at ports in the Pribilofs, and an emergency rat response team in case of a shipwreck.

Actions:

- Eradicate rats, foxes, and other predators; remove herbivores (reindeer, caribou, marmots, and European rabbits).
- Enact policies to combat climate change.

Red-legged Kittiwakes nest on sea cliffs in the Pribilofs; their wintering range is largely unknown, however.

IMPORTANT BIRD AREA

NORTHWEST HAWAIIAN ISLANDS

BIRDSCAPE: **MARINE**
BCR: **69**   STATE: **HI**
AREA: **90 MILLION ACRES**

WATCHLIST SPECIES
RED: **8**
YELLOW: **2**

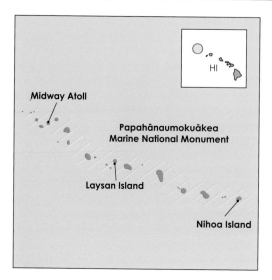

The Northwestern Hawaiian Islands form an 800-mile chain across the central Pacific. They are sparsely vegetated and vary from rocky pinnacles to low-lying atolls. They provide vital habitat for a range of seabirds and for a few endemic species.

Birds: More than 14 million seabirds breed here, with the largest colonies on Laysan and Midway. WatchList species include Laysan, Black-footed, and Short-tailed (recent breeding attempt on Midway) Albatrosses; Tristram's Storm-Petrel, Laysan Duck, Millerbird, and Laysan and Nihoa Finches. Other breeding seabirds include the Wedge-tailed and Christmas Shearwaters, Bulwer's and Bonin Petrels, White-tailed Tropicbird; Masked, Brown, and Red-footed Boobies; Great Frigatebird, Blue-gray and Brown Noddies, and Gray-backed and Sooty Terns. WatchListed Bristle-thighed Curlews winter here in small numbers.

Threats: Historically, the commercial harvest of guano (for fertilizer), eggs (albumen for the photographic industry), and feathers (for down and the millinery trade) decimated seabird colonies. Intro-

duced rabbits devoured virtually all vegetation on Laysan leading to the extinction of the endemic rail, and subspecies of the Millerbird and Apapane. Introduced rats, mice, grasshoppers and ants, and plants such as sandbur and golden crownbeard, also threaten bird habitat. Beaches (and the ocean) are littered with trash. Plastics and other garbage kill seabird chicks if ingested. On Midway, albatross chicks eat lead-laced paint chips from decaying buildings; up to 10,000 die annually as a result. Additional threats include botulism in Laysan Ducks, the potential arrival of avian diseases from the main Hawaiian Islands, sea level rise, droughts, severe storms, and grass fires.

PHOTO: BILL HUBICK

More than 90% of all Laysan Albatrosses nest in the Northwest Hawaiian Islands.

A Laysan Duck chases brine flies; once numbering fewer than 10 birds, the species is increasing and has been reintroduced to Midway.

PHOTO: M.J. RAUZON / VIREO

Conservation: Laysan and other islands were first protected in 1909, and enforcement efforts helped end the exploitation of seabirds. FWS controls invasive plants, eradicates introduced predators, and has successfully translocated Laysan Ducks and Laysan Finches between islands to establish new populations. Nesting seabirds rebounded dramatically on Midway following rat eradication in 1997, with Bonin Petrels growing from 5,000 pairs in 1979 to 130,000 pairs in 2008; both Tristram's Storm-Petrel and Bulwer's Petrel also re-colonized the island. The 90-million-acre Papahānaumokuākea Marine NM was established here in 2008.

Actions:
- Clean up lead-based paint contamination on Midway; remove floating plastics.
- Eradicate invasive species, especially rodents, sandbur, and golden crownbeard.

- Establish new populations of Laysan Duck, Millerbird, Laysan Finch, and Nihoa Finch on additional restored islands.
- Increase biosecurity to avoid introduction of additional invasive species.
- Regularly monitor the most remote islands.

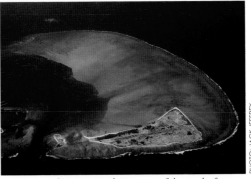

PHOTO: JACK JEFFREY

Rising seas and storm surges threaten tens of thousands of seabirds nesting on low-lying atolls.

HUMAN BIRDSCAPE

Area: **3,100,000 sq. miles**	WatchList Species
States: **All 50**	Red: **0**
BCRs: **All**	Yellow: **3**

Humans have significantly altered more than 80% of the U.S. landscape. Our impacts are most severe where people live in the highest densities—in cities and suburbs. We have also put much of the southern part of the continent to work in the production of food and timber (and increasingly for water and energy development), and this has had a major impact on bird habitats. Cities differ the most from natural habitats, although tall buildings can resemble sea cliffs and provide nest sites for Peregrine Falcons. Chimneys also resemble hollow trees, and are used by breeding and migrating Chimney Swifts. Suburbs often retain elements of more natural habitats, including parks, or artificial wetlands. Harbors and

bays that are surrounded by development (even large cities) can provide habitat for birds of conservation concern, but inevitably, the quality of this habitat is affected by pollution and disturbance. Grazing, agricultural, and timber lands retain components of the natural grasslands and forests they replaced, including some of the original birdlife. With management to mimic natural processes and habitat structure, many native birds can be maintained in these areas.

Birds: Cities provide habitat for few native birds, but often support introduced species; the Rock Pigeon, European Starling, and House Sparrow are now ubiquitous in urban areas. Some south-

Birds: 1. American Goldfinch; 2. Northern Cardinal; 3. House Finch; 4. Peregrine Falcon; 5. American Crow; 6. Chimney Swift; 7. Rose-breasted Grosbeak; 8. Rock Pigeon. **Vegetation**: 9. cropland; 10. gingko. **Threats**: 11. House Finch eye-disease; 12. West Nile virus; 13. outdoor cats; 14. House Sparrows; 15. Monk Parakeet; 16. European Starling; 17. collisions with towers; 18. Reflective glass. **Other**: 19. Lights Out!; 20. fuel-efficient car; 21. birders watching Chimney Swift.

Human Birdscape Habitats	Condition	Threat Level	Flagship Bird
Pasture	Poor	Medium	Tricolored Blackbird
Silviculture	Fair	Medium	Bachman's Sparrow
Cropland	Poor	Medium	Lark Bunting
Urban	Poor	Low	Rock Pigeon
Suburban	Poor	Low	American Robin

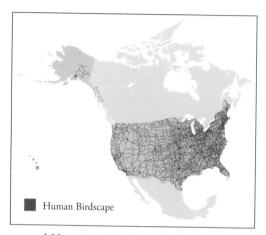

Human Birdscape

ern and Hawaiian cities have a higher diversity of introduced birds, including some parrot species, bulbuls, and mynahs. These introduced species can threaten native birds by competing for nest sites or acting as reservoirs for disease. Few WatchList species breed in cities, though the Allen's Hummingbird, Green Parakeet, and Red-crowned Parrot do, and many more pass through during migration. Agricultural areas and artificial wetlands provide habitat for a wide variety of birds depending on their location. Farmlands provide an alternative for hardier grasslands species, and reservoirs can be important for waterfowl.

Threats: Much of the Human Birdscape is itself a threat to native birds. Not only does development replace native habitat, but it can act as an entry point for invasive species, emit pollution, and present obstacles to migrating birds. Even as the Human Birdscape attracts and benefits some birds, it presents hidden dangers to others, with many falling victim to collisions with lighted structures or predation by cats. People prefer stability, and have worked to reduce the floods, fires, and populations of migratory grazing animals and large predators

that once drove the processes necessary to maintain natural ecosystems. Some of our farm, timber, and grazing lands that once supported important populations of native birds no longer do so because of intensification. People build water features in arid landscapes that serve as oases and benefit some birds, while also drawing precious water from the surrounding landscape that degrades riparian habitats elsewhere.

Conservation: Cities and other human landscapes are not all bad for birds. It is easier for many people to reduce their personal footprint on nature by living in cities, where public transportation and other services that reduce energy consumption and increase recycling are best developed. City-dwellers also typically occupy less land and consume less energy than suburbanites. Most of our conservation, research, and educational institutions including museums, zoos, universities, NGOs, and bird clubs, are located in or near large cities. Conse-

PHOTO: MIKE PARR

Agricultural lands present risks for birds, but there are also many ways farmers can enhance their land for conservation.

This looks like a safe neighborhood, but outdoor cats, pesticides, and other dangers to birds lurk in the Human Birdscape.

PHOTO: MIKE PARR

quently, we often have the best data on bird populations and trends for places where people live, and often lack this information where fewer people are available to do the work. Millions of people provide bird seed, suet, fruit, nectar, water, and nest boxes for birds. This can help birds survive the winter and even shift their ranges. Most eastern Purple Martins now nest in boxes we set out for them. Other cavity-nesters benefit from boxes, including owls, chickadees, Tree Swallows, wrens, and bluebirds. Rufous Hummingbirds appear to be wintering more commonly in the Southeast, perhaps partially due to an increase in feeders there.

We also value nature for recreation and often maintain remnants of it in our urban parks and gardens. These offer important resources to birds, and are often the places where we first experience, connect with, and learn to care for nature.

Actions:

- Reduce bird collisions with communication towers, wind turbines, glass windows, and other man-made structures through improved glass technologies, improved lighting on towers, and "Lights Out" programs in cities.
- Keep cats indoors and remove feral cat colonies.
- Reduce the negative impacts of our cities, towns, and farmlands through more efficient energy use, emissions reduction, and waste disposal.
- Determine how biofuels can benefit rather than harm natural habitats.
- Develop connections between people and birds through education and outreach.
- See ABC's Top Ten Tips for Bird Friendly Living at www.abcbirds.org/newsandreports/Toptips.pdf.

IMPORTANT BIRD AREA
CENTRAL PARK

BIRDSCAPE: **HUMAN**
BCR: **30**   STATE: **NY**
AREA: **862 ACRES**

WATCHLIST SPECIES
RED: **1**
YELLOW: **10**

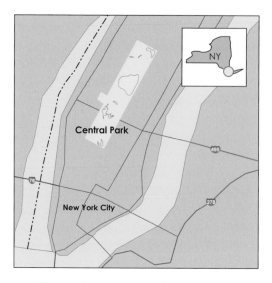

Central Park

New York City

Red-headed Woodpecker, Olive-sided Flycatcher, Wood Thrush; Golden-winged, Bay-breasted, Prairie, Cerulean, Prothonotary, and Canada Warblers; and Rusty Blackbird. Like many urban parks, some of the most abundant birds are not native, such as the Rock Pigeon, European Starling, and House Sparrow. Introduced Monk Parakeets also inhabit New York City.

Threats: Birds in urban areas such as Central Park face many threats, including feral cats, introduced rats, disease (e.g., House Finch eye disease), and collisions with windows and other structures. The first outbreak of West Nile virus began in New York in 1999, before spreading across North America and killing millions of birds and hundreds of people.

Central Park is located in the center of Manhattan, New York City. It is an oasis of habitat in a canyon surrounded by skyscrapers. Though WatchList species only occur here during migration, it provides access to birds and nature for millions of people living nearby.

Birds: Several WatchList species pass through during migration or occur in winter, including the

Conservation: Starting on Christmas day 1900, Frank Chapman began substituting Christmas bird hunts with Christmas Bird Counts, thereby establishing the longest continuously running bird monitoring program in North America and providing important information on bird population trends. Central Park was one of 25 sites where the first count took place. The Peregrine Fund and

Urban parks form vital "islands" of habitat where migrating birds, such as this Blackburnian Warbler, can rest and refuel.

PHOTO: ROBERT ROYSE

Central Park is a green oasis within an obstacle course of glass and concrete buildings that threaten migrating birds.

other groups worked to restore Peregrines to the eastern U.S., and now more than 12 pairs nest on buildings and bridges in New York City. In 1998, Eastern Screech-Owls were reintroduced to Central Park (they had been absent since the 1960s, but now breed). In 1991, Red-tailed Hawks recolonized Central Park on their own, including one dubbed "Pale Male" which became a celebrity itself after human celebrities tried to remove its nest from their building. To reduce bird mortality from collisions, New York City Audubon initiated a "Lights Out" program to encourage owners of tall buildings to turn off their lights after hours during fall migration in September and October. Turning out lights saves both birds, and money on electricity bills.

Actions:
- Expand participation in "Lights Out New York" to reduce collisions.
- Incorporate bird-safe glass and lighting guidelines as part of environment-friendly building certification schemes.

- Improve education and outreach on migratory birds in the park.

The Hooded Warbler can be hard to find in forest undergrowth, and is often easier to see in urban parks during migration.

IMPORTANT BIRD AREA
CHEASAPEAKE BAY WATERSHED

BIRDSCAPE: **HUMAN**
BCR: **30** STATES: **MD, VA, PA, NYC, DE, WV**
AREA: **64,000 SQ. MILES**

WATCHLIST SPECIES
RED: **5** YELLOW: **15**

The watershed has 16 million residents, several large cities, extensive suburbs, and large tracts of farmland and forest (58% of the land area). It is drained by more than 100,000 streams and rivers that feed into the polluted Chesapeake Bay.

Birds: Most bird species that occur in the eastern U.S. have been recorded in the watershed. Watch-Listed species that breed in the area's uplands include the Red-headed Woodpecker; Cerulean, Blue-winged, Golden-winged, and Prairie Warblers; and Henslow's Sparrow. The Bay itself supports migrant shorebirds, and up to one-million wintering waterfowl, although waterfowl numbers are significantly reduced from the past. Breeding WatchList species include the Least Tern, Black Skimmer; Black, Clapper, and King Rails; and Saltmarsh and Seaside Sparrows. The Bay area also supports the largest wintering concentration of the Bald Eagle in the "Lower 48" states.

Threats: The region's birds face the entire gamut of threats expected in a human-dominated landscape: forest fragmentation, suburban sprawl, predation by feral cats, collisions with buildings and communication towers, pesticides, and invasive species. Excess fertilizer from farms causes algal blooms and oxygen-depleted "dead-zones" in the bay; invasive *Phragmites* crowds out native plants, and introduced Mute Swans and nutria denude native vegetation. The impacts of mosquito management actions on the WatchListed Black Rail need to be determined. The most serious long-term threat to the bay likely comes from sea level rise, however.

Conservation: Most conservation focus in the region is concentrated on the Bay itself, yet the watershed as a whole is also vitally important to landbirds such as the Wood Thrush, Brown Thrasher, Prothonotary Warbler, and Eastern Towhee. Species such as the Bald Eagle and Osprey that were decimated by DDT have now almost returned to their historic population levels. A major effort to reduce pollution in the Bay from agricultural runoff is underway with a "nutrient trading program" for farmers recently being proposed to provide additional cleanup finance.

Actions:
- Reduce nutrient runoff from farmlands.
- Determine how to maximize habitat for the Black Rail.
- Manage saltmarshes to mitigate sea level rise, if and where possible.
- Remove or control invasive species.
- Restrict development around key bird sites and reduce habitat fragmentation.

The Louisiana Waterthrush, which is actually a warbler, returns to the U.S. in early spring to nest along the banks of forest streams.

PHOTO: ROBERT ROYSE

Areas surrounding the Bay provide a mosaic of bird habitats, including the various successional ecosystems shown here.

PHOTO: MIKE PARR

Although beautiful, the Mute Swan is an invasive species that consumes aquatic vegetation needed by native waterfowl.

PHOTO: MIKE PARR

CHAPTER 3

THREATS

The population size and trend of a bird species is determined by the balance between mortality and reproductive recruitment. To maintain population size, species that suffer high natural rates of mortality must compensate with high reproductive rates, whereas those that live longer lives (usually those at the top of the food chain), tend to have lower fecundity. Some species maintain consistent population levels, whereas others are adapted to boom and bust food cycles that mean their populations can fluctuate wildly from year to year. However, it is the long-term population trend that determines the fate of a species. Consistent, long-term declines occur when mortality continually exceeds reproductive recruitment. This can be due to natural events, and many species have suffered declines and eventual extinctions due to factors that were entirely independent of humans. But today, many bird populations are in decline due to an array of man-made threats. These threats have accumulated more rapidly and with greater severity than many species have been able to adapt to. Although it can be argued that most birds go through natural cycles of decline and increase, we have reached a point in history when the impacts of human activities are so profound and far-reaching that from now on, it will always be impossible to untangle the completely natural declines from those that are partially or completely anthropogenic. From a conservation point of view, it is largely irrelevant anyway. Any human-caused stress that we can alleviate from a declining species can potentially benefit its population, and we should take action to lessen that stress if we can.

The WatchList tells us that roughly one third of all bird species breeding in the U.S. are of conservation concern. Breeding Bird Surveys and Christmas Bird Counts show that some 160 species in the continental U.S. are declining, but also that we lack sufficient data to evaluate trends for roughly 165 species. Even armed with good population statistics, it is often difficult to link declines to specific causes or threats. This problem is especially challenging for migratory birds which are difficult to track and face a variety of threats on their breeding and wintering grounds, and along their migratory routes.

In the case of most WatchList species, the cumulative impact of multiple issues across their ranges and life-cycles, rather than one single threat, is affecting their populations. Therefore, it is not helpful to birds for us to simply shift blame and responsibility from one threat to another. For example, wind turbine manufacturers may be correct in asserting that more birds are killed in collisions with communication towers or windows than at wind farms, but that does not justify or excuse the bird mortality at turbines, or those that will likely occur when more turbines are constructed. Instead, we must address all of the major threats responsible for killing large numbers of birds, using the best information and research available, while promoting further research and monitoring where it is most needed.

By understanding how our activities threaten bird populations, and learning what alternatives exist to reduce our impacts, we can make changes that

PHOTO: WFVZ

Photo of an Eskimo Curlew taken in 1962 by Donald Bleitz in Galveston, Texas, held in the archives of the Western Foundation of Vertebrate Zoology. The species was doomed by multiple threats, including habitat loss and market hunting.

ANTHROPOGENIC MORTALITY EXCEEDING ONE-MILLION BIRDS ANNUALLY (estimates)

Outdoor cats	>532 million
Collisions with buildings/glass	100 million-1 billion
Collisions with power lines	130-175 million
Collisions with automobiles	50-100 million
Hunting	41.5-120 million
Collisions with communication towers	4-50 million
Pesticides and toxics	16 million
Persecution	>4 million
Oil and waste-water pits	1.4-2 million

can maintain healthy bird populations and avoid further extinctions. The accounts in this chapter explain the problems, solutions, and what you can do to minimize some of the major issues affecting birds. These include habitat loss, invasive species, collisions, pollution, and climate change.

Some of these threats result in direct mortality, while others reduce reproduction or long-term survival. Although our actions may not be intended to harm wildlife, the sensitivity of bird species to environmental change means that this is often the end result.

HABITAT LOSS

Habitat loss is widely considered the number one threat to birds, but which habitats are we losing, which birds are most affected, and what can be done about it?

We can probably all agree that "paving paradise and putting up a parking lot" qualifies as habitat loss, yet many urban environments still provide habitat for birds. In fact, what we usually refer to as "habitat loss" is actually the change from one habitat type to another. Change can occur naturally, such as with forest succession, where naturally occurring forest fires convert forests into open meadows. These meadows are then replaced by bushes and shrubs, which eventually develop into forests again. With each habitat change along this natural path of succession, some species win and others lose, but as long as some of each habitat type exists somewhere across a large enough landscape at all times, all the birds in that region will be able to persist over time. When people per-

manently convert habitats to other uses, however, or disrupt natural succession processes, many bird populations suffer. This conversion can be sudden and dramatic, such as the conversion of forests to subdivisions, or slow and subtle, such as the suppression of fire across America's forests.

Habitat conversion and disturbance can directly kill birds, especially young birds and eggs during the breeding season. For example, logging a forest or harvesting an agricultural field during nesting can destroy all the nests and young in an area. More significantly, habitat loss displaces adult birds, depriving them of breeding sites, food resources, and cover from predators. Displaced to marginal areas, birds may fail to breed, and suffer increased mortality. Although a bird may live on as an individual, without the ability to breed, it no longer contributes to the continuation of the species, and unless conditions improve later, is equivalent in evolutionary terms to an individual that has been killed. Habitat loss, then, reduces the capacity of a landscape to support birds, placing a new, lower limit on their populations.

Often, it may appear to us that some viable habitat remains after human intervention, albeit in fragmented patches, but these patches may be unsuitable for the birds that relied on the original landscape. The habitat fragments may be too small to support birds requiring large territories, such as some raptors, or birds that require multiple, neighboring territories rather than isolated ones. Birds that breed in habitat fragments may also suffer higher rates of predation or nest parasitism because the adjacent areas can support large numbers of predators and cowbirds. This is called "edge effect", but may actually extend well past the edge of a fragment itself. Young birds searching to establish territories of their own may also have trouble traveling through adjacent, unsuitable areas to find new habitat. Likewise, suitable habitat fragments may be too isolated from larger tracts and therefore be left unoccupied.

PHOTO: MIKE PARR

You should still avoid starting forest fires, but land managers are increasingly changing fire suppression policies.

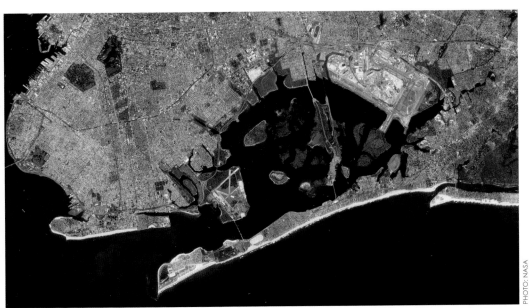

Jamaica Bay is surrounded by New York City. Sea level rise could wipe out its marshes which would be unable to move inland.

Even unfragmented habitat that appears intact can be degraded by the alteration of natural disturbance regimes, such as fire or flood, or native animals such as bison, prairie dogs, and beaver. Invasive species that alter vegetation, kill birds, or spread diseases, further impact habitat; as do air, water, noise, and light pollution.

A bird population spread out among habitat fragments of varying size and quality can show a "source-sink" relationship between habitat patches. In such a source-sink dynamic, poor-quality habitat fragments may contain birds that do not reproduce well, acting as a population sink, whereas high-quality tracts of habitat, where birds thrive, act as a source. Birds from the source habitat disperse outward to colonize other areas. If they can establish themselves in another source habitat, they can become viable contributors to the population. If all they can find is a smaller, sink patch, they disappear into a reproductive "black hole". Yet these smaller fragments, even if they cannot support the reproductive needs of adult birds, may still be valuable during migration, for non-breeding young birds, or for post-breeding adults. Some also contribute at least a few adults to the population.

Habitats are destroyed and degraded by many human activities. These include agriculture, logging, mining, and development. These activities have profoundly altered North American landscapes during the last 500 years and continue to do so today. Of course, not all of them are bad for all birds; managed forest is still better than subdivisions, even if natural forest is better than managed forest. Some species also flourish in human-dominated landscapes. Our challenge is to continue our way of life, while improving habitat conditions that maximize the abundance and diversity of birds, with a special focus on those that are currently most threatened.

The Northern Bobwhites declined 82% from 1966 to 2006 due to intensifying management of farms and pine plantations.

Development

EXTENT: C. 10% OF U.S. (URBAN AND RURAL RESIDENTIAL, ROADS, RAILWAYS, ETC.)
TREND: 400% INCREASE SINCE 1945

The average human density in the U.S. is 79.6 people per square mile. Human development replaces natural habitats with buildings, pavement, and other hard surfaces. Although some birds are able to persist in developed landscapes, these species are rarely of conservation concern because they tend to have large ranges and population sizes, can survive in a wide variety of habitat types, or are non-native.

Problems: During the 1990s, development claimed more than 11 million acres of the "Lower 48" states. Habitat loss caused by development has historically been most severe along our coasts and rivers. Linear habitats, such as beaches and saltmarsh are naturally more vulnerable to fragmentation because they are so narrow. In 2003, 53% of the U.S. population lived in coastal counties, which make up only 17% of the land area outside Alaska. Coastal development is encouraged by government insurance for beach homes and other buildings located within flood hazard areas that are considered too risky for private insurers.

Compared to logging, grazing, and farming, development occupies significantly less land, but associated habitat change tends to be more complete and irrevocable. Its footprint also extends beyond the land it occupies and the natural habitats it displaces or fragments. Construction requires lumber, energy, and other materials produced through logging and mining. Hard surfaces change water flow patterns, increasing runoff and decreasing absorption. Natural wildfires on public lands are suppressed near residential areas. Developed areas are also sources of pollution, present high risks for bird collisions, are often the entry points for invasive species, and have high concentrations of feral cats. Currently, many zoning regulations aim to protect open space by limiting housing density, but this actually facilitates sprawl by spreading it over a larger area.

Solutions: Our growing population continues to demand living and working space, so we must learn to use land more efficiently to avoid increased habitat loss. This must involve reforms to zoning regulations, which are often made by local county and municipal governments. Land-use planning should encourage high density housing near public transportation in order to maximize living units per acre and reduce the need for space devoted to development. Further regulations and conservation easements can be used to protect and buffer sensitive habitats, such as stream corridors and wetlands, from surrounding development, and to provide natural areas for recreation. Bird conservation planning at a regional scale can also help to minimize impacts on priority bird populations.

Where development does occur, energy efficiency should be maximized and environmental impacts minimized. Publicly funded building projects should meet green building standards. Corporate

Power line corridors can create early successional habitat, which benefits some birds species, but also increase "edge effect".

PHOTO: MIKE PARR

Development fragments habitat, such as the forest patch in the upper left, making it unsuitable for forest interior species.

boards and home buyers should demand similar standards for office, home, and other construction projects. However, green building standards need to include measures that minimize bird collisions (see p. 317). Meeting these standards currently costs more than not doing so, but some costs are offset in energy efficiency savings. Porous concrete and permeable pavements may help to slow storm runoff. Government subsidies (disaster insurance and aid) for new development or rebuilding in sensitive coastal habitats and flood-prone areas should end, or premiums be increased to price risk appropriately.

Actions:

- Attend local zoning meetings and encourage mass transit and high density housing projects, as well as the setting-aside of natural habitat to mitigate losses to development.
- Oppose federally subsidized development of flood-prone coastal areas.
- Ask for environmentally certified features in new home construction, including lumber and

energy-efficient appliances, as well as window features to minimize bird strikes (see p. 317). Existing homes and offices can also be upgraded with energy-efficient appliances, lighting, and windows. Tax rebates for installation may be available to offset a portion of the cost.

Some birds, such as the American Robin, thrive in developed areas, though most WatchList birds do not.

Mining and Drilling

Extent: c. 5% of U.S. (with 20% more open for leasing)
Trend: Increasing

There are more than 25,000 mines and drill sites in the U.S., and more than 500 million acres of federal land is open to oil and gas drilling, with 120 million acres in the West already degraded by well pads, pipelines, and associated roads. Drilling permits issued for BLM lands increased from 1,803 in 1999 to 6,399 in 2004.

Problems: Surface mining for coal, stone, metals, and oil from tar sands often results in the removal of native vegetation and substrate. Additional land is used to store waste (including mine rubble), as well as to construct roads and other mining infrastructure. Waste products often pollute soils and watersheds, and may include arsenic, lead, mercury, or petrochemicals.

Mountaintop removal-valley fill coal mining is impacting close to one-million acres in the southern Appalachians. In this destructive mining process, entire mountaintops are removed and the waste rock is dumped into adjacent stream valleys, leaving behind a barren moonscape. From 1985 to 2001, 724 miles of streams were buried, and water quality along a further 1,200 miles was degraded. "Edge effect" is also increased reducing interior forest habitat by up to five times the acreage actually cleared. This negatively affects species such as the Wood Thrush, Kentucky and Worm-eating Warblers, and the Louisiana Waterthrush. The Cerulean Warbler is especially hard hit as it uses both the ridge top and valley forests impacted by mountaintop mining. Phosphate strip mining also threatens to affect 400,000 acres in southwest Florida.

Mining can also directly cause bird mortality. For instance, open pits used to store waste-water and oil at production and drill sites attract and kill up to one-million birds annually. Struggling birds that are oiled or poisoned can draw in raptors, and those that escape can poison themselves further through preening, and contaminate and kill eggs back at nests. The Lesser Prairie-Chicken, Peregrine Falcon, Short-eared Owl, and Loggerhead Shrike are among the species found dead in oil pits. Noise from drilling can disturb Greater Sage-Grouse and other game birds, causing them to abandon lek sites. Offshore drilling can result in oil spills, and kill migrants through collisions with lighted platforms.

Solutions: Solar and wind energy provide cleaner sustainable alternatives to fossil fuels, but their infrastructure can also affect habitat and pose collision risks for birds. If designed and sited properly, however, these risks can be minimized. Reforming mining regulations, including the General Mining Act of 1872, could improve environmental safeguards. For example, new operations should be sited away from grouse leks. Drilling operations should also cover oil and water pits to prevent bird kills. Mined lands in the Appalachians can be reclaimed to native forest. For example, Appalachian

Mines, such as this one in Utah, often store waste-water in open pits that appear to be wetlands but are hazardous to birds.

PHOTO: ISTOCKPHOTO.COM / IOFOTO

"Mountaintop mining" in the Appalachians clears ridges and fills valleys where Cerulean Warblers breed (Bolt Mtn, WV, pictured).

Regional Reforestation Initiative partners have planted nine-million trees and reclaimed 15,000 acres to native hardwoods. Mature forest bordering mined areas can also be restored for forest interior birds by planting fast-growing trees that reduce "edge effect". Many mine-lands can also be reclaimed to grassland and shrub-land, which can benefit populations of the Golden-winged Warbler and Henslow's Sparrow. Protecting habitat on public lands mandated for multiple uses will ultimately require setting some areas aside completely for wildlife. Efforts to require royalty fees from mines on public lands to fund the cleanup of abandoned mine sites (e.g., the Hardrock Mining Reclamation Act of 2007), have not yet succeeded due to opposition from mining states.

Actions:

- Improve energy efficiency at home and work to reduce reliance on fossil fuels; drive a fuel-efficient car.
- Write to elected officials and request policies to stop mountaintop mining, mitigate mining impacts, clean up abandoned mine sites, and encourage the development of cleaner energy sources (especially in eastern coal mining states, and western states where public land is open to mining and drilling).
- In affected states, become active in your local watershed group, many of which are involved with mine-land restoration efforts.

Lekking grouse avoid noisy oil drill sites; mosquitoes breeding in waste-water from drills spread West Nile virus.

AGRICULTURE

EXTENT: **19% OF U.S.** (ONLY C. **75% OF THIS USED ANNUALLY**)
TREND: **7.5% DECLINE SINCE 1949**

Over 400 million acres of the U.S. is farmed to produce food. The vast majority of this land is privately owned, though some NWR lands are also leased for crops. Half the farmed land is made up of farms of 2,000 acres or more in size. In any given year, only a proportion of the available cropland is actually farmed, though increasing commodity prices can drive more land into production (e.g., 15 million acres were moved into corn ethanol production in 2007). When prices fall, more land is enrolled in "Farm Bill" programs such as the CRP that provide subsidies to farmers for idle land (covering c. 10% of all cropland). When prices rise again, CRP incentives become less enticing, and expired conservation contracts are less likely to be renewed.

Problems: In total, more than 100 million acres of North American wetlands and 290 million acres of grasslands have been converted to agriculture. Agricultural runoff is also the top polluter of waterways. Following European settlement much of the eastern forests were logged and then cleared for

farmland, but over recent decades, forest cover has increased as abandoned cropland has reverted to forest. In the southeast, pine lands and bottomland forests were cleared first for cotton and tobacco, and then later (in the 1970s), for soybeans. The Great Plains region has suffered the brunt of the conversion of grasslands to agriculture, but Florida, California, and Washington have also lost significant grassland acres. Prairies continue to be ploughed, with, for example, 1.1 million acres in South Dakota lost between 1982 and 1997. Much of the northern Everglades are occupied by sugar cane plantations, and Florida scrub continues to be cleared for citrus orchards. Most of the lowland forests and wetlands in Hawaii have been cleared to grow pineapples and other tropical crops. Farming has also affected much of the natural habitat in arid regions, such as the Lower Rio Grande Valley in Texas.

Some grassland birds remain in crops that resemble native grasslands, but faster growing varieties now allow farmers to harvest while birds are still breeding, destroying nests and reducing breeding success. Pesticides can poison incubating birds, nestlings, and birds entering fields to forage. Farming also takes water away from wetlands and riparian habitats for irrigation, and pollutes waterways with pesticides and excess nutrients from fertilizers. Farming has benefited some bird populations though, such as those of the Snow Goose, which has become overabundant as a result. Most Watch-List species associated with grassland and wetland habitats prefer native habitats to tilled agricultural lands, however. Pesticides also kill millions of birds in farmland each year (see p. 330 for detailed section on pesticide effects).

Solutions: The most far-reaching policies affecting conservation on farmlands are found in the federal Farm Bill which is enacted roughly every five years to set and fund USDA policy. The May 2008 version appropriates close to $300 billion over five years towards agricultural programs, including $24

PHOTO: TOM GREY

The Horned Lark has declined 56% since 1966 partly due to early and repeated harvests that destroy nests.

Mandated increases in corn-ethanol drives the conversion of grasslands to corn, reducing available habitat for grassland birds.

PHOTO: MIKE PARR

billion for financial incentives to farmers to maintain or restore wetlands, grasslands, and forests. For example, the CRP pays farmers (c. $50 per acre per year) to conserve agricultural lands as natural habitat under ten to 15 year contracts. This program started in 1986, and now enrolls more than 36.8 million acres owned by 400,000 farmers. For details on other Farm Bill conservation programs (see p. 433). Native grass biofuels may present a bird-friendly alternative to corn ethanol as technology advances and additional refineries are built.

Actions:

- Farmers interested in conserving habitat can contact the USDA or NRCS for assistance.
- Farmers can practice integrated pest management, reducing the use of pesticides.
- Delaying harvest until after nesting can help birds that breed in croplands (e.g., California compensates farmers with Tricolored Blackbird colonies on their lands for delayed harvests, saving tens of thousands of nests).
- Buy organic to reduce the use of pesticides, and eat less meat which reduces both the land required for raising cattle, and feed acreage.
- Write to your elected officials to support Farm Bill programs that encourage habitat conservation.

PHOTO: MIKE PARR

Agricultural intensification has led to substantial declines in grassland birds that were once more tolerant of farmland.

GRAZING

EXTENT: **35% OF U.S.** (INCLUDES GRAZED FOREST LAND)
TREND: **25% DECLINE SINCE 1945**

There are approximately 94 million cattle in the U.S. Millions of acres of public and private range-lands including grassland, shrub, riparian, and arid habitats, are severely damaged by over-grazing. Public grazing lands include 163 million acres managed by BLM, and 97 million acres managed by the USFS, almost all in western states. Grazing is also allowed in some NPs, NMs, and Wilderness Areas.

Problems: Ranchers using public lands must pay grazing fees, but these are priced far below those on comparable private lands, encouraging over-grazing at the expense of the environment. Great Basin sagebrush is especially incompatible with grazing, though prairie systems that have evolved with large herds of bison (that in most cases are no longer present), can benefit from well-managed grazing.

Besides degrading habitat and displacing native wildlife, over-grazing facilitates the spread of invasive plants, increases erosion, and often requires roads or fencing that restrict wildlife movements. It also leads to increased water withdrawals from local streams and aquifers, and requires supplemental

feed that is grown on agricultural lands elsewhere. Over-grazed lands are sometimes treated with herbicides and reseeded with invasive grasses. Southeastern pastures are dominated by fescue and other exotics that have little value to birds. Ranchers, at times with government assistance, eradicate woody vegetation, predators such as wolves and pumas, and competitors (e.g., prairie dogs). Cattle ranching has also allowed cowbirds to dramatically increase their range and numbers, and to expand nest parasitism to species that were not previously cowbird hosts. Livestock also emit significant amounts of greenhouse gases in the form of methane.

Solutions: Grazing can help prevent the conversion of habitat to more intensive agriculture or other land uses. Carefully managed, seasonal grazing can also reduce fire risk, control invasive plants, and maintain habitat for some priority bird species. Even moderate to heavy grazing is compatible with the needs of short-grass prairie species such as the Mountain Plover and Chestnut-collared Longspur. Impacts can be reduced through rotational grazing between seasons or between pastures, which allows more time for habitat to recover. Rotational graz-

Chestnut-collared Longspurs benefit from moderate to heavy grazing in short-grass prairies. The species winters in Texas and Mexico.

PHOTO: ROBERT ROYSE

PHOTO: D. CASEY

Little vegetation remains on the over-grazed side of Ashley Creek, Montana, compared to the right bank where cattle are excluded .

ing is more efficient on larger lands, but can also be achieved with smaller land parcels through co-ops of multiple landowners. In the northern Great Plains, private ranchers are taking down fences and restoring bison and native prairies. FWS and Ducks Unlimited work in this region with ranch owners to protect grasslands and prairie pothole wetlands from over-grazing through easements, land purchases, tax incentives and other methods that benefit ranchers financially. In Montana, the American Prairie Foundation has purchased grazing rights from ranchers on 45,000 acres of federal land in order for the FWS to manage these lands for wildlife. The above efforts, along with bison ranching, recreation, hunting, and eco-tourism offer viable alternatives to cattle ranching in many northern Great Plains communities suffering from long-term economic decline. Improving bird habitat on public grazing lands could involve removing cattle from key areas, raising grazing fees, restricting herd densities, and offering compensation in return for grazing permits. Two Congressional acts, the Voluntary Grazing Permit Buyout Act of 2003 and the Multiple-Use Conflict Resolution Act of 2005, were aimed at doing just that, but neither was passed. Opponents to changes in grazing policies argue that they would harm the economy, though in many areas ranching is already a marginal business, and losses could be offset through Farm Bill or other subsidies.

Actions:

- If you own pasture and graze livestock, contact NRCS about programs that can help you in sustainably managing your land.
- Write to your elected officials to ask that they support policies that protect public lands from over-grazing.
- Help reduce grazing pressure by eating less red meat.
- Support groups such as TNC and local conservationists that are working to restore range-lands.

FORESTRY

EXTENT: **20% OF U.S.**
TREND: **14% DECLINE SINCE 1953**

Forests in the U.S. have had a dynamic history since European settlement. An initial wave of logging and clearance for agriculture in much of the East has given way to a period of re-growth over the past century. Colonists also cut significant amounts of old-growth forest in the West for construction purposes. Though now much reduced, the cutting of old-growth continues, particularly in the Pacific Northwest, and in Canada's boreal forest. Of the remaining forested land, much is managed for periodic harvest on both public and private lands. The U.S. consumes some 15 million cubic feet of wood each year, seven million of which is used for paper production.

Problems: Although forestry lands can be valuable for birds, and can serve as an economically viable alternative to agriculture or development, these forests typically provide lower value habitat for priority forest birds, especially those that depend on old-growth stands. Plantation forests, which occupy some 40 million acres, are usually less diverse in tree species, structure, and age-classes than the forests they replace, and are often devoid of the standing dead trees (snags) needed by some birds for foraging or nesting. To maximize timber pro-

duction, commercial forests are often planted with different trees from those naturally present (e.g., loblolly pine in place of longleaf pine), and may be managed to skip the development of early-successional habitats needed by some birds. For example, some forestry operations in the Pacific Northwest apply herbicides and fertilizers to quickly establish Douglas-fir, accelerating regeneration past early shrub stages used by birds such as the Mountain Quail and Calliope Hummingbird.

Solutions: Paradoxically, many U.S. forests are in dire need of thinning, prescribed burns, and even clear-cutting in some areas, to restore more natural conditions that include snags. For instance, clear-cuts in the eastern U.S. soon provide large patches of early-successional forest needed by WatchList species such as Prairie and Blue-winged Warblers, and the thinning of ponderosa pines helps species such as the Flammulated Owl and White-headed Woodpecker. On the other hand, remaining old-growth forests need complete protection, especially those that provide habitat for endangered species such as the Marbled Murrelet and Spotted Owl.

Some public lands, including NPs, NMs, and designated Wilderness Areas provide strict protection to forests, whereas many National and State Forests mandate multiple uses for timber extraction, mining, recreation, and wildlife conservation. Management of some NFs has been attacked for not adequately protecting endangered species, yet in other cases, unfairly criticized for applying management practices such as thinning and burning that can help some priority birds.

Over half the forests in the U.S. grow on private lands, and these also play an important role in wildlife conservation. Numerous government agencies and conservation groups offer guidelines and assistance to private landowners to manage their forests for birds. These include the federal Safe Harbor Program, used widely in southeastern pine forests to help protect Red-cockaded Woodpeckers. Oth-

PHOTO: MIKE PARR

Plantations may be harvested sustainably, but offer limited habitat for birds requiring mature forests.

Clear cuts benefit few priority western birds, and usually only if snags are left (private lands near Eugene, OR, pictured).

er examples include work by ABC and its partners to help private landowners manage ponderosa pine forests for cavity-nesting birds. The USFS, American Forest Foundation, and many state wildlife and forest agencies also offer landowner assistance.

Since the early 1990s, the Forest Stewardship Council (FSC) and Sustainable Forestry Initiative (SFI) have worked to certify forestry products, including lumber and paper, to encourage sustainable forestry. Use of certified lumber is also part of many green building standards. Products bearing FSC or SFI logos can be purchased from many retailers, mills, and manufacturers.

Actions:

- Learn the distinction between different kinds of forest management and help educate others.
- If you own forest acreage, contact the USFS or American Forest Foundation to get help in managing your land for birds.
- Reduce paper consumption by removing your

address from junk mail lists and canceling catalogs; recycle paper and cardboard; choose recycled timber or paper products, or those certified by SFI or by FSC (now certifying eight-million acres); reuse scrap paper for notes or shopping lists and print double-sided at the office.

- Support the conservation of forest on public lands, particularly Important Bird Areas, old-growth forests, and roadless areas; as well as efforts to restore ecosystem function.

The Marbled Murrelet nests high in old-growth forest. Habitat loss has resulted in an annual population decline of 4-7%.

FIRE AND WATER MANAGEMENT

EXTENT: **EPHEMERAL BUT CAN AFFECT MOST OF U.S.**
TREND: **POTENTIAL FOR INCREASE, ESPECIALLY WITH SEVERE WEATHER EVENTS**

Many ecosystems in North America depend on floods and fires to remain healthy. Unfortunately, human activities often suppress, change, or compete with these processes, resulting in widespread habitat degradation for birds.

Problems: Riparian and bottomland forests, rivers, wetlands, and estuaries depend on natural floods that reset vegetation succession and deposit nourishing silt. In many parts of the U.S., we have altered these hydrological processes by damming rivers, ditching and draining wetlands, and eradicating beavers. Flood control has resulted in the loss of 1.2 million acres of wetlands in the Mississippi Delta alone, and 2.5 million dams block 20% of our river miles. Our farms and cities compete with wetlands for water resources. For example, heavy use for irrigation means that the Colorado River no longer consistently reaches the ocean, resulting in the loss of wetlands and riparian forests upon which Endangered "Yuma" Clapper Rails and "Southwestern" Willow Flycatchers depend. Beavers build dams which create ponds that later transform into meadows after the dams decay, providing important habitats for birds. There were

once some 60 million beavers in North America, but they were trapped out of much of their range by fur hunters in the early 19th Century, though they are naturally re-colonizing.

Most of our forests and grasslands depend on wildfire. Some pine forests (e.g., lodgepole, pinyon-juniper) require high-intensity fires that destroy trees, but only occur once every few hundred years. Other pine forests (e.g., longleaf and ponderosa), require low-intensity fires every five to ten years that do not harm adult trees, but kill competing brush in the understory. Similarly, many grasslands burn over three to ten year intervals, maintaining plant diversity and restricting growth of woody brush. Other habitats dependent on fires at intermediate burn frequencies include coastal chaparral in California, and sagebrush. Some birds depend on successional habitats that only develop following fires. The Kirtland's Warbler, Florida Scrub-Jay, and Black-capped Vireo each need habitats that develop a few years after burns, but that may only be suitable for these species for ten to 15 years, unless fire returns. Black-backed and Three-toed Woodpeckers are also most frequently found in recently burned forests.

In many areas, natural fires no longer occur frequently enough to maintain healthy forests and grasslands. Forest fragmentation and roads act as firebreaks, limiting the spread of natural wildfire. Government fire suppression policies for federal lands have created dangerous conditions in many places, with understory choked with dense brush, and dead, woody debris. This fuel builds up so that fires that once burned only understory, now burn more intensely, climb into forest canopies, and can kill mature trees. Adult trees located within these unnaturally dense stands may also become more vulnerable to pest outbreaks (e.g., bark beetles) and water stress during droughts. Fire suppression is common in habitats near where people live, such as in oak scrub used by Florida Scrub-Jays. The use of

PHOTO: ROBERT ROYSE

Water diversion for cities and agriculture stresses southwestern riparian forests used by the Abert's Towhee.

Fire enhances the health of longleaf pine forests by clearing out hardwoods while leaving pines unharmed.

PHOTO: ©ISTOCKPHOTO / PETE PATTAVINA

water for urban areas or crop irrigation can reduce natural flood conditions and exacerbate fire risk.

In other ecosystems, over-grazing has reduced the buildup of dry grass fuels, reducing wildfires needed to maintain grasslands and prevent the invasion of woody vegetation. Over-grazing in sagebrush has facilitated the invasion of the highly flammable cheatgrass that increases fire frequency, clearing vast areas of sagebrush which it then replaces.

Solutions: Seasonal floods can be restored on dammed rivers by either removing dams or releasing water to mimic floods. Many eastern dams have been removed to restore fish runs, beginning with the Edwards Dam on Maine's Kennebec River in 1999. Where dams cannot be removed, downstream conditions can be managed to benefit birds. For instance, sandbar nesting habitat for Least Terns and Piping Plovers on the Missouri River can be created and maintained mechanically. Reintroduction could bring back beavers to more remote watersheds, such as some in the Southwest, where they have not yet re-colonized naturally.

Prescribed or controlled burning is commonly used to mimic natural fire regimes or prevent dangerous fuel loads from building up in a variety of forest, scrub, and grassland ecosystems. Where prescribed burns are not appropriate, other techniques can help restore vegetation structure. Mechanical thinning of forests can restore the open understory in dense forests where fires have been suppressed and where prescribed fire risks killing trees or burning homes. Tree girdling can create dead snags that are important for cavity-nesting birds in younger forests unsuitable for burning. Overall though, fire suppression programs need to be changed to favor more natural fire regimes.

Actions:

- Many government agencies and conservation groups provide landowners with assistance on water management, controlled burns, and other habitat manipulations that could otherwise be difficult or dangerous to implement.

- Landowners can be more tolerant of beavers, or protect particular trees with mesh to prevent damage.

- Encourage government officials to allow more frequent use of natural wildfire in appropriate areas, protect rivers from new dam projects, and restore them by removing dams or by releasing water to mimic natural flows.

INVASIVE AND OVERABUNDANT SPECIES

Invasive species are organisms that are not native to a region, but were introduced intentionally or accidentally, mainly as a consequence of actions by people. Many of these new arrivals do not survive, but some thrive, free of the ecological constraints of their native homes. Some native species have also benefited from anthropogenic habitat changes to become overabundant, impacting birds and their habitats in similar ways to invasives.

Next to habitat loss and alteration, invasive species present the greatest threat to birds in the U.S., threatening more than one third of birds on the WatchList. Invasive species are also the greatest threat to many protected areas, such as NWRs. Invasives are responsible for the majority of bird extinctions since the 1800s, most of which have occurred on oceanic islands. On Hawaii alone, invasive pathogens and predators have contributed to the extinction of 71 bird species: 48 following colonization by Polynesians, and 23 more since the arrival of Europeans in 1778. They also cost the U.S. an estimated $137 billion annually through damage and control expenses.

Invasive and overabundant species harm birds by predating or otherwise killing them, reducing their health or reproductive rates, competing for similar resources, or by contributing to habitat destruction and degradation. The invasive species expansions that bird conservationists fear the most are the potential spread of the brown tree snake and West Nile virus to Hawaii. While the virus may be impossible to halt (though captive birds can be vaccinated), the snake should be possible to detect by carefully inspecting all cargo from Guam. The Hawaii State Legislature passed a bill to pay for increased inspections of imported freight, but this was vetoed by the Governor in 2008 due to its prospective cost. A federal program to address possible brown tree snake invasion is currently in

place, but its long-term funding remains insecure. Many established invasive and overabundant species already cause serious problems, however, and we provide examples of some of the most significant specific threats on subsequent pages with introductory texts for each category. Not all of these invasives cause direct bird mortality, and in some cases specific bird impacts may be hard to quantify, but the cumulative effect on birds and their habitats is nonetheless extremely widespread and significant.

Hybridization between overabundant species and those with smaller populations can also be a problem. This is a concern for the Hawaiian Duck and Mottled Duck that hybridize with feral Mallards. In cases where these species are both potentially of conservation concern such as with the Golden-winged and Blue-winged Warblers, solutions can be very challenging.

PHOTO: JACK JEFFREY

Mosquito-borne diseases have decimated birds such as the Hawaiian Crow which now only survives in captivity.

PHOTO: ROBERT ROYSE

Feral Mallards threaten the elegant Mottled Duck (above) through hybridization in southeastern coastal areas.

Invasive species control and eradication efforts face several challenges, including minimizing harm to non-target organisms, preventing re-invasion, and public opposition to control efforts. Unfortunately, the killing of native species by invasives typically attracts less outrage, but in contrast to invasive control, its consequences for populations of the species concerned are often far more profound.

In addition to the following categories, there are many other invasive species (e.g., invasive snails that predate native Hawaiian snails which provided food for the Poo-uli, and bullfrogs which feed on chicks of Hawaiian water-birds) that impact birds. Invasive and overabundant species can have both positive and negative (and often complex) consequences for birds. For example, the Cooper's Hawk has benefitted from hunting Rock Pigeons in cities, and its resulting population growth may be increasing its predation of the American Kestrel.

The most efficient solution to the problem of invasive species is to safeguard against future invasions and outbreaks, since prevention is easier and cheaper than removal or control. Securing our borders from biological invasions requires increased screening of imported foods, plants, lumber, and other cargo at ports of entry. Improving regulations on the dumping of ballast water from ships that can contain invasive shellfish and aquatic plants, would also help deal with the invasive species problem.

Eradicating or controlling invasive species is challenging, but organizations around the world are succeeding in restoring both marine islands and mainland reserves of increasing size. Advances in fence design allow reserve managers to exclude invasive grazing animals and rodents without having to remove them completely from the surrounding landscape. In some cases, invasive or overabundant species may not need to be eradicated, but simply reduced to levels such that they no longer pose a serious threat to native endangered species.

MAMMALS

Invasive predatory mammals are particularly harmful to beach-nesting birds, and to those that occur on islands such as Hawaii. In the accounts below, we highlight some of the most significant, including rats, cats, foxes, and mongooses.

Many native medium-sized predators ("mesopredators") including coyotes, skunks, raccoons, and opossums also predate birds and their nests. These native predators have greatly increased in number due to human activities that both increase the amount of food available to them through landscape changes and garbage, and the eradication of top predators (e.g., wolves and mountain lions) that would otherwise keep smaller predator populations in check.

Invasive and overabundant mammalian herbivores can cause severe damage to bird habitat (herbivore species accounts follow directly on from predatory mammals below). For example, rabbits were introduced to Laysan Island in 1903. Freed from natural predators they proliferated, and soon ate most of the native vegetation. In 1923, they were eradicated using poisoned alfalfa, but by then, the damage had caused the extinction of the island's Millerbird, Apapane, and Laysan Rail, and almost claimed the Laysan Finch and Laysan Duck. Island

Conservation successfully eradicated rabbits from Lehua Island, Hawaii, in 2005-2006, benefiting seabird colonies there. Today, a variety of invasive and overabundant native herbivores degrade our forests, tundra, and wetlands. Feral ungulates (sheep, goats, and deer) on the major Hawaiian Islands continue to damage native trees and plants, slowing habitat restoration efforts. Fencing and eradication has protected some sites in Hawaii, but political opposition by sport hunters has successfully hindered attempts at wider scale eradication efforts.

DOMESTIC CATS

"House" cats (*Felis catus*) were originally domesticated in Egypt 4,000 years ago and brought to the U.S. by European colonists, with most cat introductions on islands occurring during the 19th and early 20th Centuries. Cats now occur throughout the U.S., including in places where few, if any, native mammalian predators originally occurred, such as on oceanic and barrier islands. Domestic cats sometimes establish feral colonies, and are particularly common in southern residential and agricultural areas. In 2007, there were estimated to be more than 80 million pet cats in the U.S., of which roughly 43% have access to the outdoors. Estimates suggest that there are 60-120 million additional feral cats.

Problems: All outdoor cats hunt and kill birds (20-30% of cat prey) and other small animals. A study in England found that wearing bells may reduce predation by outdoor cats, but does not prevent it. A study in Wichita, Kansas, found that de-clawed cats kill as many birds as cats with claws. Studies report variable predation rates by cats, for example, 35.5 birds per cat per year in Michigan, 15 in San Diego, 9.6 in England, and between 5.6 and 109.5 in Wisconsin. Multiplying the most conservative predation rate (5.6 birds per cat per year) by the most conservative estimate of outdoor

Hundreds of millions of birds become the victims of cats in the U.S. each year.

PHOTO: GAÉTAN PRIOUR

Managed feral cat colonies are a hazard to birds and attract the dumping of additional unwanted cats.

and feral cats in the U.S. (95 million) yields 532 million birds killed annually by outdoor cats, with the actual number likely being much higher.

Aided by food provided by people, cats often occur at much higher densities than native predators. Birds that nest or feed near the ground, exhausted migrants, seabird colonies, and endemic Hawaiian birds are the most vulnerable, and may be severely affected by domestic cats. Cat predation has helped cause the extinction of at least 33 island bird species worldwide, including the Stephens Island Wren in New Zealand, and the Guadalupe Storm-Petrel in Mexico. Cats predate many birds listed under the ESA, including the Hawaiian Petrel, "California" and "Light-footed" Clapper Rails, Piping Plover, Florida Scrub-Jay, California Gnatcatcher, and Palila. Many well-intentioned but misguided people exacerbate feral cat problems through a technique called Trap, Neuter, Release (TNR), whereby colony cats are caught, spayed or neutered, and then returned to colonies to be fed by volunteers. Despite its apparent appeal, few colonies managed under this system shrink, as the program takes more time than most volunteers are able to give, some cats are never caught, and the colonies often become dumping grounds for more unwanted cats.

Solutions: Cats cannot be blamed for their predatory nature; instead people must take responsibility for them. Keeping pet cats indoors or within

The flightless chicks of beach-nesting birds, like this endangered Piping Plover, have little defense against feral cats.

Rats were recently eradicated from the 6,871-acre Rat Island in Alaska; even larger-scale eradications could follow.

RATS AND MICE

Several species of rats and mice originally native to Asia and Oceania were introduced (largely accidentally) to the U.S. by European colonists, to the Hawaiian Islands by both ancient Polynesians and European settlers, and to the Aleutian Islands during World War II. These include the house mouse (*Mus musculus*), black rat (*Rattus rattus*), brown or Norway rat (*R. norvegicus*), and the Polynesian rat (*R. exulans*). Introduced rats now occur throughout forests on the major Hawaiian Islands, as well as in most U.S. cities and agricultural areas.

Problems: Rats and mice have contributed to numerous bird extinctions, particularly on oceanic islands. Predominantly seedeaters, these rodents can severely alter vegetation on islands, and reduce food supplies for insect- or seed-eating birds. Rats

screened enclosures outdoors, is the simplest and best solution for preventing pet cats from killing wildlife. Keeping cats indoors also helps protect them from disease, predators such as coyotes, and collisions with cars. Neutering and spaying pet cats is a humane method for reducing cat over-population. Numerous studies have shown that TNR fails to eliminate cat colonies or address problems of wildlife mortality caused by cats, however.

By 2004, feral cats had been successfully removed from at least 48 islands worldwide. Since 1815, cats had decimated seabirds on the 24,000-acre Ascension Island in the South Atlantic, reducing seabird numbers by 98% from 200 million to 400,000. The remaining birds survived on offshore islands or sheer cliffs that were inaccessible to cats. Ascension was declared free of feral cats after a multi-year eradication effort that began in 2001. This is one of the larger islands from which feral cats have been removed and the first with residents who keep pet cats. Five species of seabirds have since returned to nest on the island's mainland, and this success may serve as a model for other large islands with human populations, such as Juan Fernández, Chile, where ABC is working to protect the Juan Fernández Firecrown and other native birds.

Actions:

• Expand education campaigns so that owners keep their cats indoors; make TNR and the feeding of feral cat colonies illegal.

Crested Auklets and other seabirds are re-colonizing islands where rats have been eradicated in the Aleutians.

and mice also predate eggs, nestlings, and adult birds opportunistically on the ground or in trees. Ground-nesting birds, seabird colonies (in Hawaii, California, and Alaska), and endemic Hawaiian land-birds are the most severely impacted.

Solutions: Rat populations are most effectively controlled using bait stations, or through the aerial broadcast of rodenticides. This is most straightforward when there is a low risk of harming non-target animals, such as on islands without native mammals, though secondary effects on raptors can be significant and must be addressed prior to the implementation of control programs. Rodents have been successfully eradicated from at least 284 islands globally, including Anacapa Island off California in 2001 (already benefitting the Xantus's Murrelet), with larger scale efforts recently completed on the 6,871-acre Rat Island in Alaska (aimed at benefitting the Whiskered Auklet among other species). Rodent-proof fences are being used in New Zealand and Hawaii, allowing rodents to be eradicated from within reserves. Inspecting cargo at ports can reduce the chance of rat introductions, and monitoring islands can alert response teams quickly in case of a rat invasion. Rodent eradications are sometimes opposed by animal rights groups, despite the harm that rats and mice do to native birds and other wildlife.

Actions:

- Eradicate rats and mice from uninhabited Aleutian Islands (e.g., Kiska) and other U.S. coastal islands (e.g., Farallons).
- Control populations in Hawaii while investigating methods for larger scale control or eradication efforts.

FOXES

Both the red fox (*Vulpes vulpes*) and arctic fox (*V. lagopus*) were introduced by fur traders to many of Alaska's Aleutian Islands starting in the mid-

Birds are rebounding following the eradication of introduced Arctic Foxes from many of the Aleutian Islands.

The "Aleutian" Cackling Goose now exceeds 70,000 birds; introduced foxes had reduced them to hundreds.

1700s, and peaking between 1913 and 1940. By 1940, around 100 of the islands had invasive foxes.

Problems: Foxes eat eggs, nestlings, and adult birds, severely impacting ground-, marsh-, and island-nesting species (particularly colonial seabirds). Many bird populations on the Aleutian Islands were decimated by introduced foxes including the "Aleutian" Cackling Goose which was at one point feared extinct as a result of fox predation. The loss of seabirds was also severe enough to reduce guano deposits by birds, reduce soil fertility, and transform lush grasslands to tundra.

Solutions: FWS began fox eradications in the Aleutian Islands in 1949 (Amchitka Island), and increased its efforts in the 1970s. By 2002, foxes had been removed from 40 Aleutian Islands and had died out on most of the rest after exhausting the food supplies, with foxes now remaining on only a handful of islands. Fox eradication is credited with successfully boosting native bird populations, including local populations of the Cackling Goose, Rock Ptarmigan, and many seabirds. Tufted Puffins have increased in abundance and re-colonized many islands where they were extirpated by foxes; Whiskered Auklets increased from 25,000 in 1974, to 116,000 in 2003.

Actions:

- Eradicate foxes from remaining Aleutian Islands.
- Consider controlling foxes in the San Francisco Bay area.

Small Indian Mongoose

The small Indian mongoose (*Herpestes javanicus*) is native to tropical Asia. It was intentionally introduced to most major Hawaiian Islands (Hawaii, Maui, Molokai, and Oahu, but not Kauai) in 1883, and to many Caribbean Islands (e.g., Cuba, Jamaica, Puerto Rico, Hispaniola) around the same time in a failed attempt to control black rats in sugar cane fields and other areas.

Problems: Mongooses eat eggs, nestlings, and adult birds. Ground and marsh-nesting birds, seabird colonies, and other island birds have been severely impacted or extirpated by mongooses (e.g., Hawaiian Goose, Hawaiian Petrel, Jamaica Petrel, and the Jamaican Poorwill). Mongooses also predate poultry causing economic damage, and can spread disease (e.g., rabies) to people and other mammals.

Solutions: Trapping has been used to control mongooses in many nature reserves, including Haleakala NP (Hawaii), and on small Caribbean islands, such as Buck Island in the U.S. Virgin Islands.

PHOTO: BILL HUBICK

Mongooses kill both wild and domestic birds, causing significant damage to poultry in Hawaii and Puerto Rico.

Actions:
- Expand mongoose control programs in Hawaii.
- Safeguard against the accidental spread of mongooses to Kauai through increased biosecurity efforts at ports.
- Research additional control methods.

Feral Pigs

Feral pigs (*Sus scrofa*) are often mixes of domestic pigs and wild boars native to Europe. Pigs have been introduced to many parts of the U.S. (e.g., Hawaii, Texas, Tennessee, California, and Florida). Small Polynesian pigs were introduced to the Hawaiian Islands around the year 400, but were probably restricted to pens. These pigs were supplanted by larger feral hogs in 1778.

Problems: Feral pigs have voracious appetites and consume native plants, nuts, and fruits, causing extensive damage to bird habitats. Pigs also predate ground nests, compete with birds for food, and disperse invasive plants. In Hawaii, pigs wallowing and biting open hollow tree fern stems creates areas where water can collect and mosquitoes can breed, aiding the spread of avian malaria and pox. Feral pigs can also impact habitat for ground nesting seabirds in Hawaii, and compete for food with Whooping Cranes at Aransas NWR in Texas.

Solutions: The successful eradication of 5,036 feral pigs from California's 61,830-acre Santa Cruz

PHOTO: JACK JEFFREY

The Small Indian Mongoose feeds on the eggs of ground- and marsh-nesting birds, such as the Hawaiian Coot.

Island in 2006 to reduce competition for acorns with the Island Scrub-Jay demonstrates that these animals can be controlled on large islands. Controlling pigs in Hawaii is challenging due to dense forest cover and steep terrain, however, but it can be done. Areas in Hawaii have been cleared of pigs and fenced to prevent re-entry. Fencing is also used in Puerto Rico, and pig numbers have been reduced by hunting with dogs, and setting snares and traps. The main challenge facing larger scale pig eradication in Hawaii comes from the political opposition of sport-hunters.

Pig wallows facilitate the spread of mosquito-borne diseases that can kill the Akiapolaau and other Hawaiian birds.

Feral pigs were eradicated from California's Santa Cruz Island, but sport hunters oppose control efforts in Hawaii.

Actions:

• Continue and expand efforts to fence forest areas and remove feral pigs.
• Build grassroots support for broader pig eradication efforts in Hawaii.

DEER

White-tailed deer (*Odocoileus virginianus*) are native ungulates in North America, and range south through Mexico and Central America to South America. Unregulated hunting reduced white-tailed deer from more than 20 million animals to fewer than half-a-million by 1900. The U.S. population now numbers around 30 million due to protective game laws, changes in habitat, and the eradication of large predators such as wolves and mountain lions. White-tailed deer are particularly overabundant in many portions of the eastern and Midwestern states. Black-tailed or mule deer (*O. hemionus*) are also native to western North America, but have been introduced to several Hawaiian Islands, and the Queen Charlotte Islands off western Canada. Axis deer (*Axis axis*) are native to south Asia and were introduced to Hawaii. Reindeer or caribou (*Rangifer tarandus*) have been introduced to islands in the Aleutians, and several other deer species (such as sika, and fallow) have been introduced to ranches, barrier islands, and other parts of the continental U.S.

Problems: Overabundant white-tailed deer over-browse the forest understory, creating browse lines up to six feet high, below which little leafy vegetation remains. Removing understory harms native herbaceous plants, prevents tree regeneration, and impacts birds that nest near the ground, such as the Ovenbird and Kentucky Warbler. Deer introduced to islands that would otherwise lack large native herbivores can severely damage native vegetation and reduce bird abundance. Deer will also predate nests opportunistically, and compete for acorns with native mice that in turn help regulate gypsy moth numbers through predation. Overabundant deer also damage crops and gardens, reduce regeneration of commercially valuable trees, host ticks that spread Lyme disease to people, and collide with automobiles.

Solutions: Reintroducing natural predators is the ideal method of controlling deer, but political resistance over safety concerns to livestock, pets, and people make this difficult. Wolves are re-colonizing some areas in the northeastern U.S. on their own, but could do so faster with our help. Hunters currently offer the best available management tool for reducing deer populations in rural areas, but this is not an option in most urban environments, though some culling has been carried out. Reducing deer numbers through hunting is sometimes opposed by animal rights activists as well as by a minority of hunters who want more deer. Relo-

Vegetation damage by deer degrades understory nesting habitat for species such as the Ovenbird.

cating deer simply moves problems elsewhere and stresses animals. Contraception is expensive and requires repeated treatments to maintain infertility. Fences 7.5 feet in height or taller are used to exclude deer from commercial timberlands, and vineyards. Fences constructed in Hawaii have typically not been high enough to exclude deer or sheep.

Actions:
- Restore diversity of forest ecosystem structures through increased hunting and culling, and where possible, by restoring native predators.
- Expand hunter access to lands, and build political support for reduced deer populations. Where numbers of hunters are declining, deer hunting restrictions may need to be relaxed in order to maintain hunting pressure.
- Research use of large (>100-acre) deer exclusion areas on a rotational basis (five to ten years) as a potential management tool for forest restoration in heavily impacted forests.

Hunting pressure is insufficient to control most white-tailed deer, which can also leap fences less than 7.5 feet tall.

BIRDS

Invasive birds are well-established in cities and suburban areas across the U.S. Species include House and Eurasian Tree Sparrows, the European Starling, three mynah species, almost 30 species of parrots and parakeets, the Rock Pigeon, and the Eurasian Collared Dove. Birds such as the Eastern Bluebird, Northern Flicker, and Red-headed Woodpecker have suffered from nest site competition with sparrows and starlings. Introduced birds such as the White-rumped Shama and Japanese White-Eye are likely impacting Hawaiian forest birds through food and nest site competition, and by acting as disease reservoirs, but no hard data yet exist to show population-level effects. In addition to the species highlighted here, some larger gull species have become overabundant due to the availability of food at land-fills and fish docks, and are generating unnatural levels of predation pressure on species such as the Roseate Tern in the Northeast, and on seabirds such as the Ashy Storm-Petrel on California islands. Populations of species such as the Blue Jay and American Crow may also be somewhat inflated due to human activities, allowing increased nest predation of other native birds in populated areas.

Forest clearance and cattle farming enabled the Brown-headed Cowbird to expand its range and parasitize new host species.

COWBIRDS

Several species of cowbird are native to the U.S., the Caribbean, and Mexico. The Brown-headed Cowbird (*Molothrus ater*) was originally native to the Great Plains where it foraged alongside migratory bison herds. Today, it is mainly found alongside livestock, in feedlots, and in other agricultural areas. It has greatly expanded its range across the "Lower 48" states and north into much of Canada, thanks to forest fragmentation and the expansion of livestock into new areas. The Breeding Bird Survey shows that Brown-headed Cowbird populations in North America number between 20 and 40 million birds, but indicates that despite some local increases, overall numbers declined by 1.5% annually (45% total) from 1966-2006, perhaps due to a reduction in open habitats in the East. The Bronzed Cowbird (*M. aeneus*) is native to arid regions of the U.S. along the southern border from southeast California to the Texas Coast, and the Shiny Cowbird (*M. bonariensis*) is native to the Caribbean and Central and South America, but now also occurs in southern Florida. Both Bronzed and Shiny Cowbirds benefit from human activities (irrigation, pastures) and are expanding their ranges in the U.S.

European Starlings and House Sparrows compete for nest cavities with native species such as the Eastern Bluebird.

Problems: Cowbirds are nest parasites (also known as brood parasites). Female cowbirds lay single eggs (up to 40 in a season) in nests belonging to other bird species, which then rear the cowbird chicks at the expense of their own. Cowbird chicks hatch earlier than their hosts' eggs, are larger than host chicks, and out-compete host nestlings for food and attention from the unsuspecting foster parents. Cowbird nest parasitism often results in the death of the host eggs or nestlings, and can greatly reduce the reproductive success of some host species. Brown-headed Cowbirds have been recorded parasitizing the nests of more than 220 species, and have been successfully reared by over 150, including WatchList species such as the Wood Thrush. Forest species that build open cup nests (e.g., vireos and warblers) and have only been exposed to cowbird parasitism during the last few hundred years are particularly vulnerable. Nests close to cowbird foraging areas are more vulnerable to parasitism. Deforestation by humans frequently results in small remnant forest patches with a high perimeter to interior area ratio. Brown-headed Cowbirds thrive along these edge habitats, causing dramatically increased rates of parasitism. Shiny Cowbirds also parasitize a wide variety of host species (including Endangered species in the Caribbean such as the Yellow-shouldered Blackbird), but Bronzed Cowbirds are more specialized, favoring nests of orioles, thrashers and mockingbirds, Northern Cardinals, and Green Jays.

Solutions: Live-trapping of cowbirds, which are later humanely killed using carbon monoxide or other methods, has been used locally. Trapping cowbirds in critical breeding areas for Endangered species (e.g., Kirtland's Warbler, Golden-cheeked Warbler, Black-capped Vireo, and "Least" Bell's Vireo) successfully reduces parasitism and increases reproductive success of hosts near traps, but is insufficient to impact cowbird numbers in subsequent years. Reducing fragmentation and maintaining habitat cover over large intact landscapes may be the best solution for forest birds vulnerable to cowbird parasitism.

Actions:

- Continue cowbird control programs to maintain populations of endangered species. Consider expanded local cowbird control that might aid other species, such as the Audubon's Oriole along the Rio Grande, and the ESA-listed Florida population of the Grasshopper Sparrow.
- Larger-scale Brown-headed Cowbird control at winter roosts requires further research, caution to avoid harming non-target birds, and cost-benefit analysis in light of overall declining population trends.

Geese

Snow Geese (*Chen caerulescens*) and Canada Geese (*Branta canadensis*) are both native to North America. Certain populations (*C. c. atlantica*, *C. c. caerulescens*, and *B. c. maxima*) of both species have experienced rapid growth over the last 40 years and become overabundant in eastern and central North America, thanks in part to abundant food supplies provided by agricultural areas, man-made lakes and ponds, cattle feed lots on their wintering grounds, and protective hunting regulations. There are presently more than ten-million Snow and Canada Geese in North America.

PHOTO: ROBERT ROYSE

Brown-headed Cowbirds lay their eggs in the open-cup nests of other birds, such as this Bell's Vireo.

Overabundant Snow Geese are eating away saltmarshes, and tundra vegetation in the Canadian arctic.

PHOTO: ROBERT ROYSE

Problems: Recovery of North American goose populations from historical over-hunting is a great conservation success, but we now have too much of a good thing. Captive-raised Canada Geese released for hunting have remained as year-round residents far to the south of their normal breeding range. Canada Geese graze on grasses and Snow Geese dig up roots, tubers, and rhizomes. Overabundant Snow Geese have over-grazed saltmarshes along the Arctic and Mid-Atlantic Coasts, and also impacted tundra sedge meadows. Tundra vegetation recovers slowly (15 years for some tundra grasses) following damage by geese. Local declines in other birds, such as Semipalmated and Stilt Sandpipers, have occurred in areas occupied by Snow Goose breeding colonies. "Giant" (*maxima*) Canada Geese also compete for food with other smaller-bodied (*canadensis*) populations.

Solutions: Problem geese populations could be reduced by relaxing restrictions on hunting. About 2.6 million Canada Geese are annually harvested

in the U.S. and Canada, however, this is insufficient to reduce numbers, and does not address landscape changes that are the cause of overabundance. Lethal control of geese is controversial and sometimes opposed by animal rights groups. Efforts to scare Snow Geese off wildlife refuges using propane cannons, scarecrows, and planes have not proven effective. Trained dogs are used to scare Canada Geese away from some golf courses, but total goose numbers continue to rise.

Actions:
- Increase harvest rates of problem Snow Geese and Canada Geese.

MUTE SWANS

Native to Eurasia, Mute Swans (*Cygnus olor*) were intentionally introduced to multiple North American locations between the mid-1800s and 1912, mainly as ornamental birds in city parks, gardens, private bird collections, and zoos. They principally inhabit coastal areas from Maine to South Carolina, and the Great Lakes, with populations rapidly increasing since the 1970s. The Atlantic Flyway population totaled 14,300 feral birds in 2002; numbers in the Chesapeake Bay increased from 324 birds in 1986 to more than 4,000 today.

Problems: Despite their romantic symbolism, Mute Swans behave aggressively towards other birds, displace native waterfowl (e.g., the Tundra Swan) and scare off beach-nesting birds (e.g., Least Tern and Black Skimmer). They consume submerged aquatic vegetation, reducing these important plants for other birds and wildlife. Mute Swans also benefit from marshes invaded by *Phragmites* reeds.

Solutions: Lethal control (via hunting or capture and euthanasia) is the best solution to reduce or eradicate Mute Swans, but often faces opposition from local residents and animal rights activists. Washington and Montana have removed Mute Swans to help reestablish Trumpeter Swans. Mute Swans are not hunted in most states. A 2001 court ruling conferred protection to Mute Swans under the Migratory Bird Treaty Act, hampering control efforts until 2004, when the species was explicitly excluded from the Act and authority over swan control was returned to states. Destroying eggs, by addling or shaking at nests in Rhode Island failed to stem population growth there, but may be a useful control tool in urban parks where lethal control is not permitted.

Actions:
- Maintain "swan-free zones" near sensitive habitats and Important Bird Areas using lethal control.
- Eradicate all feral Mute Swans, while maintaining legally-permitted swans in captivity (in same sex pairs to prevent breeding).

PHOTO: KERRY HOLLEY

Mute Swans are aggressive and will chase native waterfowl, such as this female Mallard, off their territories.

REPTILES

The brown tree snake represents a serious threat to birds, but thankfully it remains a potential invasive to the U.S., for now. Introduced Burmese pythons also represent a potential threat to birds in the Everglades.

BROWN TREE SNAKE

The brown tree snake (*Boiga irregularis*) is native to New Guinea, northeastern Australia, and some western Pacific islands, but arrived in the U.S. territory of Guam after World War II (presumably in cargo). Snake numbers may now exceed two-million there, with densities up to 12,000 snakes per square mile.

Brown tree snakes decimated 17 of 18 native land-birds and seabirds in the U.S. territory of Guam.

PHOTO: BJORN LARDNER / USGS

Problems: Brown tree snakes eat eggs, nestlings, and adult birds. They decimated Guam's birds, causing the extinction of the Guam Flycatcher, which was endemic to the island, and the near-extinction of the Guam Rail, which now survives only in captivity, and in a small managed area on Guam. In the absence of avian pollinators, trees and other plants on Guam now no-longer produce normal quantities of seeds. These aggressive and mildly venomous snakes also attack people, and like to hide in transformer boxes causing frequent power outages on Guam. They have also reached Rota, Saipan, and Tinian via air and sea transport, but have not become established as yet. Ships and planes travelling between Guam and the Hawaiian Islands now threaten to spread snakes to Hawaii, where they could devastate endangered birds and cost the state $28-40 million annually in electrical problems. Eight snakes have already been detected and intercepted at the Honolulu Airport on Oahu, thus-far narrowly averting disaster. Planned expansion of military infrastructure on Guam could increase air and sea traffic to Hawaii necessitating expanded biosecurity efforts.

Solutions: Effective control methods have yet to be developed, but poisons and biocontrol are be-

ing investigated. Smooth, vinyl-surfaced barriers can keep snakes out of fenced areas, but are vulnerable to damage by rats and storms, whereas molded concrete designs are more durable, but also more expensive. Preventing spread to Hawaii depends on stringent inspections of cargo using trained dogs at Guam's departure ports and on arrival in Hawaii, and the use of snake traps. Traps are also used to suppress snake numbers near electrical lines. Bait (baby mice) laced with poison (acetaminophen) is also being dropped into forest areas in an experimental control program. A 56-acre snake exclusion zone was constructed at Anderson Air Force Base on Guam to safeguard Guam Rails.

Actions:
- Secure stable funding for thorough cargo screening in Guam and Hawaii.
- Build snake-proof fences around Hawaiian ports.
- Develop an emergency plan to respond to the possible arrival of tree snakes in Hawaii.
- Research methods to control or eradicate snakes on Guam and other islands.
- Re-authorize and fund the Brown Tree Snake Control Act of 2004.
- Continue monitoring for outbreaks, treating new infestations to help prevent spread, and research new biocontrol methods.

DISEASES

Parasites, diseases, disease vectors, or pests can harm bird populations by killing birds, or by reducing survival and reproductive success. Particularly devastating examples include avian malaria and pox, West Nile virus, and botulism. New strains of avian influenza (e.g., H5N1) could also potentially spread from Asia to North America. "Avian flu" is transported primarily by infected poultry (outbreaks have been shown to follow poultry transport routes). Wild birds can be flu vectors, and many regularly carry mild forms, but in the case of H5N1, they usually die before the disease spreads along migratory routes. Some people have called for the culling of wild birds during flu scares, but evidence suggests that this would be ineffective in limiting the spread of disease. Increased biosecurity for imported poultry is a more effective tool and can benefit both humans and wild birds.

Invasive pathogens also affect trees, insects, and other components of ecosystems, thereby indirectly affecting birds, though the direct impacts on birds can be difficult to measure. Waves of invasive pests and diseases have harmed trees in eastern forests since European colonization. Between 1900 and 1940, American chestnuts were largely wiped out by introduced chestnut blight. Several fungal agents of Dutch elm disease were introduced to North America beginning in 1928, and devastated most American elms. Beech bark disease was introduced to Nova Scotia around 1920 and has since spread south to North Carolina and west to Michigan. More recently, anthracnose fungus, introduced in the mid-1970s, has attacked dogwoods; sudden oak death syndrome has killed many western oaks since 1995; and since 2005, a fungus has attacked ohia trees in the Hawaiian Islands.

AVIAN MALARIA AND POX

Avian malaria (*Plasmodium relictum*) is a protozoan parasite that is transmitted to birds by female *Culex* mosquitoes. Avian pox (*Poxvirus avium*) is a virus that infects birds through the ingestion of contaminated food or water, or through skin abrasions (also including mosquito bites). Various forms of avian malaria and pox (which do not infect people) occur naturally in most continental bird populations, where birds have evolved resistance. However, they are a major threat to birds in Hawaii. Neither disease was known in the Hawaiian Islands before 1826 when mosquitoes were introduced. The diseases then spread from domestic poultry and introduced birds (both resistant hosts), to non-resistant native birds.

Problems: Avian malaria can infect the blood, liver, spleen, bone marrow, and brains of birds, cause anemia, and can ultimately kill non-resistant individuals. Avian pox can cause dry lesions on bare skin around the eyes, beak, legs, and feet, or wet lesions inside the mouth, throat, trachea, and lungs. These symptoms can impair breathing, feeding, vision, and perching, and can ultimately kill infected birds. Having evolved in isolation from these diseases, most native Hawaiian birds lack resistance, and infections cause high mortality rates (60-90+% depending on the species). These diseases pose the most serious current threat to native Hawaiian songbirds, and likely were the most important factors in the disappearance of more than 20 bird species there since 1778.

PHOTO: JACK JEFFREY

An introduced mosquito bites the skin around an Apapane's eye possibly infecting it with avian malaria or pox.

Watering tanks for cattle and the wallowing holes created by invasive feral pigs provide mosquito breeding areas, helping to spread disease. Many native Hawaiian songbirds are now limited to high-altitude forests where cooler temperatures (<55° F) hamper disease development and transmission, but climate change threatens to warm temperatures so that even birds surviving in high-altitude forests may eventually become vulnerable.

Solutions: Few solutions currently exist. Spraying pesticides to kill adult mosquitoes is largely ineffective, harms native insects, and could compromise birds' immune systems. Reducing the small pools of still water required by mosquitoes for breeding is the best solution currently available to reduce mosquito numbers, including removing feral pigs, cattle tanks, and other objects or trash that hold standing water near or within key bird sites. The use of larvicides in certain areas may also prove worthwhile. Encouragingly, some species, including the Hawaii Amakihi, appear to be developing natural resistance to these diseases.

Actions:

- Increase research efforts aimed at fighting mosquito-borne diseases in Hawaii. Recently, researchers have infected *Aedes* mosquitoes in labs with *Wolbachia* bacteria, which have rapidly spread throughout mosquito populations. Eggs laid by uninfected female mosquitoes are killed if fertilized by infected males. These bacteria also reduce the lifespan of infected mosquitoes enough so that mosquito-borne diseases, such as dengue fever (and potentially malaria), have insufficient time to develop within the mosquito and be passed onto new human (or bird) victims. Additional funding is required to expand research and apply this cutting edge technology to *Culex* mosquitoes to aid Hawaiian birds.
- Until such solutions can be developed, efforts to protect habitat for Hawaiian songbirds, combat global warming, and reduce other stresses to Hawaiian bird populations affected by avian malaria and pox need to be increased.

West Nile virus likely killed 49% of all Yellow-billed Magpies between 2004 and 2006.

West Nile Virus

West Nile virus (WNV) was first identified in 1937 in Uganda, and is likely native to Africa and parts of Eurasia. WNV is transmitted to birds by mosquitoes, and was accidentally introduced to the New York City area in 1999. From there it has since spread across the "Lower 48" states and into Canada, Mexico, and south into South America.

Problems: The virus attacks the nervous system, with symptoms including brain inflammation (encephalitis), weakness, inability to fly or walk properly, tremors, and death. Hundreds of bird species in the U.S. have been infected with WNV. Most birds in North America recover from WNV infections, but crows, jays, magpies, raptors, and some grassland birds appear to be particularly vulnerable, suffering high mortality rates. WNV reduced Greater Sage-Grouse summer survival rates by 25%, killed 40-68% of infected American Crows in marked populations, and likely killed 49% of California's Yellow-billed Magpies from 2004-2006. Expansion of coal-bed methane activities that create waste-water pools likely increases local mosquito populations and facilitates the spread of WNV in the arid regions where Greater Sage-Grouse occur. Other endangered species with small populations may also be vulnerable to this disease and require more time to develop resistance.

PHOTO: ROBERT ROYSE

Solutions: Few solutions currently exist. Spraying pesticides to kill mosquitoes in residential areas has little effect on the spread of WNV, and can harm non-target insects, birds, and people. Some mosquitoes in North America have also developed resistance to insecticide sprays. Any mosquito control efforts should follow Centers for Disease Control and Prevention models for similar diseases such as St. Louis encephalitis, which use less toxic methods to target mosquitoes at aquatic larval stages. Mosquitoes need small pools of standing water to breed, and reducing these is the best solution to keeping mosquito numbers low in residential areas. Regularly changing water in birdbaths and disposing of water-catching trash (e.g., plastic cups, old tires) helps. Attracting bats that eat large numbers of mosquitoes by erecting bat boxes may also be beneficial. Large bird populations should eventually develop resistance to WNV naturally, but smaller populations are at risk. ABC helped fund the development of a vaccine for captive populations of endangered species, which has since been used successfully on California Condors and other vulnerable species.

Actions:

- Continue monitoring efforts to measure the effects of WNV on WatchList species, particularly the Greater Sage-Grouse, Yellow-billed Magpie, Pinyon Jay, Florida Scrub-Jay, Island Scrub-Jay, and Oak Titmouse. Protect habitat for these species and reduce other threats and stresses to populations affected by WNV.
- Increase biosecurity efforts in Hawaii to prevent the spread of WNV to the Hawaiian Islands.

Botulism can threaten water-birds, and shorebirds, such as these Red-necked Phalaropes, on saline lakes in the West.

PHOTO: ROBERT ROYSE

BOTULISM

Botulism is caused by the consumption of botulinum toxin, produced by bacteria (*Clostridium botulinum*) that thrive in oxygen-depleted waters worldwide.

Problems: Water-birds are most susceptible to botulism outbreaks, and become infected after eating contaminated fish or other aquatic organisms. Affected species include loons, grebes, pelicans, ducks, gulls, and shorebirds. Botulism kills birds by paralysis. Mortality rates depend on the dosage of toxin ingested. Botulism outbreaks kill fish and thousands of birds annually in the U.S. (e.g., 25,000 birds were killed at Lake Erie in 2002). Most water-birds killed by botulism drown before being detected. Although botulism-causing bacteria are a ubiquitous and natural part of the environment, human activities (eutrophication through excessive nutrient runoff from farms) create conditions where outbreaks are more frequent and severe. In the Great Lakes, overabundant invasive zebra mussels deplete oxygen levels and the invasive round goby fish helps move the bacteria up the food chain towards birds. In the U.S., outbreaks are most common during peak summer temperatures in alkaline lakes in western states (Great Salt Lake, Salton Sea), and at managed waterfowl refuges, where ponds frequently have algal blooms. In 2008, an outbreak killed at least 134 Laysan Ducks on Midway Atoll.

Solutions: Government agencies, universities, and conservation groups (e.g., Ducks Unlimited) work to monitor, research, and manage botulism outbreaks in birds across the U.S. and Canada.

Actions:

- Continue and expand research into botulism outbreaks, and water management techniques to minimize risk of outbreaks.
- Increase biosecurity efforts to minimize introduction of invasive species (zebra mussels, gobies) that facilitate outbreaks in birds.
- Reduce runoff that leads to eutrophication.
- Increase freshwater in-flow into saline lakes.

INSECTS

In the Hawaiian Islands, a variety of insects (e.g., wasps and beetles) have been introduced and now prey upon or parasitize other native invertebrates upon which forest birds depend for food, while themselves being unsuitable as a food source for birds. Aphid-like insects, the balsam wooly adelgid (introduced in 1900) and hemlock woolly adelgid (introduced in 1924) kill firs and hemlocks in high-elevation Appalachian forests. Foresters in the Northeast fear that the Asian long-horned beetle will become the next invasive pest to damage their industry. These beetles arrive from China and Korea in wood used for packing crates and in bonsai trees, and have been intercepted at ports in 17 states. Their larvae damage maples, poplars, willows, elms, mulberries, and black lo-custs. Localized infestations in the late 1990s in New York and Chicago resulted in the destruction of many trees. Since 1980, gypsy moth outbreaks have defoliated over one-million acres of forest in North America. Fire ants also predate nestlings of species such as the "Attwater's" Greater Prairie-Chicken, Golden-cheeked Warbler, and Black-capped Vireo.

Warm winters and drought contribute to beetle outbreaks that affect millions of trees in the West, including the ponderosa pine.

BARK BEETLES

There are many species of native bark beetles that attack conifers in the western U.S., including the mountain pine beetle (*Dendroctonus ponderosae*), Douglas-fir beetle (*D. pseudotsugae*), spruce beetle (*D. rufipennis*), and others in the same genus.

Problems: Adult female beetles tunnel into bark to lay eggs and the hatched larvae feed on the wood before eventually emerging as adults. Bark beetles transmit fungus that helps kill infested trees and stains wood blue. Periodic outbreaks are capable of killing millions of trees. Tree kills by beetles benefit some birds in the short-term, such as the Black-backed Woodpecker, but degrade habitat for other forest species. Bark beetles often attack trees previously injured by spruce budworms or that are otherwise stressed. Outbreaks are often most se-vere in dense stands of trees during droughts when trees are less able to produce resin defenses against the beetles. Human-induced changes to forest structure through fire suppression combined with climate change (particularly warmer winters) fa-cilitate larger and more severe outbreaks, as well as range expansion of bark beetles in western forests.

Solutions: Freezing temperatures can kill larvae and pupal stages and serve as natural controls for bark beetles, along with native predators (clerid beetles, woodpeckers). Forest management, in-cluding thinning susceptible stands to leave trees well spaced, is the best long-term remedy to reduce outbreak damages. Spraying pesticides and cutting infected trees is expensive and kills natural insect predators that help control beetles, but has been used successfully to save uninfected trees over small areas.

Actions:
- Combat global climate change and monitor the spread and severity of bark beetle outbreaks.
- Manage healthier forests to be more resistant to beetles.

PHOTO: ©ISTOCKPHOTO.COM

PLANTS

Historically, plants have been intentionally introduced for animal forage, pasture improvement, erosion control, wind breaks, ground cover, shade, or as ornamentals, or accidentally introduced in contaminated soil or feed. Once here, roads, grazing, off-road vehicles, water projects, and other human activities that disturb landscapes and natural processes often help to spread invasive plants. Invasive plant damage and control costs the U.S. more than $34 billion annually.

Invasive plants replace or degrade native habitats, particularly grasslands, sagebrush, wetlands, and island systems. For example, Florida's Everglades are infested with paper-bark trees (*Melaleuca quinquenervia*) from Australia, Old World climbing fern (*Lygopodium micropyllum*), and Brazilian pepper (*Schinus terebinthifolius*). Meanwhile, invasive plants such as purple loosestrife (*Lythrum salicaria*), hydrilla (*Hyrdilla verticillata*), and Eurasian common reed (*Phragmites australis*) have impacted freshwater marshes and brackish wetlands throughout the "Lower 48" states. Other habitats are also impacted. For instance, kudzu (*Pueraria montana*) is an invasive vine that infests forests in much of the southeast, and Chinese tallow (*Triadica sebifera*) is invading "Attwater's" Greater Prairie-Chicken habitat in Texas. Hawaiian forests are battling invasive Miconia trees, strawberry guava

trees (*Psidium cattleianum*), banana poka (*Passiflora mollissima*), Lantana flowers, and some grasses and ornamentals. Buffelgrass (*Pennisetum cilare*), cheatgrass, leafy spurge (*Euphorbia esula*), yellow starthistle (*Centaurea solstitialis*), spotted knapweed (*C. maculosa*), and other invasive weeds have displaced native plants and altered fire regimes on our western range-lands, prairies, and deserts. Although many invasive plants come from other continents, some arrive from different parts of the U.S. For instance, smooth cordgrass (*Spartina alterniflora*), native to East Coast saltmarshes, is invading West Coast mudflats and reducing habitat for shorebirds in the Pacific Flyway.

Controlling invasive plants often requires using multiple techniques, including a variety of physical methods (grazing, burning, cutting, pulling), herbicides, and biocontrol agents. Using biocontrol agents involves releasing non-native insects that evolved alongside the invasive plant and can effectively attack it. This is risky due to the possibility that the released insects may themselves become a problem, and must be done very cautiously. Ideal candidates include insects with specific host requirements that are unlikely to attack non-target plants. If successful, biocontrol agents can be the cheapest, most effective and sustainable solutions to the problem of invasive plants and other pests. Finally, if an area is dominated by an invasive plant and that plant is removed, care must be taken to replant the area with native vegetation and prevent the exotic from returning. You can help reduce the impact of invasive plants by planting native species in your own yard or garden, by removing any invasive plants that may appear, and by participating in efforts to remove invasive species from local parks and refuges.

CHEATGRASS

Cheatgrass (*Bromus tectorum*), also known as drooping brome, is an annual grass that typically

PHOTO: ©ISTOCKPHOTO.COM

Cheatgrass is highly flammable and rapidly regrows in burned areas, replacing less fire-tolerant sagebrush.

Over 50% of sagebrush is invaded by cheatgrass, which now covers more than 100 million U.S. acres (Wyoming sagebrush pictured).

PHOTO: ©ISTOCKPHOTO.COM

grows 19-23 inches tall and likely originated in southwestern Asia. It was first found in New York and Pennsylvania in 1861. It now covers more than 100 million acres, occurs in all 50 states, and grows along roadsides, on eroded lands, grazing lands, prairies, and in sagebrush. It is most abundant in the Great and Columbia Basins.

Problems: Fires and over-grazing help cheatgrass spread. Its extensive root systems reduce soil moisture and help it out-compete native grasses. Dead cheatgrass is highly flammable and fire return frequencies are less than five years for areas invaded by cheatgrass, compared to 32-70 years or longer in Great Basin sagebrush. Cheatgrass is now likely a permanent habitat component in the Intermountain West, where it dominates 100 million acres. Over 50% of sagebrush is likely invaded by cheatgrass, including 36% of the Greater Sage-Grouse's range. Sagebrush requires at least 15 years (and up to 50) to reoccupy burned sites. Cheatgrass invasion costs North American wheat farmers over

$350 million annually in control and lost yields, and also increases the costs of fire suppression and restoration activities for federal, state, and local governments.

Solutions: Cheatgrass can be reduced through herbicides and mechanical methods (burning, grazing, disking, and plowing). Planting strips of fire-resistant vegetation ("greenstripping") can reduce the spread of wildfires. Herbicides (sulfometuron methyl) that kill cheatgrass but spare most native perennials are often the most effective control option, but are expensive. Control efforts must be carefully planned and timed, and follow-up treatments are often necessary. Native perennials should be seeded on lands following control efforts to prevent cheatgrass recovery from the seedbank.

Actions:
- Dealing with cheatgrass is very challenging. Reducing cheatgrass in areas close to sage-grouse habitat is a top priority for bird conservation.

Salt Cedar

Salt cedar or tamarisk (*Tamarix spp.*) are trees and shrubs native to Eurasia that grow five to 20 feet tall, frequently at the edge of streams and springs. Eight species were planted in the Southwest U.S. during the 1800s as windbreaks, for shade, and as erosion control.

Problems: Salt cedar had infested 3.6 million acres across 17 western states by 2003, including many important riparian habitats for birds. Although a few bird species can survive in salt cedar (e.g., the "Southwestern" Willow Flycatcher), it supports fewer insects, less mistletoe, and fewer bird species than native willow and cottonwood riparian woodlands. Southwestern populations of the Yellow-billed Cuckoo, Bell's Vireo, and Summer Tanager are among the birds that are nearly excluded. Salt cedar consumes large quantities of water compared to native vegetation, costing irrigators and hydropower producers $54-164 million annually in

Salt cedar has infested 3.6 million acres and exudes salt, which helps inhibit the growth of native plants nearby.

PHOTO: STEVE HILDEBRAND / FWS

lost water. It also increases the frequency of fire in southwestern riparian woodlands. Its leaves exude salt, which accumulates in the soil beneath trees and inhibits the growth of native plants nearby.

Solutions: Once salt cedar is well established, it is very difficult and expensive to eradicate. Monitoring, prevention, early detection, and local eradication are the most effective means of control. Biocontrol agents include the salt cedar leaf beetle

(*Diorhabda elongata*), a mealybug species (*Trabutina mannipara*), and a weevil (*Coniatus tamarisci*), but so far, only the leaf beetle has been approved for use and released in nine states. Sawing and immediately applying "eco-friendly" herbicide such as Garlon to the stump is effective. Burning and manipulation of water levels to mimic natural flow regimes can favor native species. Bulldozing and plowing up roots has also been used. Aerially spraying herbicides in areas with little or no native vegetation can reduce salt cedar by 90%. Once it is removed, it may be necessary to actively replant native vegetation.

Actions:
- Reduce salt cedar over large areas using biocontrol methods. In riparian woodlands, replace it with native willows and cottonwoods.

Miconia

Miconia (*Miconia calvescens*) is a tree native to Central and South America which may grow up to 50 feet tall. It has been introduced to many Pacific islands as an ornamental tree, including to the Hawaiian Islands in the 1960s.

Problems: Miconia's shady foliage inhibits regeneration of native Hawaiian trees, depended on by native honeycreepers. Its shallow root system does not hold soil as well as those of displaced native forest trees, resulting in soil erosion that can damage nearby coral reefs. Miconia produces large amounts of fruit, which is then dispersed by non-native birds. By the late 1980s, miconia dominated 60% of Tahiti causing landslides and endangering native plants and animals.

Solutions: Controlling miconia in Hawaii is important to prevent damage similar to that which occurred in Tahiti. Trees can be cut and their stumps treated with herbicides to prevent resprouting. Seedlings and small saplings can be uprooted by hand. Repeated treatments may be required to remove new seedlings that germinate following the removal of adult trees. Government

Invasive miconia shades out native trees in Oahu's Koolau range where the "Oahu" Elepaio can still be found.

PHOTO: ©ISTOCKPHOTO.COM

agencies, private organizations, and volunteers have helped eradicate miconia on Oahu and Kauai, and are working to eradicate it from Maui and the Big Island. Monitoring cleared areas for seedlings is important so they can be quickly removed before they can re-establish. Miconia sightings can be reported to the Hawaii Department of Land and Natural Resources. Research into potential biocontrol agents (fungal pathogens and insects) for miconia is ongoing.

Actions:

• Continue miconia control efforts to stamp out new infestations, and complete miconia eradication from remaining Hawaiian Islands where it occurs.

Golden Crownbeard

Golden crownbeard (*Verbisina encelioides*) typically grows one to five feet in height and is native to the western U.S. It was introduced to Hawaii prior to 1871, where it dominates some coastal areas. It likely arrived as seed in soil brought by the military to Midway and Kure Atolls,

where it now forms dense stands. In 2000, it occupied 13% of Midway's 1,200-acre Sand Island and 39% of the 334-acre Eastern Island.

Problems: Crownbeard displaces native vegetation (bunchgrass) on Midway. The Laysan and Black-footed Albatrosses, whose global population stronghold is on Midway, are unable to nest in dense stands of crownbeard. This rapidly growing plant can also enclose existing nests, trapping albatross chicks that eventually die of starvation. Crownbeard also supports invasive ants that may harm ground-nesting birds.

Solutions: FWS started a control program on Midway in the late 1990s that includes handpulling and mowing to prevent crownbeard from flowering, removing stems and roots, treating with herbicide, monitoring, and replanting areas with native bunchgrasses and sedges.

Actions:

• Complete removal of crownbeard and continue habitat restoration efforts on Midway. Monitor for potential return of crownbeard.

• Expand crownbeard control to Kure Atoll and elsewhere in the Northwest Hawaiian Islands.

PHOTO: MICHAEL LUSK

Invasive golden crownbeard can prevent Laysan Albatrosses from gaining access to their nests, and starving their chicks.

COLLISIONS

Birds accidentally collide with a variety of man-made structures and very often die as a consequence. Collisions impact birds regardless of sex, age, or species, and contribute to declining populations around the globe. People are often unaware of collisions, as bird carcasses can be hard to see and are quickly scavenged. Despite this, staggering numbers of birds are known to be killed in collisions annually.

Hundreds of species have been recorded as collision victims, but nocturnal migrants such as rails, the American Woodcock, cuckoos, vireos, thrushes, warblers, and sparrows appear to be particularly vulnerable. Night-flying migrants use stars to navigate and can be attracted by artificial lights. In North America, 41% of all north-south migration routes cross at least one major urban area that is artificially illuminated at night. The lights disorient birds, especially on cloudy and rainy nights when they travel at lower altitudes. Many then fatally collide with buildings, towers, and other structures. Places where migrants concentrate, such as the Florida coasts, the Gulf states, and Great Lakes, are notorious for large numbers of collisions among nocturnal migrants. More than 30,000 Endangered Newell's Shearwaters have been recorded colliding with towers and power lines illuminated at night on the Island of Kauai, and Endangered Hawaiian Petrels have also been recorded in collisions there. Daytime collisions with glass are also a problem, for both resident birds and migrants.

Cranes and other water-birds can be fatally injured by flying into power lines or telephone wires near wetlands (e.g., at least 46 Endangered Whooping Cranes have been killed in collisions with power lines since 1956). Prairie-chickens and sage-grouse suffer high mortality rates in collisions with barbed wire fences, but most of these collisions can be avoided by using small, visible markers on the wires.

While large numbers of birds are killed by vehicles each year, effective solutions to this problem are still hard to determine. Suggestions include avoiding the planting of berry bushes that can attract birds to highway medians, but while potentially useful, this is likely to provide only localized benefits. The authors look forward to hearing additional suggestions from readers on this topic. Collisions with planes are virtually insignificant from a bird population perspective, but obviously can be devastating for aircraft (though are not always so). If large-scale programs were developed to eradicate birds from around major airports in an attempt to reduce collisions, then it could become a major bird conservation issue, however.

If you discover an injured bird that has been the victim of any type of collison, contact your local wildlife rehabilitation center. Birds that survive the initial impact can often be saved.

This Wood Thrush is one of perhaps a billion birds killed annually in collisions with glass windows and other structures.

PHOTO: MIKE PARR

BUILDINGS AND GLASS

There are billions of square feet of window glass in the U.S. Unfortunately much of this presents a deadly hazard for birds. Both residential homes and tall office buildings can be problems, and few buildings are immune to occasional collisions and fatalities. The sheer number of homes and glass windows makes this one of the biggest killers of birds in the U.S., with estimates ranging from 100 million to one-billion birds killed each year.

Problems: Birds don't see glass, and perceive reflections as real. They may try to fly to sky or vegetation reflected in windows, or to plantings seen through glass in building lobbies or across glass building corners. This happens in backyards as well as in city centers. An average building kills from one to ten birds per year, but some buildings lit at night, or with large expanses of glass, can kill far more. In addition, birds attracted to feeders may collide with windows during panicked attempts to escape predators, such as hawks or house cats.

Solutions: Turning off lights in office buildings at night, especially during spring and fall migration, can greatly reduce collisions while also saving energy. Recent legislation in Minnesota mandates this for all buildings owned or leased by the state. Shields above outside lights block light radiating upwards, which is the most dangerous to birds. The city of Tuscon, Arizona, has strict lighting codes to preserve visibility of the night sky for citizens and astronomy observatories, benefiting migrant birds in the process.

Anything that can be done to make glass more visible is likely to reduce daytime bird strikes. Decals are popular and readily available, but not effective if gaps larger than two inches remain between decals. External screens (or better, vegetation screens outside windows), decorative window films, or exterior tape stripes are other options. Situating bird feeders within three feet of a window can reduce mortality, as birds cannot generate sufficient velocity when flying away to injure themselves. Netting can be stretched in front of windows that are particularly problematic. Films that differentially reflect or absorb ultraviolet light (which birds can see, but people can't), non-reflective coatings (such as CollidEscape), glass with thin black lines, and fritted glass, can all serve as effective deterrents to prevent bird strikes at windows. Such deterrents should be included in green building standards.

PHOTO: CHRISTINE SHEPPARD

Birds often just see vegetation reflected in glass, and are unaware of the danger from collisions.

Actions:
- Turn off your office lights after work; ask your office manager to do the same for the whole building.
- Modify the placement of bird feeders and the visibility of your windows (see p. 416).
- Volunteer to help monitor collision mortality at city buildings (and rescue survivors).
- Write to your mayor and governor to suggest "Lights Out" policies for city and state buildings.
- Green building standards should include bird collision avoidance.

COMMUNICATION TOWERS

The federal government requires night lighting on approximately 80,000 communication towers that are 200 feet or more in height, or located near airports. This number is increasing, with 7,000 new towers constructed annually to meet the demand for cellular telephone and digital television networks. Estimates suggest that these towers kill between four and 50 million birds annually.

Problems: Tower lighting, especially red light, disrupts birds' ability to navigate, resulting in collisions with towers and guy wires. Total annual mortality at communication towers is particularly significant for the Ovenbird, Red-eyed Vireo; and Tennessee, Bay-breasted, Blackpoll, and Cerulean Warblers.

Solutions: Communication tower collisions can be reduced by putting multiple antennae on existing structures to avoid constructing new towers, removing inactive towers, constructing towers less than 200 feet tall to avoid lighting requirements, using strobe lighting rather than continuous lights, eliminating guy wires where possible, and using visual markers on wires to prevent daytime collisions involving waterfowl and raptors. Thanks to a lawsuit won by ABC and the Forest Conservation Council in 2008, the Federal Communications Commission must now assess the potential environmental impacts, including bird mortality, of new towers prior to construction, and all new permits for towers must comply with existing environmental laws, including the ESA and National Environmental Policy Act.

Actions:
- The Federal Communications Commission (which licenses towers) must require white strobe lights and un-guyed tower structures.
- Avoid constructing towers near Important Bird Areas.
- If you are opposing a specific tower, ABC has a number of pre-written ordinances and supporting documents that can help you appeal a tower project (see www.abcbirds.org). Contacting a local bird club, state Audubon chapter, and the media may also help build support for your campaign.

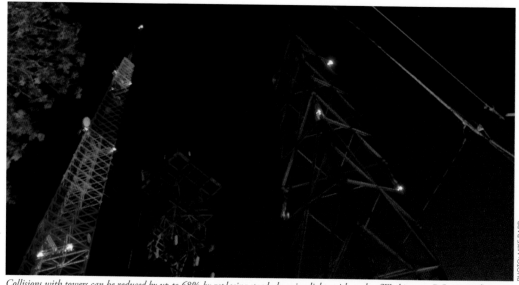

PHOTO: MIKE PARR

Collisions with towers can be reduced by up to 68% by replacing steady-burning lights with strobes (Washington, DC, pictured).

WIND TURBINES AND POWER LINES

At the present time, there is approximately 31,000 megawatts of installed wind generation capacity in the U.S., with some 5,000 additional megawatts currently under construction. This generation is still equivalent to less than 1% of U.S. energy consumption needs, though. New construction is expected to reach 16,000 megawatts per year by 2018, however, and continue at that rate or faster until 2030 when wind would supply 20% of the nation's energy needs. At that point, the U.S. would be able to produce approximately 300,000 megawatts of wind-generated electricity, equivalent to some 200,000-300,000 large industrial-scale turbines (assuming that turbine size continues to grow), killing close to one-million migratory birds per year, and occupying some 19 million acres. The first offshore wind farm in the U.S. was recently approved for the Cape Cod area by the Department of the Interior, though it is expected to face ongoing opposition. The first offshore wind farm in the U.S. was recently approved for the Cape Cod area by the Department of the Interior, though it is expected to face ongoing opposition.

Problems: Raptors, seabirds, songbirds, and bats collide with the rotating blades of wind turbines when these structures are built near their hunting grounds, nesting colonies, flight paths, or migration routes. Studies suggest that wind turbines are each capable of killing 2.3 birds per year (3.1 birds per megawatt of electricity generation). In addition to direct mortality, wind farms have the capacity to seriously disrupt sensitive breeding species such as the Greater Sage-Grouse through disturbance and habitat fragmentation, and to impact Endangered species such as the Whooping Crane. The crane already has a history of collisions with power transmission lines, and thousands of new miles of these lines will be needed (many expected to traverse the Whooping Crane's migration route) to carry wind-generated power to consumers. Birds such as eagles, hawks, and owls can also be electrocuted when they land on power lines. Estimates suggest that annual mortality could be in the tens of thousands.

Annual bird mortality from wind turbines will likely rise to almost one-million birds by 2030 (Altamont, CA, pictured).

PHOTO: MIKE PARR

Solutions: New wind farms should not be constructed within Important Bird Areas or at migratory hotspots, but should use already fragmented habitats and agricultural areas. Wind-wildlife maps have been produced to help the wind power industry avoid areas with high risks for bird and bat collisions. Siting turbines away from ridge crests, canyons and other areas frequented by resident raptors and migrating birds will also reduce collisions. Radar can be used to detect flocks of birds at risk so that wind projects can be temporarily shut down to avoid collisions. FWS has produced guidelines for improved pole configurations to avoid raptor electrocutions on power lines. Federal prosecutions of power line companies, such as the Moon Lake Electric Association and Pacifi-Corp, for bird electrocutions have alerted industry to the need to halt this mortality. Visible markers can also help reduce crane fatalities.

Actions:
- Wind turbines should be sited away from sensitive locations such as sage-grouse lek sites.
- Federal guidelines for bird collision avoidance including temporary shutdowns when needed must be made mandatory.
- Revoke federal tax credits for operators who do not comply with regulations.
- Power lines that traverse Whooping Crane migration routes must be prominently marked or buried where possible.

EXPLOITATION AND PERSECUTION

Historically, market hunters exploited waterfowl, shorebirds, seabirds, and some other species, for their feathers, eggs, and meat; yet more species, such as hawks, were persecuted as pests. These threats contributed to the extinction of several bird species in the U.S., including the Great Auk, Carolina Parakeet, Eskimo Curlew, Passenger Pigeon, and probably the Ivory-billed Woodpecker. During the last century, however, the passage of hunting regulations allowed many bird populations to recover, and taxes on guns and ammunition helped fund habitat protection and restoration efforts for both game and non-game species. Direct persecution of birds within the U.S. does still occur, however, with several species (e.g., terns, cormorants, vultures, and Red-winged Blackbirds) killed because they are regarded as pests. Fish-eating birds are most frequently targeted because they are thought to contribute to declines in fish stocks. The real culprits are principally over-fishing, pollution, and dams, however. Much of this killing is regulated (and sometimes initiated) by state game managers, the USDA, or other government agencies with permission from FWS. The USDA Animal and Plant Health Inspection Service killed four-million birds in control programs in 2008, including 1.6 million cowbirds (over and above the targeted control used locally to protect endangered birds), 1.5 million starlings, and 900,000 blackbirds. Killing programs are expensive, rarely reduce damages over the long-term, and in many cases, birds are not even the primary cause of the damage they seek to redress. Alternative strategies, including government or private insurance programs that compensate for damages caused by birds, can resolve conflicts more effectively, and without needlessly killing target and non-target animals or wasting taxpayer dollars. Some species, such as the Painted Bunting, are still exploited for the pet trade in Latin America, and this threat is further discussed in the International chapter.

PHOTO: JAMES DEAN / SMITHSONIAN INSTITUTION

Once the most abundant bird in North America, three to five billion Passenger Pigeons were wiped out by over-hunting.

HUNTING

Before 1900, over-hunting decimated many bird populations in the U.S. and contributed to the extinction of several species. Laws passed in the early 1900s (the Lacey Act and Migratory Bird Treaty Act), in combination with a new conservation ethic among hunters and wildlife management agencies, largely stopped this slaughter and allowed many species to recover. Today, more than 41.5 million birds are killed by hunters annually in the U.S. Most hunted birds are waterfowl, doves, quail, grouse, American Woodcock, and Wild Turkey, but smaller numbers of prairie-chickens, ptarmigan, American Coot, Wilson's Snipe, rails, and Sandhill Cranes are also harvested. Most game species in the U.S. are well-managed and are not suffering population declines due to hunting. This is demonstrated by the continued abundance of Mourning Doves, and dramatic increases in populations of well-managed game species, especially waterfowl. For game species that are declining, habitat loss rather than hunting is usually the cause, but hunting regulations need to be driven by science. The hunting of rare, rapidly declining species, such as prairie-chickens, should be constantly evaluated to determine whether shooting is contributing to declines. In countries such as Barbados and Brazil, shorebirds such as the American Golden-Plover and Semipalmated Sandpiper are also still overharvested in large numbers.

Problems: Toxic lead ammunition was banned for waterfowl hunting in the early 1990s, but is still widely used by dove and big game hunters, placing scavenging birds such as the California Condor at risk from lead poisoning when they feed on abandoned carcasses. Subsistence hunting by native Alaskan communities may contribute to declines in eider populations through lead poisoning acquired from shot used over wetlands. Some hunters oppose reducing populations of invasive or overabundant game species (e.g., deer in the eastern U.S.; feral pigs, mouflon, and other species in Hawaii) and help maintain and stock natural areas

Hunters have contributed over $10 billion to government wildlife conservation programs since 1934.

with introduced game species that can harm native birds.

Solutions: Conservation organizations, such as Ducks Unlimited, Quail Unlimited, the National Wild Turkey Federation, and others, draw support from hunters, and are leaders in acquiring, restoring, and managing land and water for both game and non-game species. Hunters and fishermen also fund habitat restoration and NWRs through taxes on hunting equipment and licenses (e.g., Duck Stamps), generating a cumulative total in excess of $10 billion for wildlife conservation since 1934. Hunting is also the best management tool available for controlling destructive and overabundant populations of geese and deer.

Actions:
- For game species that are declining, hunting should be regularly evaluated to determine whether it is contributing to declines. If so, hunting regulations should be adjusted accordingly.
- The effect of subsistence hunting in Alaska requires careful monitoring.
- Hunters should end the use of lead ammunition that contaminates the meat of game animals, and threatens birds and the environment.

CROP-EATING BIRDS

Many birds benefit from feeding on agricultural crops. These include Snow Geese that have become over-populated by taking advantage of southern farm fields during the non-breeding season, and other species such as Blue Jays and Cedar Waxwings that feed opportunistically in orchards. It is widely believed that blackbirds (primarily Red-winged) cause the most substantial economic losses by feeding on corn and sunflower seeds.

Problems: Hundreds of thousands of Red-winged Blackbirds are killed with poison baits (DRC-1339) each fall in North and South Dakota in an effort to reduce damage to crops. This also places non-target birds such as the Western Meadowlark, Mourning Dove, and the WatchListed Baird's Sparrow in jeopardy.

Solutions: In 2001, the USDA proposed killing up to two-million additional blackbirds over five years in the Dakotas, but an economic review found that the benefits to sunflower growers would be negligible, and the proposal was not approved by FWS. The USDA also provides free herbicide spraying to remove cattails on private lands to discourage flocks of blackbirds.

Actions:
- Rather than poisoning blackbirds, non-target birds, and wetland plants, USDA funds would be better spent by directly compensating the few farmers who suffer sunflower crop losses due to blackbirds.

PHOTO: ROBERT ROYSE

Government programs kill hundreds of thousands of Red-winged Blackbirds annually to protect crops, but with questionable benefits.

VULTURES

Members of the public frequently complain about vulture roosts (and droppings) in the vicinity of their properties. Although Black Vultures are primarily scavengers, they do occasionally kill newborn livestock (though far more cattle are killed by dogs).

Problems: The USDA Animal and Plant Health Inspection Service killed more than 2,400 Black Vultures in the U.S. in 2003, as well as around 400 Turkey Vultures, which do not kill livestock at all. These killings are unlikely to have any significant effect other than allowing APHIS to say that they are doing something to respond to public complaints.

Solutions: Lethal control at current levels may help appease local concerns, but will not impact the overall increasing vulture population trend. Suspending a dead vulture or effigy at roosts may disperse birds away from conflict areas, but the efficacy of this method requires further study.

Actions:
- Research non-lethal methods to resolve conflicts between ranchers and vultures.
- Ensure that Turkey Vultures or other raptors are not indiscriminately killed.
- Evaluate the benefits to livestock of lethally controlling Black Vultures where conflicts occur.

PHOTO: ROBERT ROYSE

Government programs needlessly kill hundreds of Turkey Vultures each year in response to public complaints.

FISH-EATING BIRDS

Thanks to pesticide regulations, Double-crested Cormorant populations have increased from recent lows in the 1950s-1970s. Cormorants and other fish-eating birds are often perceived to be a threat to commercial and recreational fisheries, but no good scientific evidence exists that they cause significant damage except at certain aquaculture and fish hatchery fa-

Caspian Terns are made scapegoats for salmon declines; dams, pollution, and over-fishing are the real culprits.

cilities. Evidence also suggests that cormorant populations are still below their natural carrying capacity, and that far more cormorants were historically able to coexist with healthy fisheries.

Problem: In response to public pressure from fishermen and the aquaculture industry, cormorant kills have occurred in both Canada and the U.S., with more than 45,000 cormorants killed legally in the U.S. annually from 1998-2002. Similarly, fish-eating birds in the Pacific Northwest are blamed for declines in salmon stocks, despite larger threats to salmon from dams, pollution, and fisheries. During the late 1990s and early 2000s, Caspian Terns, Double-crested Cormorants, and other fish-eating birds were killed, harassed, or displaced from breeding colonies in the Columbia River estuary in Washington State. In 2008, the U.S. Army Corps of Engineers began implementing a plan to reduce Caspian Tern populations in the estuary from 10,700 nesting pairs to as few as 3,125

pairs by redistributing a portion of the tern colony on East Sand Island to alternative sites in Oregon and San Francisco Bay. As breeding habitat on East Sand Island is reduced from six acres to two acres or less, the Corps plans to create new habitat at alternate sites where decoys and acoustic signals will be used to attract terns to nest (some proposed alternate sites are more than 500 miles away and social attraction techniques are unlikely to work).

Solutions: Funding would be better spent investigating and addressing the real causes of declining commercial and game fish populations, such as pollution, dams, and over-fishing; and developing methods to non-lethally exclude fish-eating birds from aquaculture ponds and hatcheries. Cormorant killing programs are likely ineffective at reducing damages at such facilities, and lack a clear definition of success or adequate monitoring to evaluate their impacts on either fish or cormorants.

Actions:

- Stop federal subsidies for the killing of fish-eating birds.

Thousands of Double-crested Cormorants are killed each year to appease complaints from fishermen and fish farmers.

FISHERIES

The U.S. consumes more than 15 billion pounds of fish each year. Commercial fisheries cause two major problems for bird populations. Firstly, seabirds, particularly albatrosses, can be accidentally drowned in certain types of fishing gear (primarily longlines and gillnets, but also some trawl fisheries). The slow rate of reproduction of many seabird species makes their populations particularly sensitive to this adult mortality. Secondly, overharvesting of fish and other sea creatures (e.g., horseshoe crabs) can cause bird population declines.

Fisheries are declining globally. One widely used management tactic is that of setting fishing quotas, but it has proven difficult to regulate an industry that once had unlimited access to resources. Fisheries regulation also tends to be dominated by industry representatives, and it is understandably difficult for them to voluntarily reduce quotas if some captains may need to exceed them just to meet their costs. Regardless of the economic complexities, the present system has led to the commercial exhaustion of multiple fisheries, and few can be regarded as truly sustainable. Purchasing seafood responsibly can encourage sustainable fisheries, however, as well as helping to protect healthy marine ecosystems that benefit birds, and reducing bycatch of non-target fish, birds, sea turtles, whales, dolphins, sea lions, and other marine life. A variety of organizations provide free online guides to purchasing seafood. The Marine Stewardship Council (MSC) certifies fisheries, and we recommend buying MSC certified seafood in most cases. There is debate, however, concerning their certification of some Chilean sea bass (also known as Patagonian toothfish), as some scientists believe that this fishery needs to be closed completely to allow stocks to recover. Despite attempts at regulating the fishery, much of the world's Chilean sea bass is also caught by pirate longliners that kill thousands of seabirds.

Colored streamers scare albatrosses and other seabirds away from the hooked lines until the baits sink safely out of reach.

PHOTO: WASHINGTON SEA GRANT. ©COPYRIGHT UNIVERSITY OF WASHINGTON BOARD OF REGENTS

LONGLINES

Longlines are fishing lines up to 60 miles in length, baited with as many as 30,000 hooks set to catch tuna, swordfish, cod, and other fish. Seabirds accustomed to scavenging fish offal thrown overboard from fishing boats are attracted to the baited hooks as the longlines are reeled into the water. Birds that take the baits are often impaled by the hooks, pulled under water by the sinking line, and drowned.

Problems: Longline vessels from many countries ply the world's oceans killing hundreds of thousands of seabirds including albatrosses, petrels, shearwaters, boobies, and gulls. Birds that forage near the surface are most vulnerable, though shearwaters and other diving birds can become hooked far below. Eighteen of the world's 22 species of albatross are threatened with extinction, primarily due to high mortality on longlines. During the 1990s, U.S. longline fisheries killed more than 3,800 albatrosses annually.

Solutions: Simple remedies exist to reduce seabird bycatch on longlines. These include streamers (see photo on previous page) that flap above the longlines, and keep birds at a distance long enough to allow the hooks to sink out of reach. Setting hooks at night when most birds are less active, discarding offal when hooks are not being set, removing hooks from offal, adding weights to quickly sink lines, and even dying bait blue to make it less visible to birds can help avoid accidental seabird mortality, though not all methods are equally effective in all fisheries. If used appropriately, however, these bycatch avoidance measures can nearly eliminate seabird mortality.

In 2004, Alaskan longliners were required to use bird-scaring streamer lines (paired lines for boats over 55 feet long) and over 5,000 lines were given free to fishermen by the federal government. Seabird avoidance measures were also mandated for Hawaiian longliners in 2004, including setting lines at night, setting lines off the sides of boats

Wandering Albatross hooked and drowned on a longline in the southern oceans.

PHOTO: GRAHAM ROBERTSON / AAD

(instead of at the stern) to allow baits more time to sink, using tubes extending below the water through which the lines run, and offal discharge regulations. These measures have greatly reduced seabird bycatch in both fisheries (see table below). The Convention on the Conservation of Antarctic Marine Living Resources (known as CCAMLR) has also mandated strict seabird bycatch avoidance policies for vessels in the southern oceans, and succeeded in recording no albatross deaths from 2005-2006.

Actions:

- Replicate successful seabird avoidance measures in fisheries across the world.

ALBATROSS DEATHS 1990-2006

Fishery	1990	2006
Alaska	1,308	291
Hawaii	2,563	88

GILLNETS

Gillnets are vertical mesh curtains, often made of nylon, that typically hang from floats on the ocean surface. They can be set at variable depths and with different mesh sizes to target specific fish, and are used to catch salmon, squid, cod, flatfish, and other species.

Problems: Nets that are discarded, damaged, or lost during storms can continue to "ghost-fish", trapping ocean wildlife for years. Many diving birds swim at high speeds and to great depths in search of prey, and often fail to see gillnets until it is too late. U.S. birds affected include ducks (such as scaup, goldeneye, mergansers, and scoters), and alcids (murres, murrelets, and puffins). The full extent of bird mortality in gill nets is poorly known due to a lack of consistent monitoring. Outside the U.S., gillnets are a problem for several species of flightless grebes that are endemic to high Andean lakes (Junin Grebe, Titicaca Grebe), as well as many other marine birds.

Solutions: Ocean driftnets were extensively used in the 1980s to catch tuna, and severely impacted whale, dolphin, and seabird populations. The United Nations banned their use in international waters in 1993, though at least 11 driftnet vessels still operated in the North Pacific in 1999, and the ban did not extend to near-shore national waters.

Several tools exist for reducing seabird bycatch in coastal gillnet fisheries. Modifying gear with visual markings (e.g., using white-colored mesh in the upper portions of nets) or acoustic alerts ("pingers"), avoiding fishing at dawn and dusk when birds are most active, and timing fishing for peak salmon abundance could reduce bycatch of Common Murres and Rhinoceros Auklets in parts of the Pacific Northwest by up to 75%. Restricted use of gillnets off Newfoundland during the spawning period for capelin would also reduce bycatch of Common Murres.

Actions:
- Require that gillnet fisheries implement seabird bycatch avoidance measures.
- Limit the use of gillnets in sensitive bird areas.

Implementing bycatch avoidance techniques in gillnet fisheries could substantially reduce mortality of vulnerable seabirds.

PHOTO: MIKE PARR

HORSESHOE CRAB OVERHARVEST

Along the Mid-Atlantic Coast, millions of horseshoe crabs come ashore each spring to spawn, depositing their eggs on the region's beaches. Particular concentrations of spawning crabs occur in the Delaware Bay, where huge numbers of shorebirds, including Red Knots, Ruddy Turnstones, Dunlin, Sanderlings, Semipalmated Sandpipers, and Short-billed Dowitchers stop to feed on their eggs, refueling for the final leg of the northward migration to their breeding grounds.

Problems: Since the 1980s, horseshoe crabs have been harvested at unsustainable levels for use as bait in eel and conch traps. As a result, the number of eggs available for shorebirds has been dramatically reduced. Studies of Red Knots have shown that without an abundance of this high-protein food source, many birds have been unable to gain sufficient weight to successfully complete their migration and breed.

Between 1990 and 1998, horseshoe crab egg densities declined by up to 90% in the Delaware Bay, and the number of days that eggs were available declined from 70 to 30. This correlates with steep population crashes in the *rufa* subspecies of the Red Knot (>50%) and Semipalmated Sandpiper (70-80%). Roughly 13,000 Red Knots were

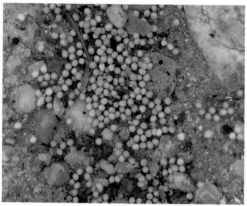

Horseshoe crab eggs are a crucial food source for migrating shorebirds needing to refuel along their northward journey.

counted in the Delaware Bay in 2004, where historically more than 100,000 are estimated to have occurred. Without a reversal in this trend, some conservationists fear that the *rufa* Red Knot could soon become extinct.

Solutions: Thanks to efforts by ABC and other groups, Delaware and New Jersey declared statewide moratoriums on horseshoe crab harvests in 2006. This moratorium has been maintained in New Jersey, but was overturned in Delaware in 2007. In 2009, Maryland began requiring fishermen to harvest two male crabs for every female crab, in order to leave more females and eggs on the beaches. Despite these policy gains, however, shorebird numbers continue to decline. Horseshoe crabs take nine to ten years to reach sexual maturity, and so further restrictions on their harvest in Delaware Bay are necessary to give crab numbers, crab egg densities, and shorebird populations the chance to recover. Reusable bait bags cut the amount of crab needed in half; a synthetic alternative to horseshoe crab is also still in development.

Actions:
- A moratorium is needed on horseshoe crab harvests until crabs and shorebirds recover.
- Finish developing an artificial bait alternative.
- List the *rufa* Red Knot under the ESA.

A population crash among Red Knots followed a 90% drop in horseshoe crab egg densities in Delaware Bay during the 1990s.

POLLUTION AND TOXICS

Large oil spills are dramatic and well-publicized examples of how pollution causes harm to birds. Less well-publicized, however, is how we pollute our environment every day, and the impacts of this pollution on bird populations. Pesticides, heavy metals (such as lead and mercury), oil, toxic waste, and other forms of pollution contaminate our air, land, and water, contributing to bird mortality and population declines. Approximately 15 million birds are killed each year by pesticides in the U.S. The sensitivity of birds to pollutants can indicate environmental threats that may also harm people.

Pollution and toxins can also cause sub-lethal effects that do not directly kill birds, but reduce their longevity or reproductive rates. In addition to pesticides, other contaminants, including heavy metals and plastic garbage, can cause mortality and health problems, reducing birds' life spans and reproductive success.

Light and noise pollution can also cause problems for birds. Lights on communication towers can disorient nocturnal migrants on cloudy nights, directly killing numerous birds, and exhausting many others during their already challenging migrations (see p. 318). Drill noise can disturb displaying sage-grouse, although the negative impacts of oil and gas development also extend well beyond this.

Pollution threatens even the most remote places where birds live. Air pollution from coal-fired power plants is deposited in the form of acid rain, affecting birds in mountaintop forests far away from the pollution's source. Acid rain damages trees in eastern forests and reduces healthy snail populations required by birds such as Wood and Bicknell's Thrushes. Plastic garbage can float thousands of miles from its source over the open ocean, takes centuries to degrade, and is sometimes mistakenly consumed by seabirds and fed to their chicks. Excess nutrient runoff (e.g., nitrogen and phosphorus) from farms or developed areas

Banning harmful pesticides such as DDT has helped birds such as the Brown Pelican to recover from past population crashes.

PHOTO: ROBERT ROYSE

can lead to the eutrophication of water far downstream, degrading wetland and marine habitats, favoring invasive plants, and sometimes leading to algal blooms, dead zones, and botulism outbreaks that can kill thousands of birds. For example, periodic algal blooms along the coast of California caused by nutrient-rich river outflows from agricultural areas and coastal cities result in die-offs of pelicans, murres, and marine mammals. Natural leaching and evaporation from agricultural runoff can also lead to potentially harmful concentrations of elements such as selenium in estuaries and saline lakes, which may be further exacerbated by mining and oil refinery effluents.

PESTICIDES

Worldwide, five billion pounds of pesticides are applied annually, 10% of which are used in the U.S. Agricultural lands occupy 19% of the U.S. and account for 70% of our pesticide use (average of 2.4 pounds per acre). Residential yards also occupy approximately 50,000 square miles of the U.S. Much of this is maintained as well-manicured but barren lawns that use ornamental plant varieties and offer little in the way of bird habitat, while requiring significant amounts of pesticides (and water). Ten to 20 percent of pesticide applications are for cosmetic purposes to produce blemish-free produce for consumers, with little effect on crop yield or taste. Far from declining since the publication of Rachel Carson's book *Silent Spring* in 1962, annual pesticide use has continued to increase in terms of pounds applied and number of chemicals registered. Millions of birds are still killed each year by these substances.

Fish-eating birds, such as the Osprey, and other raptors atop the food chain can accumulate dangerous levels of pesticides.

PHOTO: BILL HUBICK

Problems: In the 1950s and 1960s, organochlorine insecticides including DDT and dieldrin were used widely. These pesticides are fat-soluble and resistant to degradation, and therefore tend to accumulate in animals and the environment. They become particularly concentrated in the bodies of predatory bird and fish species that consume prey animals with high organochlorine residues. Problems with these chemicals and other pesticides inspired the publication of *Silent Spring*, which raised awareness of pesticide threats and helped start the environmental movement in the U.S. DDT, dieldrin, and other organochlorine pesticides were banned in the U.S. and Canada in the 1970s, but they degrade very slowly in the environment, and as a result, trace amounts are still found in the blood of over 90% of migrant songbirds today. Some organochlorines are still used in Latin America, where migratory birds from North America winter. As organochlorine pesticides were phased out, they were replaced by organophosphates and carbamates. These insecticides break down much faster in the environment, but are also much more toxic to birds at lower doses. For instance, carbofuran (a carbamate), is linked to the largest number of direct bird fatalities caused by any pesticide used in the U.S., and also to population declines in the Horned Lark, Western Meadowlark, and other grassland birds. Fortunately, thanks to efforts by ABC and its partners, carbofuran has now been cancelled. Monocrotophos (an organophosphate), was withdrawn from use in the U.S. in 1989, accidentally killed tens of thousands of Swainson's Hawks in Argentina during the 1990s. Argentina, along with most other South American countries, banned this pesticide in the late 1990s, but it is still used to deliberately kill Dickcissels in Venezuela and Bobolinks in Bolivia, where these migrants feed on rice and other crops.

Agricultural pesticide use is rising rapidly in Latin America where many U.S. breeding birds winter, with over 100 million pounds imported into the region each year. As our economy becomes more

PHOTO: BILL HUBICK

Rodenticides used outdoors can poison birds such as this young Great Horned Owl that might be fed contaminated rodents.

globalized, Latin American farmers are also supplying more and more fruits and vegetables to the U.S. market, particularly during the northern winter. Pesticides that are restricted or banned in the U.S. are often still used in great quantities in parts of Latin America, including on produce sold to the U.S. Weak regulations and uneven enforcement can result in high pesticide applications that expose birds. Pesticide use in Latin America causes health problems for resident as well as migrant birds – and for farm workers.

In total, the EPA and ABC have documented more than 2,500 incidents of bird kills attributed to pesticide use, many of which killed hundreds or thousands of individual birds. In most cases, the pesticides were used legally and according to label instructions. The effects of individual pesticides on birds can be difficult to predict in advance of widespread use, as toxicity can vary among species, or depend on interactions with other pesticides in the environment. Furthermore, the impacts of pesticides on birds are difficult to detect. All but the largest-scale bird kills caused by pesticides tend to

go unnoticed by people because, for the most part, we are not out actively looking for dead birds. Birds also often hide when they are sick, their small size makes them hard to find, and scavenging predators remove most carcasses within 24 hours. Pesticides can also take many hours to kill a bird, by which time it may have flown far away from the source of its poisoning, making correlations with specific pesticide applications difficult. Nevertheless, more than 670 million birds were estimated to be directly exposed to pesticides each year in the U.S. during the 1990s, with 67 million of these probably dying as a result. However, recent pesticide cancellations due to ABC action have likley reduced this mortality to 15 million birds annually.

Pesticide exposure also often harms birds without killing them. For example, DDT contamination caused many water-birds and raptors to lay eggs with thin shells that broke before the chicks could hatch. Populations of the Osprey, Bald Eagle, Peregrine Falcon, Brown Pelican, and other birds declined rapidly in the 1960s in the U.S. as a result. Other sub-lethal effects include disorientation dur-

PHOTO: ROBERT ROYSE

Spraying pesticides to fight spruce budworm in Canada reduces food supplies for Bay-breasted Warblers, and can poison birds.

ing migration, reduced ability to maintain weight, impaired territory defense, decreased ability to avoid predators, and reduced care of young. Organophosphate and carbamate pesticides are neurotoxins that paralyze the nervous system of target insects, but they can have the same effect on birds. In a study of White-throated Sparrows, 20% of birds living in forests sprayed with the insecticide fenitrothion died or disappeared. Survivors showed a 20-50% reduction in cholinesterase, an enzyme critical to the nervous system. Affected breeding White-throated Sparrows also produced 75% fewer young in sprayed forests compared to unsprayed forests. In another example, birds exposed to the insecticide acephate (four-million pounds of which are applied in the U.S. each year to control aphids on crops) have been shown to lose their ability to properly orientate during migration. Other pesticides can also act as synthetic hormones, causing health problems in birds, other wildlife, and people.

In the hands of professional applicators, rat poisons and other rodenticides can be effective tools in the battle against household pests, but rodenticides that leave toxic residues in the carcasses of poisoned rodents pose a very high risk to raptors and scavengers when used outdoors or in agricultural settings. If control of prairie dogs, ground squirrels, or meadow mice is truly required, only those rodenticides that do not leave a toxic residue should be used. Rodenticides can also be an important tool for eradicating invasive rodents from seabird islands, but should be used with great caution to minimize the exposure of non-target birds.

Pesticide applications often kill non-selectively, and as a result, beneficial, non-target insects that are key food sources for birds may also be killed on a large scale. This can cause population-level effects in some bird species. Herbicides tend to have low direct toxicity to birds, but their use may end up destroying bird habitat and potentially reducing populations. The aerial spraying of South American forests to reduce cocaine and marijuana cultivation is a prime example.

Solutions: Following the establishment of the EPA in 1970, the system for regulating pesticides in the U.S. improved and most of the worst pesticides have since been banned. Recently, this included fenthion and carbofuran, two insecticides that are particularly harmful to birds. Bird populations can successfully recover if threats from pesticides are reduced. For example, with intensive management, Bald Eagles and other raptors have largely recovered in North America since DDT and dieldrin were banned. Unfortunately, in an effort to promote agricultural and commercial interests, the EPA has encouraged pesticide use, which has unintentionally been harmful to the environment, in spite of extensive testing and review of products prior to registration. The ongoing review and assessment of pesticides requires the careful attention and engagement of NGOs to ensure that science and the environment do not take a back seat to profit or convenience. In the cases of fenthion and carbofuran, ABC led concerted campaigns to counteract pressure from applicators and the manufacturer to continue using the pesticides in the face of irrefutable evidence that they were hazardous and had killed hundreds of thousands to millions of birds.

To reduce the negative impacts of pesticides on birds, we need a more open registration process with improved transparency, monitoring, and ongoing field studies, to ensure that wildlife impacts are fully taken into account. When necessary, we also need to block or restrict the use of the most harmful pesticides. Assembling, maintaining, and publicizing information on the impacts of pesticides on birds is critical to this effort. ABC curates a database of bird poisoning incidents called the Avian Incident Monitoring System (AIMS), which is available online at www.abcbirds.org, but data on bird kills are poorly documented by the government and agricultural operators, and this must be dramatically improved. See chapter five for more on how to influence regulatory policy.

Protecting boreal songbirds, such as the Bay-breasted, Cape May, and Blackburnian Warblers, should be part of the solution to biologically controlling outbreaks of the native spruce budworm, rather than using pesticides. Songbirds typically eat 80% of budworm caterpillars, likely lengthening the time between outbreaks. This "service" is valued at $5,000 per square-mile in increased tree growth from natural pest protection. Healthy forests and bird populations depend on each other, and, therefore, conventional pesticide use to control spruce budworm should be reduced, and carefully managed biological controls should be used instead to help maintain boreal songbirds as natural pest control agents, as well as for the birds' own sake.

Actions:
- The U.S. should not allow the import of farm produce that is sprayed with pesticides that have been cancelled in the U.S. due to their negative effects on birds, other wildlife, and people.
- Reduce or eliminate pesticide uses in and around the home (including the lawn)—planting native trees and bushes instead provides backyard bird habitat and requires little or no pesticide maintenance; drain standing water containers on your property where mosquitoes may breed; if you must use pesticides, always follow label instruction carefully; only dispose of pesticides or their containers at hazardous waste collection sites, never down household or storm drains.
- Avoid the use of rat poisons by sealing access points for rodents, and cleaning up food sources.
- Where possible, purchase foods grown organically, particularly corn, alfalfa, wheat, and potatoes, which are often grown using large amounts of pesticides that can be harmful to birds; buy clothing made with organically-grown cotton. Supermarkets respond to consumer demand, so if your local store does not stock organic food items, petition them to do so, and engage your neighbors in the effort.
- Avoid buying non-organic fruits and vegetables from Latin America until pesticide use in these countries is changed in favor of protecting birds and the environment. More pesticides are used per acre to grow bananas than just about any other crop, and residues of more than 50 pesticides are allowed on imported bananas. Organic bananas and organic, shade-grown coffee are widely available. These are bird-friendly alternatives that reduce pesticide exposure for tropical and migrant birds (see International chapter).

ACID RAIN

Most of our energy is still generated by burning fossil fuels, resulting in significant emissions of pollutants, especially from coal. These include greenhouse gases (see Climate Change section p. 342), mercury, arsenic, and the oxides of sulfur and nitrogen (c. 20 million tons of sulfur dioxide are discharged each year). These oxides combine with water in clouds, falling as acid rain, snow, or fog. Nitrogen oxides also react with other compounds in the air to produce smog, and can contribute to eutrophication.

Problems: Acidic precipitation damages vegetation and alters the chemistry of soil and wetlands, negatively impacting birds in several ways. Damage to trees degrades forest habitat. Acid rain caused more than a 50% die-back of red spruce during the 1970s and 1980s in the Adirondack and Green Mountains of the northeastern U.S.; important habitat for the Bicknell's Thrush. Acid rain also leaches nutrients (calcium, magnesium, potassium) from soils and increases the availability of toxic aluminum. Calcium deficiency is detrimental to birds and can increase absorption of other toxic metals, such as lead and cadmium. Songbirds require calcium to lay eggs and grow, and food sources such as snails and other soil invertebrates become rare in acidified soils. Increased acidity of rain has been shown to reduce the probability of Wood Thrushes breeding.

The loss of insects (including dragonflies, damselflies, caddis flies, and mayflies), fish, and other aquatic organisms from acidified waterways also reduces food supplies for birds. For instance, Common Loons are less likely to nest on acidic lakes in Canada, and those that do also show lower reproductive rates, likely due to reduced fish and other food supplies. Louisiana Waterthrushes in Pennsylvania suffer reduced food supplies on acidified streams, resulting in delayed egg laying, smaller clutch sizes, and the need for twice the foraging range compared to birds on healthy streams.

PHOTO: DAVID MATHEWS

Acid rain kills mountain top forests in the Appalachians and New England, reducing important breeding habitat for birds.

PHOTO: ©ISTOCKPHOTO.COM

Acid rain kills fish in many northern lakes, reducing breeding success among Common Loons.

Solutions: To reduce local air pollution problems, power plants initially built taller smokestacks, but these chimneys just transported gaseous emissions higher and further away from their source. Power plants can reduce acidifying emissions by installing scrubbers to chemically remove sulfur dioxide from gases leaving smokestacks, and by using low-sulfur coal or natural gas for fuel. Catalytic converters, required by federal law on cars and trucks, remove nitrogen oxides from exhaust. The Clean Air Act first began regulating emissions that cause acid rain in 1970, but mandated further reductions in 1990 that included a cap-and-trade program to reduce sulfur dioxide (but not nitrogen oxide) emissions to 50% of 1980 levels by 2010. Acid rain decreased by approximately 30% between 1989 and 2006 as a result. Thanks to more stringent regulations, Canada's acid rain program reduced sulfur dioxide emissions there by 58% by 1999 compared to 1980 levels, showing that even deeper cuts are possible. Loopholes in the Clean Air Act have allowed older, dirtier power plants to continue operating without installing scrubbers and other pollution control technologies. Strengthening emissions caps and accelerating the phase out of the oldest and dirtiest coal-fired power plants would further reduce acid rain problems.

Experimental liming of soils in Pennsylvania increased soil calcium, reduced soil acidity, and increased snail and Ovenbird abundance. This technique may help mitigate acid rain problems in forests until emissions are reduced. Liming is used in Europe to counter the effects of acid rain, but is expensive and must be done repeatedly as long as acid rain continues. Ultimately, using energy efficiently and producing energy with alternative technologies (wind, solar etc.) that do not produce air pollution will reduce reliance on burning coal.

Actions:

- Write to your representative to request that they push for even more stringent regulations on sulfur dioxide emissions.
- Most consumers now have a choice concerning what type of electricity generation is used to power their home. See the Department of Energy website, or contact your utility company about purchasing "Green Power" from sources that do not emit sulfur dioxide (and are also climate-friendly).

LEAD

Lead is an inexpensive and malleable metal used in ammunition, fishing weights, and historically in paint, water pipes, and as a gasoline additive. Like people, birds suffer health problems or death from ingesting lead. Over the past three decades the Clean Water Act and Clean Air Act have helped reduce lead exposure for people. We still pump thousands of tons of it into the atmosphere each year, however.

Problems: Birds will accidentally ingest lead shotgun pellets, ammunition fragments, or fishing sinkers while foraging on lake bottoms, scavenging on carcasses left behind by game hunters, or consuming grit to aid digestion. Once digested and absorbed into the blood, as little as a single lead pellet can be lethal to a bird, but sub-lethal amounts also compromise health and cause neuro-logical dysfunction. Birds are particularly susceptible to lead poisoning, because of their small size and fast metabolic rates. Lead poisoning killed two to three percent of North America's waterfowl between 1938 and 1954, and is still the leading cause of death for adult loons in the Northeast. It also continues to hamper California Condor recovery, and kills other scavenging birds including crows, ravens, eagles, and hawks. Lead contamination in game meat poses a health risk to people as well. As many Mourning Doves may die annually in the Midwest from consuming lead shot as are killed by hunters. Finally, Laysan and Black-footed Albatross chicks suffer lead poisoning from ingesting contaminated paint chips at their main breeding colony on Midway Atoll, where abandoned military buildings remain in deteriorating condition. The effects of atmospheric lead on birds are harder to ascertain and need further research.

Solutions: The U.S. began restricting the use of lead in the 1970s, banning lead paint on toys and furniture in 1977 to safeguard human health, and banning lead in gasoline in 1996 under the Clean Air Act. Lead water pipes were banned under the Clean Water Act of 1984. Lead shot was banned for use in waterfowl hunting in 1991, though it is still permitted for use on upland game birds. Canada banned lead shot completely in 1999. In 2007, California announced a ban on lead ammunition in the state within the range of the California Condor. Arizona currently gives away non-toxic copper ammunition to hunters, and although voluntary participation in this program is high, it is not enough to prevent all lead poisoning of condors in the state. Meanwhile, many poisoned condors have been captured, treated to reduce blood lead levels, and re-released into the wild.

Canada and some New England states have also restricted the sale and use of lead sinkers and jigs in fishing gear below certain weights (as has Great Britain). The use of lead ammunition and sinkers was scheduled to be phased out by 2010 on all Na-

California Condors are poisoned by lead bullet fragments ingested while scavenging animal carcasses.

PHOTO: FWS

tional Park Service lands, but the Park Service has backed off on ammunition due to complaints from pressure groups. It is estimated that banning lead shot for waterfowl hunting in the U.S. may have prevented the deaths of more than one-million ducks per year, since ducks often ingest spent shot that falls into wetlands. Tens of thousands of shot pellets are found per acre in heavily hunted areas, and bird fatalities can result from the ingestion of just one or two pellets. This ban has also saved many Bald Eagles, which were being poisoned by eating crippled ducks. Lead poisoning among raptors and other land-birds will not end until all lead ammunition is phased out and replaced by non-toxic alternatives. ABC is working with government, manufacturers, retailers, hunters, and private land-owners to reduce lead contamination from ammunition, promote non-toxic alternatives, and clean up lead contamination on Midway.

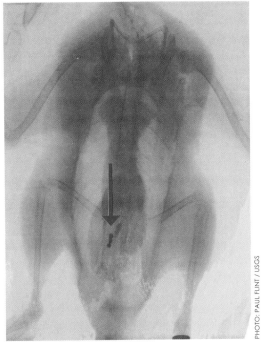

PHOTO: PAUL FLINT / USGS

This X-ray shows three lead pellets in the digestive tract of a live Spectacled Eider ingested while the bird was foraging.

Actions:

• Sport hunters and fisherman can use lead-free ammunition and tackle; non-toxic shot made of steel, tungsten, and other alloys is available to hunters as an alternative to lead shot. Winchester recently began manufacturing lead-free .22 bullets, and many lead-free rifle bullets are also available for large game hunting. As the variety and quality of non-toxic ammunition and fishing tackle improves, excuses for continuing to contaminate the environment with lead become less and less justifiable.

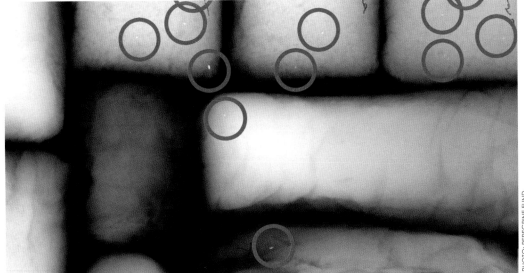

PHOTO: PEREGRINE FUND

Lead fragments are circled in this X-ray of packed venison; some would likely be too small for a person to notice when eating.

OIL

The impact of oil is probably one of the best-known threats to birds due to media coverage of high-profile spills such as the *Exxon Valdez*. In addition to these large, well-publicized events, there are hundreds of small spills each year which affect smaller numbers of birds, but which can total more than one million gallons annually. Fortunately, oil spills have been reduced over past decades (c. 70% reduction in industry-related spills since the 1970s) and regulations requiring double-hulled tankers should reduce this even further after 2010. A potential major threat from oil well leaks is exemplified by the recent Deepwater Horizon spill in the Gulf of Mexico, however. A potential major threat from oil well leaks is exemplified by the recent Deepwater Horizon spill in the Gulf of Mexico, however.

Problems: Oil and other petrochemicals are toxic to birds. Because oil floats on the surface of water,

This dead, oiled King Eider washed up on a beach in the Pribilofs, but most birds killed in oil spills are never found.

PHOTO: FWS

seabirds are particularly vulnerable to this form of pollution. Birds with severely oiled plumage lose their ability to keep warm, often dying of hypothermia before they are poisoned. Quantifying numbers of birds killed by oil pollution is difficult as most deaths are never detected. Large oil spills can occur as a result of oil tanker accidents, and can kill tens of thousands of birds in just a few days. The risk of bird mortality from oil spills is greatest where large concentrations of breeding seabirds or staging shorebirds occur near major oil transportation ports. These include Delaware Bay, the Texas Coast, San Francisco Bay, and the Gulf of Alaska, among others.

The worst oil spill in U.S. history occurred in 1989, when the *Exxon Valdez* leaked 10.8 million gallons of crude oil into Prince William Sound, Alaska. This spill oiled 1,300 miles of shoreline and killed 250,000 birds (along with many other marine animals). Three years after the spill, Black Oystercatchers on oiled shores continued to lay smaller eggs, suffered greater chick mortality, and had stunted growth. Wintering Harlequin Ducks suffered higher mortality rates and population declines in previously oiled areas compared to uncontaminated areas nine years after the spill. Oil still remains in the environment, and Barrow's Goldeneyes and Pigeon Guillemots continued to be contaminated by oil into the late 1990s. Kittlitz's Murrelets declined 84% in Prince William Sound between 1989 and 2000 (though recent surveys suggest a possible rebound). Sub-lethal effects of residual oil from this spill continue to harm wildlife 20 years later, and many seabird populations (e.g., Pigeon Guillemot, Marbled Murrelet) have still not recovered to pre-spill levels. In 2004, the *Athos I* spilled 265,000 gallons of heavy crude into the Delaware River, oiling 115 miles of shoreline. The timing (November) and location (well upriver of key shorebird sites) of the spill helped to avert what could have been a major disaster for horseshoe crabs and shorebirds. More recently, in 2007, the *Cosco Busan* cargo tanker spilled 58,000 gallons of fuel into San Francisco Bay. After this spill, at least 1,514 birds were found dead and 1,052 were rescued to be decontaminated. Surf Scoters were hit particularly hard, as 78% of their Pacific Flyway population winters in the Bay.

Oil also chronically leaks into the environment from pipelines, storage tanks and pits, rigs, automobiles, boats, and landfills (and through natural seepage). Chronic and illegal oil discharges at sea may kill more than 100,000 birds annually in the U.S., and more than 300,000 Thick-billed and Common Murres, and Dovekies annually in Newfoundland, Canada. Terrestrial oil drilling and tar sand operations routinely store oily waste-water in open pits, which can kill birds attracted to the open water. Oil pits are estimated to have killed up to two-million birds annually in the U.S. during the 1970s, and still kill between 500,000 and one-million per year today.

Solutions: In response to the *Exxon Valdez* disaster, Congress passed the Oil Pollution Act in 1990, which greatly improved oil spill cleanup response and planning, and called for a gradual phase-out of single-hulled tankers, and replacement with double-hulled vessels that are much less likely to leak oil if damaged. Meanwhile, U.S. Coast Guard regulations require all tankers in U.S. waters to be double-hulled by 2011. Europe has also banned single-hulled tankers from using its ports, and the United Nations called for a phase out of single-hulled tankers by 2010. Oil pollution has been further reduced through stricter controls on bilge washing for ships. The U.S. still needs to require open oil pits to be covered or waste to be stored in closed containment systems. However, Exxon-Mobil was recently convicted under the Migratory Bird Treaty Act of failing to protect birds from dying in open oil tanks at production fields in five states between 2004 and 2009. The company will have to pay $600,000 in fines and spend $2.5 million modifying the facilities to prevent future injury to birds. Oil extraction should be further restricted from within Important Bird Areas. For example, certain species, such as the Yellow-billed Loon that nest in the National Petroleum Reserve, Alaska,

PHOTO: ROBERT ROYSE

Over 400,000 seabirds, such as this Common Murre, are killed annually by oil discharge at sea in North America.

are particularly susceptible to disturbance on their nesting lakes. Emergency response plans, including oil recovery and wildlife rescue operations, are also helpful to quickly contain and minimize damage from oil spills when they inevitably occur.

Actions:
- Converting the global oil tanker fleet to double-hulled vessels should be accelerated.
- Ensure that your car, motorbike, or boat does not leak any oil into drainage systems; properly dispose of oil.
- In the event of an oil spill near you, volunteer with a local rescue center to care for birds and other wildlife.
- Reduce your gasoline (and therefore crude oil) consumption and save money by improving your automobile's mileage efficiency (regular services, changing and properly inflating tires) and by using public transportation or other transportation alternatives (biking, walking, carpooling).

MARINE DEBRIS

It can take more than one-hundred years for a floating plastic cigarette lighter to decay in the ocean. In parts of the North Pacific, there is six times as much floating plastic by weight as there is naturally occurring plankton. Three-hundred-million tons of new plastic products are manufactured each year, yet only 4% of all plastic is recycled. Floating plastics pose a major threat to albatrosses and other seabirds.

Problems: Floating plastic and discarded or lost fishing gear kill birds and other marine life. Marine debris originates from many sources on land and at sea: beach-goers, sewage overflows, runoff from landfills, illegal dumping, and accidental industrial spillage are all land-based sources for marine trash, including plastics, Styrofoam, metal, glass, rubber, wood, and other items. This debris includes plastic food and beverage containers, six-pack rings, cigarette lighters, beach toys, plastic bags, toothbrushes, and small pellets used in plastics manufacturing. At sea, ships and oil platforms may illegally dump trash overboard or lose fishing gear.

The amount of marine trash is both staggering and increasing, with 6.4 million tons of trash entering the oceans annually. Approximately 70% of marine debris is comprised of plastics, which are durable and can persist for hundreds of years, slowly breaking down into smaller pieces. Micro-particles of plastic are now ubiquitous in marine sediments and beach sands throughout the world, and can enter the food chain when consumed by filter feeders such as barnacles, as well as worms, amphipods, or fish. Floating plastics can travel great distances and accumulate in ocean currents, on the sea floor, and along shorelines. In the North Pacific, currents of the Great Pacific Gyre create a swirling mass of debris called the Great Pacific Garbage Patch, where net sweeps can collect hundreds of thousands of plastic items per square mile. Hormone disrupting chemicals released from degrading plastics at sea have been detected in marine mammals, commercial fish, and seabirds in surprisingly high concentrations.

Nearly half the world's seabird species are known to have ingested plastic debris, and many regurgitate these items along with food to feed to their chicks. For instance, 97% of dead Laysan Albatross chicks found on nesting islands contain plastics. Ingested plastics can block seabird digestive tracts, impair regurgitation, fill stomachs, impede digestion, pierce internal organs, leak toxic chemicals, and ultimately kill seabirds. Ingested plastics may also contain other toxins such as lighter fluid, as well as DDE, nonylphenols (a by-product of detergent manufacturing), fire retardants, and other toxins that bind to and accumulate on floating plastics. Ingested plastics are the likely source of many toxins found in seabirds, and PCB concentrations have been correlated with amounts of ingested plastics in Greater Shearwaters. Marine debris can also form rafts, helping to disperse exotic organisms across oceans.

Solutions: Reducing, reusing, and recycling plastics and other items will decrease the amount of trash entering the oceans and harming birds. Enforcement of waste disposal policies and regulations can help ensure that trash is properly disposed of. In 1988, the International Convention for the

Adult Black-footed Albatrosses often mistake floating trash for food, potentially killing themselves or chicks back at nests.

PHOTO: GLEN TEPKE

PHOTO: © CHRIS JORDAN, COURTESY OF KOPEIKIN GALLERY

This Laysan Albatross carcass has a stomach full of plastic debris; unfortunately this is common sight on Midway and Laysan islands.

Prevention of Pollution from Ships imposed a complete ban on the dumping of all forms of plastic at sea, as well as restrictions on dumping damaged fishing nets. Unfortunately, the convention's rules are widely ignored, and millions of tons of plastic are annually dumped overboard from ships worldwide. Noncompliance stems from a lack of enforcement, including inspections and penalties. More waste bins and educational outreach are also required to reduce plastic litter on beaches and in other coastal areas.

Removing ocean trash is another solution. Numerous volunteer-based beach cleanups occur worldwide, including efforts organized by the United Nations and groups such as the Ocean Conservancy. Government agencies can also take a more active role in cleaning up ocean trash. For instance, The Marine Debris Research, Prevention, and

Reduction Act of 2006 provided $10 million to the National Oceanic and Atmospheric Administration, and $2 million to the U.S. Coast Guard over four years to identify trash accumulations in ocean currents and to send ships to clean them up. Between 1996 and 2008, over 579 metric tons of derelict fishing gear was removed from the Northwestern Hawaiian Islands alone.

Actions:
- Significant additional funding is needed for cleanup operations; write to your elected officials requesting that they support this effort.
- Do not litter; recycle your plastics (including bags) instead of throwing them away.
- Use reusable cloth bags at the grocery store and reusable beverage containers to reduce your consumption of plastics.
- Participate in local beach and river clean-ups.

CLIMATE CHANGE

The overwhelming body of independent scientific evidence shows that the recent warming of the global climate is caused by greenhouse gas emissions. The effects of climate change on birds could be wide-ranging, challenging, and in many cases, unique when compared to other pollution issues. Climate change threatens to exacerbate threats from habitat loss and invasive species, as well as creating new challenges that birds must overcome. This includes changing habitat distributions and a shift in the timing of peak food supplies such that traditional migration patterns may no longer put birds where they need to be at the right time.

Problems: Our planet's climate systems are driven primarily by energy from the sun. Sunlight enters our atmosphere where it can either be absorbed, or reflected back into space. Greenhouse gases, including carbon dioxide (CO_2), nitrous oxide, methane, and ozone, trap solar energy in the atmosphere, much like glass traps heat in greenhouses.

PHOTO: ROBERT ROYSE

If warming temperatures cause habitats to shift upslope, Black Rosy-Finches that inhabit alpine areas may have nowhere to go.

Earth's atmosphere is vulnerably thin, however, and greenhouse gases emitted from the burning of oil, coal, gas, and other fossil fuels have raised the atmospheric concentrations of carbon dioxide by 36% over pre-industrial levels. Significant amounts of greenhouse gases are also released by agriculture, cattle, and the burning or clearing of forests, because much of the carbon stored within the wood of trees and the soil escapes into the air. Each year, the U.S. emits more than 1.5 billion tons of CO_2 in total.

Average global temperatures increased 1°F during the 20th Century, and are projected to rise by another 2.0-11.5°F by 2100. The most severe changes in climate are expected to occur near the poles and at high elevations. An average global increase of 5°F means equatorial areas would warm 1-2°F, but polar areas could heat up by 12°F.

Already glaciers on mountaintops, ice sheets in Greenland and Antarctica, sea ice at both poles, and tundra permafrost are melting. Water expands when heated. Although sea levels have been rising since the last ice-age, the pace of this rise is accelerating well beyond background rates. Sea levels have risen about eight inches since 1870, and are projected to rise 11-13 further inches by 2100 if current trends continue. Scientists also worry that ocean currents such as the Gulf Stream could be slowed or disrupted by reductions in salinity caused by an influx of fresh water from melting glaciers and ice caps. Disrupting these currents could have profound impacts on marine habitats, and on the climate of adjacent continents. Oceans also absorb much of the atmospheric CO_2 we emit, making sea water more acidic. Many of the corals, mollusks, plankton, and other small creatures that comprise the base of marine food chains can't build shells or develop properly in water that is too acidic.

Climate change threatens to impact bird habitats over immense areas of land and sea. Its effects will

PHOTO: PETER ZWITSER

Ivory Gulls scavenge polar bear kills on the pack ice, but their icy world is rapidly melting ("Kumlien's" Iceland Gull at front).

be different for each bird species. Some may benefit, but many will suffer. Melting sea ice, glaciers, and permafrost alter habitats depended on by WatchList birds. For example, melting arctic pack ice is reducing foraging opportunities for the Ivory Gull, as well as for polar bears whose kills are scavenged by the gulls. Sea level rise is predicted to inundate and fragment coastal habitats, such as saltmarshes that are important for WatchList species including the Black Rail, and Saltmarsh and Seaside Sparrows. Changes in sea temperatures or shifts in ocean currents are blamed for reducing food supplies near seabird colonies causing widespread reproductive failure. As climate change progresses, some habitats are predicted to shift pole-ward or up mountain slopes. Unfortunately, tundra and alpine habitats located in the far north or at the tops of high mountains have nowhere to shift to. Habitats might also not shift fast enough in the face of rapid climate change. For instance, sea level rise may occur too fast for saltmarshes to shift inland, and roads and coastal developments now surrounding many saltmarshes present ba-

rriers to habitat movement. Range shifts could also cause problems for migrants, requiring them to travel further to reach breeding areas if wintering areas do not shift as well.

Some migratory birds are already arriving at their breeding grounds and laying eggs earlier to adjust to advancing spring times. Short-distance migrants appear to adjust to such changes more rapidly than long-distance migrants. Migrant birds that do not adjust may arrive at their nesting grounds to find food in short supply. For example, egg laying dates for Thick-billed Murres in Canada are not advancing as fast as sea-ice melt, putting breeding time out of step with peak food availability and reducing chick growth rates. Milder winters are also allowing the spread of forest pests such as bark beetles that are responsible for killing spruce trees over millions of acres in Canada and Alaska this decade. Many of Hawaii's native songbirds find their last refuges in high-elevation forests where temperatures are cooler than the average minimum threshold for malaria

Warming will accelerate malaria transmission, though the Kauai Amakihi appears to be develping some natural resistance.

PHOTO: PETER LATOURETTE

transmission (55 °F). If temperatures increase by 3.6°F in this region, then many of these forests will become warm enough for malaria to cause infections at the highest altitudes, which could threaten the extinction of species that cannot develop resistance.

Solutions: To reduce future climate instability, we need to reduce our greenhouse gas emissions now. The U.S. represents only 5% of the world's population, but produces 25% of global greenhouse gas emissions. Therefore, reducing emissions is disproportionately incumbent upon U.S. citizens.

Alternative sources of energy can reduce greenhouse gas emissions. Renewable and low-emission energy sources such as solar, wind, and geothermal power offer the best potential solutions, though placement and design of solar and wind farms need to minimize the risk of bird collisions and the loss of sensitive habitats. The perceived lack of a safe means for radioactive waste disposal, combined with risks of accidents, high costs, and the potential for nuclear weapons prolifera-

tion continue to prevent universal acceptance of nuclear energy. Biofuels produced from corn are renewable and can be produced in the U.S., but under current mandates may actually emit more greenhouse gases than fossil fuels, after taking into account the energy required for planting and harvesting equipment, and the carbon lost from lands cleared to plant biofuel crops. However, "next-generation" biofuels, based on sources including bacteria, algae, and agricultural waste, are under development and have the potential to be more efficient and offer partial solutions in the future. The planting and harvesting of native prairie grasses, if harvest timing does not destroy nests, could be a win-win for the climate and birds.

Cap-and-trade systems offer economic solutions to curbing emissions. In a cap-and-trade system, total emissions are limited (the cap), below which utilities such as coal power plants can buy or sell emission credits (the trade) as they retool their production facilities. Such a system has already helped to reduce acid rain pollution in the U.S. Carbon credits could also be used to finance land conservation to reduce emissions caused by deforestation. Tropical deforestation contributes

PHOTO: MIKE PARR

Rising sea-levels threaten to inundate coastal saltmarshes used by Seaside Sparrows.

Avoiding tropical deforestation can reduce greenhouse emissions, and can be paid for with carbon credits (Carpish, Peru, pictured).

PHOTO: MILKE PARR

20-25% of global CO_2 emissions. Wealthy countries or utilities that emit large quantities of CO_2 by burning fossil fuels could offset a portion of these emissions by offering other countries or projects financial incentives to protect their forests (called avoided deforestation). Funds generated by carbon credits can also be used to carry out projects to help species adapt to climate change and to ensure that endangered species do not go extinct.

Successfully conserving birds in the face of climate change makes addressing today's threats even more important. Ensuring that bird populations have sufficient habitat to breed, refuel during migration, and survive the non-breeding season requires us to protect and manage habitat now to increase the chances that more will be left intact in the future. Protecting or restoring areas into which endangered species might shift could also be an important strategy. For instance, beginning forest restoration now on cattle pastures at the highest elevations in Hawaii could succeed in creating additional habitat into which Hawaiian birds could later move if climate change makes lower-elevation forests uninhabitable for them due to the spread of malaria.

Actions:
- Write to your elected representatives to ask for rapid and significant action on climate change in Congress.
- Calculate your carbon footprint and learn where you can reduce emissions.
- Reduce energy use at home by using energy-efficient appliances (with the Energy Star label), lighting (such as compact fluorescent bulbs), by adjusting thermostat temperatures (e.g., at night or while you are out at work), insulating ceilings and attics, and sealing drafty windows and doors.
- Make your next vehicle a fuel-efficient one; walk, or ride a bike when possible; choose public transportation or car-pooling over driving.

CHAPTER 4

INTERNATIONAL BIRD CONSERVATION

Many of the threats to U.S. birds are mirrored—and often amplified—in other countries around the world. Birds do not recognize political borders, and their ranges and migratory routes often cross international boundaries. Many of the most endangered bird species in the Americas occur in tropical countries far to our south. Canada is also the nursery to a major proportion of North America's songbird population. To address the full spectrum of threats to U.S. birds, therefore, we cannot limit conservation efforts to the U.S. alone.

Most bird species native to the U.S. are also shared with our neighboring countries of Mexico, Canada, Russia, and the Caribbean nations. Many of the migratory seabirds, waterfowl, raptors, shorebirds, and songbirds that we enjoy during breeding and migratory periods depend on staging and wintering areas far beyond our borders. Only 45 U.S. bird species (representing roughly 5% of the total) occur in no other country. Of these, 28 are restricted to Hawaii. The remaining 95% of U.S. bird species are shared internationally, conferring on us a responsibility to work cooperatively with international partners.

As much as the large-scale logging and agricultural expansion of past centuries has slowed in the U.S., it continues to accelerate to our south. The type of dramatic changes seen by our great grandparents across the North American landscape can be witnessed today by anyone who takes a daytime flight across Latin America. South American forest is primarily being cleared for agriculture and cattle pasture, and the region accounted for more than one third of annual global forest loss from 2000 to 2005. While some neotropical migrants such as the Cerulean Warbler have specialized winter habitat requirements, many can at least adapt to fragmented habitats outside the breeding season, so long as there is still some cover and sufficient food available. Unfortunately most endemic birds cannot, and urgent action is needed to avert a bird extinction crisis throughout the region. Of the approximately 4,400 bird species known from the Americas, 498, or a little over 11% are considered to be threatened with extinction according to IUCN criteria. Of these, 86 are Critically Endangered, and 155 are Endangered. Thirty-two of these occur in the U.S. with most found in Hawaii (some of which are likely already extinct). This leaves 209 species of serious concern in the rest of the Americas. These species need immediate conservation assistance to ensure their survival.

BIRDS

If we leave the U.S. and travel south towards the equator, we find a tremendous increase in the diversity of breeding bird species. Global bird diversity peaks in the lowland neotropical forests of the western Amazon, near the base of the Andes. For

PHOTO: FUNDAÇÃO BIODIVERSITAS

The Lear's Macaw population has increased more than tenfold to 900 birds since its nesting grounds were discovered and protected.

example, Manu NP in Peru boasts a bird list topping 1,000 species. Excluding seabirds, more than 3,750 bird species (37% of the total global bird diversity) breed in the Neotropics, making this the richest realm for birdlife on Earth.

How so many species can coexist intrigues scientists. Bird diversity peaks near the equator, where rainfall and energy from sunlight also peak, driving the engines of life to maximum productivity. Although lowland tropical forests have undergone cycles of expansion and contraction during past climate shifts, they have a relatively stable history, persisting for millions of years and allowing species to evolve and accumulate. By comparison, boreal forests have receded and advanced hundreds of miles during alternating glacial cycles. Compared to temperate regions, birds in lowland tropical forests can also rely on continuous supplies of fruit and insects to eat, and never face the stress of harsh winters. Not only are various food supplies available year-round, but a broader diversity of types

and sizes of fruits, insects, and other food resources are found. In response to this smorgasbord of opportunity, many birds have evolved to specialize in exploiting particular habitat amd vegetation types (e.g., hummingbird bills adapted to different flower shapes), foraging substrates, elevations, or forest strata (e.g., canopy, understory, ground). Although the frequency and degree of specialization among neotropical birds is fascinating to the scientist, more specialized species are also often less numerous, may occur at lower densities, tend to be more restricted in their ranges, and can be more vulnerable to changes in the environment. For all these reasons, specialist species are often intrinsically more vulnerable to population declines and ultimately to extinction due to human activities.

The rich avifauna of the Neotropics presents a tremendous opportunity for researchers as little is known about most species. Every year sees new species described, as well as first nest descriptions, and range extensions. Furthermore, many tropical birds have very different behavior regarding breeding, migration, and communication compared to temperate species. Studies of temperate birds outnumber those of tropical species by more than one-hundred to one. North America has been blessed with a plethora of popular field guides to bird identification during the last 50 years, but some neotropical countries still lack a single field guide dedicated to their birds. There are also many more ornithologists in the U.S. and Canada than in Latin America, although Latin Americans are rapidly catching up in training and numbers.

We have not included accounts for threatened species found exclusively in Latin America and the Caribbean due to space considerations (though see the BirdLife Data Zone at www.birdlife.org).

Resident Species

Parrots, hummingbirds, trogons, vireos, and fly-catchers are familiar to North American birders, but these bird families are represented by relatively few species in the U.S. compared to the high diversity in the Neotropics. The Neotropics also harbor birds that are unlike any species found in North America, belonging to completely distinct bird families. These include jacamars, barbets, toucans, woodcreepers, antbirds, tapaculos, cotingas, manakins, and other birds that entice birders from all over the world to visit and enjoy.

Migratory Species

Birds move for many reasons, and during different times in their lives. The most dramatic movements, however, are migrations, during which birds move between breeding and non-breeding areas, sometimes over great distances. During migration, some species may gather in large flocks, congregate at staging areas, or use habitats that differ from those on their breeding or wintering grounds.

Direction of migration also varies between species. In North America, most migrants travel from northern breeding areas to southern non-breeding areas, where they spend the winter. Many of these

migrants winter within North America and are called Nearctic-Nearctic migrants (the Palm and Yellow-rumped Warblers are good examples). There are 420 species of migrants that breed in the U.S. and Canada and migrate to wintering areas in Latin America and the Caribbean, and are thus called Nearctic-Neotropical migrants, though we commonly call them simply neotropical migrants. Many such birds follow elliptical routes, flying south along an easterly or coastal path and returning along a more westerly path through the center of North America.

Migration is an evolutionary strategy for many species that allows them to live in favorable climates year-round or exploit seasonal peaks in food or other resources, often allowing greater rates of reproduction. These benefits often come at high costs, however, as birds must change their physiology for migration, must evolve the morphology

Altitudinal migrants, such as the Resplendent Quetzal, must be protected at both breeding and non-breeding sites.

PHOTO: MIKE PARR

Shorebirds such as the Dunlin congregate in large flocks during migration and are dependent on a relatively small number of key sites.

to engage in migration (longer wings, powerful muscles), and may suffer increased mortality due to hazards along the way. Natural hazards during migration include oceans, mountains, and other natural barriers, severe weather events, and native predators. Habitat loss and hunting, and collisions with buildings, towers, electrical lines, and other human structures present additional risks to birds during migration.

Most migrants appear to use a wider range of habitat types than most resident species, and therefore as a group they tend to be less vulnerable to extinction than non-migratory birds. However, migration does elevate extinction risk for some species. Many shorebirds depend on a relatively small number of discrete and irreplaceable staging areas that, if destroyed by development, agricultural expansion, or disasters such as oil spills, could have devastating effects on those bird populations. Other migrants depend on multiple habitat types during their life cycles, and may be at risk if any one habitat type is severely reduced (e.g., the Bicknell's Thrush uses

northern firs for breeding and Caribbean broadleaf forests in winter). Habitat loss, degradation, and fragmentation can make migration more difficult for birds. As a result, migrants may be forced to travel further between staging areas than they once did when habitat was more widespread, and may suffer reduced survival due to the extra distance or time spent wintering or staging in degraded areas.

Migrants may travel through areas where differences in laws pose additional risks. For instance, almost all shorebirds are protected from hunting in the U.S., but are targets for unregulated killing when they pass through Barbados. Pesticides banned in the U.S. may still kill birds in South America. For example, monocrotophos has not been sold or used in the U.S. for many years, but is still used in Bolivia where it poisons Bobolinks.

Finally, some migrant species are not adjusting their migration patterns fast enough to keep pace with rapid climate change, leading to asynchrony between arrival times, reproduction, and peak food availability. Also, more severe storms are predicted to result from climate change, potentially making oceanic crossings by migrants more dangerous.

Migrants often cross country boundaries, and therefore conserving migrant species requires international collaboration among conservationists, scientists, policy-makers, and governments. Fortunately, migrant birds can unite such stakeholders and facilitate cooperative conservation efforts, improving international relations along the way. For instance, the Neotropical Migratory Bird Conservation Act authorizes the U.S. government to fund habitat protection and conservation outside the U.S. for migratory birds. Such cooperative efforts often benefit resident species as well as migrants. In another example, La Amistad International Park spans the border between Costa Rica and Panama, which jointly manage this protected area, thereby safeguarding critically important forests for species such species as the Three-wattled Bellbird, Bare-necked Umbrellabird, Resplendent Quetzal, and wintering migrants from North America.

HABITATS

While most of the major bird habitats found in the U.S. are shared with Canada and northern Mexico, Latin America and the Caribbean host a dazzling array of unique ecosystems and bird species. Unfortunately, many of these habitats face significant threats.

In the Caribbean many of the island specialists are surviving thanks to aggressive management programs. These birds include some of the Lesser Antillean parrots, such as the St. Lucia Amazon, which depend on the island's interior forests. Those species that live close to the coasts, such as the White-breasted Thrasher and Grenada Dove, face a long struggle against the tourism and residential developments that are gradually eroding their habitat, however.

Though forest destruction in Mexico and Central America has taken place on a massive scale,

the region's endangered endemic species (e.g., the Townsend's Shearwater, Socorro Mockingbird, and Cozumel Thrasher) have suffered more on marine islands than they have on the mainland. The Imperial Woodpecker and Thick-billed Parrot from the mountains of western Mexico are two mainland species that have also been severely impacted, however. Unfortunately, commercial logging rotations there have prevented trees for maturing sufficiently to provide nesting hollows for them. The parrot has survived (just) but the woodpecker is apparently extinct as a result.

Further south, the Andes have been heavily populated for more than 500 years, though their topography is such that some areas are still relatively inaccessible (conferring a degree of natural protection). Prior to colonization, the Incas operated terraced farming systems that occupied relatively little land, but habitat fragmentation accelerated from the 1500s when European farming systems were imposed. This ultimately led to a pending extinction crisis among endemic birds that evolved in habitats ranging from the highlands, to dense cloud forests, and the Amazonian lowlands. The Amazon itself harbors hundreds of bird species, and fortunately

Semipalmated Sandpipers nest on the tundra, but require mudflats for feeding during migration and winter.

PHOTO: MIKE PARR

The Marvelous Spatuletail is perhaps the most spectacular of the hummingbirds. It is restricted to a tiny area in northern Peru.

still has some large wilderness areas capable of supporting its characteristic bird communities.

In northern South America, the seasonally flooded savannas of Venezuela, Brazil, and surrounding countries provide some of the most spectacular birding in the Americas. These systems, known as the Llanos and Pantanal respectively, are relatively resilient to land conversion because of their hydrology, and are still comparatively intact (though the hydrology of the Pantanal in particular is coming under increasing threat). Some characteristic birds include the stunning Hyacinth Macaw, Scarlet Ibis, Jabiru, and Roseate Spoonbill.

Lying across the Brazilian coastal belt, the Atlantic forest is among the world's most threatened ecosystems. These forests have extremely high bird endemism, and it is essential that large numbers of the remaining forest fragments are protected. Numerous species of parrots, antbirds, and tanagers now occur in just one or a handful of the tiny forest patches that are scattered up and down the coast alongside some of the world's most densely populated areas.

Paraguay and Bolivia host a variety of landscapes in common with their surrounding countries, including the cerrado (a dry savanna of thorn scrub and arid grassland), seasonally flooded grasslands with palm groves, and parts of the southern Pantanal ecosystem. Birds such as the Greater Rhea and Red-legged Seriema are characteristic of the drier grassland habitats. The southern Andes also reach into Bolivia from Peru to the west; and the Atlantic forest reaches Paraguay from Brazil in the east. The Atlantic forest of Paraguay is also highly threatened with most of its remaining extent occurring within San Rafael NP.

The Argentine pampas and wet grasslands in the north of the country are being cleared for agriculture, grazing, and for timber plantations, and this is where some of the country's most threatened birds occur. Several migratory species that spend the austral winter in northern Paraguay and eastern Brazil, such as the Marsh Seedeater are considered globally Endangered as a result of these land use changes. In southern Chile and Argentina there is a temperate beech forest that hosts such spectacular birds as the Magellanic Woodpecker and Black-throated Huet-Huet. Some lakes in this region also provide habitat for the fabulous Hooded Grebe.

In the south of Argentina and Chile, key wintering sites for vulnerable shorebirds such as the Red Knot and Hudsonian Godwit are found in Tierra del Fuego. Large breeding colonies of pelagic seabirds, some of which are threatened by longline bycatch, nest on islands around the coasts.

We have grouped the bird habitats of Latin America and the Caribbean into seven "Birdscapes". In the title bar of each Birdscape account we provide a summary that gives its extent, the primary countries in which it occurs, and how many bird-triggered AZE sites and threatened birds are found in it. We also provide a status assessment for key bird habitats within each Birdscape, along with a representative Important Bird Area (IBA) for each. In the title bar of each IBA we provide the Birdscape and country in which it is found, its area, and the number of threatened birds found there. Additional information follows in the body of each account.

PHOTO: FUNDACIÓN PROAVES

The Endangered Recurve-billed Bushbird was not seen between 1965 and 2004, but is now protected in two ProAves reserves.

SETTING PRIORITIES

One of the most important tools for setting bird conservation priorities outside the U.S. is the IUCN Red List, which provides a threat ranking system for all bird species in the world. Categories include (in descending order of severity) Extinct, Extinct in the Wild, Critically Endangered, Endangered, Vulnerable, Near Threatened, and Least Concern. While this list provides a useful framework for prioritizing species, fortunately it is not necessary to address the conservation of each one independently, since many of the most threatened birds co-occur at sites that can be protected for multiple species. At the most basic level, it is possible to identify Important Bird Areas (IBAs) for neotropical species using the same approach we use in the U.S. A comprehensive approach to identifying IBAs in the Neotropics is hampered somewhat by a lack of data, though, and a complete inventory has yet to be completed. Perhaps more importantly, there simply are not sufficient resources in bird conservation to protect all of the Latin American and Caribbean IBAs immediately. As a result, conservationists have developed methods of identifying the most critical sites rapidly so that conservation of the most irreplaceable ones can go ahead, while additional data and resources are mustered for the next set of priorities.

Alliance for Zero Extinction

The Alliance for Zero Extinction (AZE) is a partnership of biodiversity conservation organizations that has created a global map of key sites for Endangered and Critically Endangered species that are confined to single primary locations. If lost, these last refuges would almost certainly each result in the extinction of at least one species. Conversely, their protection can create a global front-line of defense against species extinctions. These sites are therefore the highest priorities for conservation action if they are not already protected. Of the 217 AZE bird "trigger" species worldwide, 91 (42%) occur within 77 sites in the Neotropics (ten sites have two or more trigger species). Another 18 AZE

The Jabiru stork, one of the most spectacular water-birds in the Americas, sometimes feeds in groups to herd fish in the shallows.

PHOTO: ©ISTOCKPHOTO.COM

bird species (8%) occur within nine sites in the U.S. and Canada, six of which are in the Hawaiian Islands. All North American AZE sites except one (on Oahu, Hawaii, for the Oahu Alauahio which unfortunately is likely now extinct) are currently protected, but only 58 (75%) of the neotropical AZE sites are even partially protected. One of ABC's major goals is to help build and sustain a system of private and public nature reserves for all 77 bird-triggered AZE sites in the Neotropics. Towards that end, ABC has already collaborated with international partners to establish or expand protected areas in 17 AZE sites, protecting species such as the Marvelous Spatuletail hummingbird, and the spectacular Lear's Macaw. Not only are these AZE sites critically important to help prevent the extinction of resident restricted-range birds, but their conservation also benefits numerous other resident species as well as some North American migrants. For instance, the Cerulean Warbler occurs at an AZE reserve in Colombia that is triggered by the Gorgeted Wood-Quail.

Strongholds

Species on the IUCN Red List that are found at two or more sites can be as close to extinction as those found at just one site if they lack protection and face more immediate threats. To address this issue, ABC has identified species "Strongholds"— areas that contain significant populations of all non-AZE Endangered or Critically Endangered bird species (and Vulnerable species that occur at just a single site). These sites can be regarded as the next level of priority after AZE. By protecting AZE sites and Strongholds, we can ensure that all the most threatened species are protected in at least one location. In doing so, we will also protect habitat for a large percentage of all other birds too. The next level of priority will be to identify and protect all the key sites for migratory species including all major staging, and wintering areas. The Western Hemisphere Shorebird Reserve Network (WHSRN) has already identified multiple such sites for shorebirds.

High Altitude

Area: **242,953** sq. miles	AZE Sites: **1**
Primary Countries: **Argentina,** **Bolivia, Chile, Colombia,** **Ecuador, Peru, Venezuela**	Threatened Species: **17**

High-altitude habitats occur throughout much of Latin America, but are most extensive in the Andes. Cloud forests typically grow up to elevations of 10-11,000 feet. Above this, forests are mostly replaced by tussock grassland, boggy areas, and barren rock. Páramo is a scrubby grassland that grows in the Northern Andes from Venezuela to northern Peru, as well as in Costa Rica and Panama. It is characterized by tussock grasses and small shrubs. Puna is an open grassland dominated by bunchgrass tussocks that occurs from central Peru to northwestern Argentina, and throughout the high Andean plains.

Woodlands dominated by *Polylepis* trees form above 13,000 feet within both puna and páramo ecosystems, and are among the highest woodlands in the world. *Polylepis* favors sheltered valleys and is a giant member of the rose family, characterized by its gnarled trunks covered with flaky red bark. Wetlands, including marshes, lakes, and streams, also occur at high altitudes.

Birds: Bird diversity declines with increasing altitude, but high-altitude habitats often support specialized, restricted-range species. Birds characteristic of the high altitudes include tinamous,

ILLUSTRATION: C. VEST

Birds: 1. White-tufted Sunbeam; 2. Ash-breasted Tit-Tyrant; 3. White-browed Tit-Spinetail; 4. Royal Cinclodes; 5. Andean Condor; 6. Gray-breasted Seedsnipe; 7. Streaked Tuftedcheek; 8. Golden-plumed Parakeet; 9. Andean Cock-of-the-Rock; 10. Golden-headed Quetzal; 11. Flame-faced Tanager; 12. Golden Tanager; 13. Masked Flowerpiercer; 14. Purple-throated Sunangel; 15. Barred Antthrush; 16. Sword-billed Hummingbird; 17. Green Jay; 18. Barred Fruiteater; 19. Pale-naped Brush-Finch. **Vegetation:** 20. *Polyepis* woodland; 21. puna bunchgrass; 22. cloud forest. **Threats:** 23. deforestation.

High Altitude Habitats	Condition	Threat Level	Flagship Bird
Puna	Fair	Low	White-tailed Shrike-Tyrant
Páramo	Poor	High	Bearded Helmetcrest
Polylepis woodland	Poor	Critical	Giant Conebill
Arid montane scrub	Fair	Medium	Giant Hummingbird
Humid montane scrub	Fair	Medium	Marvelous Spatueltail

High Altitude Birdscape

rheas, snipe, seedsnipe, caracaras, hummingbirds, flickers, miners, canasteros, tapaculos, ground-tyrants, pipits, wrens, seedeaters, finches, and siskins.

Polylepis woodlands also support a highly specialized avifauna including cinclodes, tit-tyrants, tit-spinetails, and conebills.

Threats: Many high-altitude habitats have been cultivated or severely impacted by over-grazing and burning. Harvesting peat for fuel and to supply mushroom farms damages important habitats for the White-bellied Cinclodes in central Peru. Rheas and tinamous suffer from hunting. *Polylepis* forest is threatened due to use for fuel, and supports some of the world's rarest birds such as the Royal Cinclodes. High-altitude habitats are also likely among the most vulnerable to global warming. As Andean glaciers melt or global warming changes precipitation patterns, more birds may suffer population declines.

PHOTO: MIKE PARR

Vicuña are slender camelids related to llamas that graze the bunchgrasses in the puna of the high Andes (nr. Lake Junin pictured).

PHOTO: ©ISTOCKPHOTO.COM

The Andean Condor is significant to many Andean cultures and is a spectacular sight for any birdwatcher.

Conservation: Fundación ProAves created the 8,300-acre Colibrí del Sol Bird Reserve in Colombia's western cordillera in 2005 in partnership with ABC. This site protects páramo and high-altitude woodlands for the recently rediscovered, Critically Endangered Dusky Starfrontlet, as well as the Endangered Chestnut-bellied Flowerpiercer, and Vulnerable Moustached Antpitta and Rusty-faced Parrot. An international assessment of *Polylepis* forests across Venezuela, Colombia, Ecuador, Peru, Bolivia, Chile, and Argentina will soon be published, helping conservation planners direct conservation attention to remaining patches of this habitat type. Projects to protect *Polylepis* forests led by ECOAN are underway in Peru (e.g., the Vilcanota Cordillera and Huascarán) and Bolivia (e.g., upper Madidi, Apolobamba, and Tunari).

Actions:
- Identify locations for potential new high-altitude protected areas at AZE sites and Strongholds.
- Protect *Polylepis* forests from overexploitation

for firewood, uncontrolled burning, and livestock grazing.
- Marcapomacocha (and nearby Ticlio) the AZE site for the White-bellied Cinclodes, and other key sites should be protected.
- Puna Rhea populations are small and require conservation attention.

PHOTO: FUNDACIÓN PROAVES

Espeletias are distinctive plants found in the páramo, a humid habitat found above the treeline high in the Andes.

IMPORTANT BIRD AREA
VILCANOTA, PERU

BIRDSCAPE: **HIGH ALTITUDE**
AREA: **5,600 ACRES**

THREATENED SPECIES: **3**

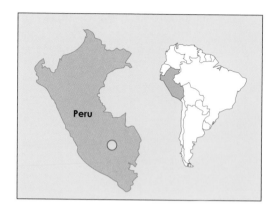

Peru

Birds: Royal Cinclodes, White-browed Tit-Spinetail, Tawny Tit-Spinetail, Ash-breasted Tit-Tyrant, and other imperiled birds of the *Polylepis* woodlands.

Threats: *Polyelpis* forest was likely never very extensive, though it coexisted with humans until the time of Spanish conquest when extensive pasture systems, burning, and highland deforestation began on a large scale. High Andean communities must endure the cold, and have little in the way of resources, so they must rely on *Polylepis* which is typically the only fuelwood available to them.

*P*olylepis forests occur throughout the Andes but have been severely depleted by people over the centuries, through the cutting of trees for firewood, and the grazing of sheep and other livestock that eat saplings and prevent forest regeneration. Only two to three percent of the original closed-canopy *Polylepis* cover remains in Peru, and even that is scattered in small (mostly shrinking) patches.

Conservation: Asociación Ecosistemas Andinos (ECOAN) and ABC have been working with indigenous communities in the Vilcanota mountains of Peru to protect and restore *Polylepis* forests since 2000. Some 1,775 families in 18 farming communities and the town of Yanahuara participate in the project. ECOAN has helped these communities by distributing nearly 6,000 high-efficiency clay

PHOTO: VALERE CLAVERIE

The Vilcanota is a stronghold for the Critically Endangered Royal Cinclodes, which may number fewer than 250 individuals.

View from Abra Malaga towards Tastayoc in the Vilcanota; Polylepis *forest cloaks the slopes, and a glacier can be seen top center.*

PHOTO: MIKE PARR

stoves that reduce firewood demand by c. 25%, planted more than 100,000 plantation trees for fuelwood, provided supplemental firewood until fuel trees reach harvestable size, mapped and secured title to the lands for indigenous communities, and provided community support through health care and education. In return, the communities have helped ECOAN by planting 350,000 *Polylepis* saplings and 30,000 other native trees. Forest restoration also helps protect valuable water supplies for local people. This partnership is now establishing private conservation areas on community lands, and providing a new model for protected areas described as "conservation by and for local people". These communally owned conservation areas are fenced off to protect trees from grazing livestock. Two private conservation areas have already been established, including one at Abra Malaga, a popular birding tourism destination, with five more awaiting official recognition. Together they form the Vilcanota Reserve Network.

Actions:
- Expand the project to additional high Andean communities in adjacent areas, as well as expanding actions in Huascarán and northern Bolivia.
- Expand energy saving programs by installing additional solar-powered lighting, improved stoves with chimneys, and better insulation.
- Continue and expand reforestation and fencing of grazing animals; expand fuelwood plantations.
- Develop ecotourism as an alternative to the overcrowded Inca trail to Machu Picchu.

MONTANE AND TEMPERATE FORESTS

AREA: **755,192** SQ. MILES
PRIMARY COUNTRIES:
ARGENTINA, BOLIVIA, CHILE, COLOMBIA, ECUADOR, GUATEMALA, MEXICO, PERU, VENEZUELA

AZE SITES: **28**
THREATENED SPECIES: **156**

Montane and Temperate Forests Birdscape

Montane forests generally occur from three to ten-thousand feet in altitude and are found in the Andes, the Tepuis of the Guianan Shield in northern South America, the mountains of southeastern Brazil; and the highlands of Mexico, Central America, and the Caribbean. Temperate forests occur at lower elevations in southern Chile and Argentina, where cool, moist climate conditions prevail. Montane and temperate forests are quite variable, including cloud forests festooned with epiphytic mosses, orchids, and bromeliads; pine-oak forests; and temperate forests dominated by southern beeches, or monkey puzzle trees.

Birds: Montane forests tend to have fewer overall bird species than adjacent lowland forests, but two to four times as many endemic species. As a result, the tropical Andes has slightly more total forest bird species (791) than the Amazon (788). A rich diversity of guans, quail, parrots, owls, hummingbirds, barbets, woodpeckers, ovenbirds, antpittas, tapaculos, tyrant-flycatchers, wrens, thrushes, tanagers, and many other genera inhabit montane and temperate forests. Montane forests in Mexico, Central America, the Caribbean, and the northern

Andes are also crucial for migrant birds during North American winters. Species of conservation concern include the Olive-sided Flycatcher, and Cerulean and Golden-cheeked Warblers.

Threats: The high concentration of range-restricted birds in montane forests makes habitat loss a particular concern. Montane forests grow on some of the richest soils in the Neotropics, and therefore have been widely cleared for agriculture and grazing. Crops such as coffee and cacao are often cultivated on mountain slopes. Conversion of shade coffee plantations to higher-yield varieties that grow in full sun, and herbicide spraying to combat coca production underneath the forest canopy severely degrades bird habitat and harms native and migrant birds. Logging has removed much old-growth forest in the highlands of Mexico and Chile. Forest fragmentation is also a significant problem, as many species are restricted to narrow

PHOTO: FUNDACIÓN PROAVES

The Critically Endangered Dusky Starfrontlet inhabits the upper cloud forest, and was recently rediscovered in Colombia.

elevational belts that easily become disconnected by small-scale land-clearance.

Conservation: Chile and Argentina have established many national parks to protect their southern temperate forests, and most neotropical countries with montane forests also have parks protecting at least some areas. Venezuela's many protected areas include montane forests in the tepuis, coastal mountains, and the Andes. For instance, Guácharo Cave NP near Caripe, Venezuela, protects a cave and surrounding forests containing one of the best-known Oilbird colonies. Most of the remaining Atlantic forest in eastern Brazil is protected in parks such as Itatiaia NP and Serra do Mar State Park. Conservation groups are also actively protecting montane and temperate forests. Private protected areas have been established by local conservation groups in Costa Rica, Colombia, Ecuador, Peru, Bolivia, and elsewhere. Fundación ProAves and Fundación Jocotoco have provided leadership in establishing networks of private reserves in Colombia and Ecuador respectively, erecting nest boxes to boost breeding opportunities for some endangered parrots, and educating communities about bird conservation issues. Several AZE species inhabiting montane forest currently lack protected areas, however.

Clouds form as air rises up mountain slopes, delivering moisture to montane forests such as these at Tapichalaca, Ecuador.

PHOTO: MIKE PARR

Actions:
- Establish protected areas for all AZE and Stronghold species in montane forests currently lacking protected habitat.
- Expand habitat protection for AZE and Stronghold sites where current habitat protection is insufficient, such as in the mountains near Caripe and the Paría Peninsula of Venezuela, the Unchoq area of central Peru, and many others.
- Search for "lost" species, such as the Táchira Antpitta.
- Explore carbon finance programs to fund forest protection and restoration efforts.
- Expand nest box programs for endangered parrots that are limited by nest site availability.
- Promote birding ecotourism at appropriate sites, to provide sustainable funding for reserve management.

Montane and Temperate	Condition	Threat Level	Flagship Bird
Hill forest	Poor	High	Rufous-webbed Brilliant
Montane forest	Poor	High	Jocotoco Antpitta
Elfin forest	Poor	Medium	Elfin-Woods Warbler
Southern temperate	Fair	Medium	Magellanic Woodpecker
Pine	Poor	High	Hispaniolan Crossbill
Pine-oak	Poor	High	Pink-headed Warbler

IMPORTANT BIRD AREA
EL DORADO, COLOMBIA

BIRDSCAPE: **MONTANE AND TEMPERATE FORESTS**
AREA: **1,600 ACRES**

THREATENED SPECIES: **11**
AZE SITE

Zero Extinction
ALLIANCE FOR

Colombia

The Sierra Nevada de Santa Marta is a Colombian mountain range perched on the edge of the Caribbean Sea. It is the highest coastal range in the world, with peaks reaching more than 16,000 feet in altitude. It supports 70 species and subspecies of birds that are found nowhere else. El Dorado is the key bird reserve within the massif. The reserve has a bird list of 364 species, 17 of which are endemic, and a lodge and trail system for birding visitors.

PHOTO: FUNDACIÓN PROAVES

El Dorado protects 17 bird species endemic to Santa Marta including the recently discovered Santa Marta Screech-Owl.

Birds: Endangered endemics include the Santa Marta Parakeet, Santa Marta Sabrewing, and Santa Marta Bush-Tyrant. The discovery of a new screech-owl and the recent elevation of the Santa Marta Foliage-Gleaner to full species status hint at the biodiversity that likely still awaits discovery. North American migrants, including the Blackburnian, Cerulean, and Golden-winged Warblers, also use this area during the northern winter.

Threats: Deforestation in the Santa Marta mountain range has been extensive and few significant tracts of habitat remain. Those which do are not only important for birds, but also act as the watershed for surrounding farmland and populated areas.

Conservation: In 2006, Fundación ProAves established the El Dorado Bird Reserve, protecting 1,600 acres of montane forest in the Santa Marta mountains with ABC's support. The reserve holds most of the threatened species in the area. ProAves built the Sierra Nevada EcoCenter, Jeniam Ecolodge, and Blue Moon Restaurant at the reserve to provide sustainable financing for reserve protection and restoration through ecotourism. The lodge also displays attractive CollidEscape window film featuring images of local birds on its windows to prevent bird collisions (see p. 317). Other conservation activities at the reserve include removing invasive Mexican pines and reforestation with native trees, the provision of artificial nest boxes for the Santa Marta Parakeet, and the monitoring of threatened species including birds, frogs, and plants on the reserve.

Actions:
- Expand the reserve to 7,000 acres to include all surrounding key areas of forest.
- Promote birding tourism to make the reserve financially sustainable.

In addition to AZE birds, this reserve also protects endangered amphibians such as Cryptobatrachus boulengeri *above.*

The Santa Marta Brush-Finch is endemic to Santa Marta and can be seen foraging near the ground at El Dorado.

LOWLAND TROPICAL FORESTS

AREA: **3, 201,194** SQ. MILES

PRIMARY COUNTRIES: **BELIZE, BOLIVIA, BRAZIL, COLOMBIA, COSTA RICA, ECUADOR, GUYANA, MEXICO, PERU, SURINAME, VENEZUELA**

AZE SITES: **20**

THREATENED SPECIES: **133**

Tropical lowland forests occur up to elevations of roughly 3,000 feet. They are found in the Amazon and Orinoco basins, from southeast Brazil into Paraguay and Argentina, from southern Mexico through Central America, along the coasts of Colombia and Ecuador, and on Caribbean islands. Tropical lowland forests vary from drier semi-deciduous forests to wet evergreen rainforests. They often contain micro-habitats that support specialized bird species, these habitats include bamboo thickets, palm swamps, and successional vegetation along larger rivers. Some forests (known as varzea) are flooded for many months of the year.

Birds: Single location bird diversity reaches its global maximum in the rainforests of the western Amazon basin; some sites boasting over 500 species. Species groups include guans, raptors, hummingbirds, woodpeckers, toucans, puffbirds, jacamars, ovenbirds, antbirds, ant-thrushes, tyrant-flycatchers, manakins, cotingas, tanagers, and oropendolas, among many other genera. The lowland forests of Central America and the Caribbean are also important for migrant birds from North America; whereas fewer northern migrants winter in the Amazon.

ILLUSTRATION: C. VEST

Birds: 1. Black-fronted Piping-Guan; 2. Red-breasted Toucan; 3. Mantled Hawk; 4. Bare-throated Bellbird (pair); 5. Three-toed Jacamar; 6. Blue-chested Parakeet; 7. Spot-breasted Antvireo; 8. Stresemann's Bristlefront; 9. Pink-legged Graveteiro; 10. Fork-tailed Pygmy-Tyrant; 11. Crested Oropendola; 12. Red-browed Parrot; 13. Great Kiskadee; 14. Plush-crested Jay. **Mammals:** 15. jaguar; 16. golden lion tamarin. **Vegetation**: 17. *Cecropia*; 18. *Erythrina*; 19. *Guadua* bamboo. **Threats**: 20. habitat loss to slash and burn; 21. grazing; 22. plantations.

LOWLAND TROPICAL FORESTS	CONDITION	THREAT LEVEL	FLAGSHIP BIRD
Amazonian-Guyanan forests	Fair	Medium	Red-and-Green Macaw
Central American forests	Poor	High	Black-cheeked Ant-Tanager
Caribbean forests	Poor	Critical	Blue-headed Quail-Dove
Atlantic forests	Poor	Critical	Stresemann's Bristlefront
Gallery forests	Poor	High	Rio Branco Antbird
Flooded forests	Poor	High	Silvered Antbird
White-sand forests	Fair	Medium	Iquitos Gnatcatcher
Palm swamps	Fair	High	Point-tailed Palmcreeper

Lowland Tropical
Forests Birdscape

Threats: The main threats facing lowland tropical forests are clearance for pasture, agricultural development, and timber extraction. The most severely affected regions are Brazil's Atlantic forests and the forests of the West Indies. Between 1500 and 1970, 80% of the Atlantic forest was cleared,

Lowland tropical forests are famous for colorful and charismatic birds, such as the Citron-throated Toucan above.

PHOTO: FUNDACIÓN PROAVES

and less than 10% remains today. Much of Central America's lowland forests have also been cleared. For example, Costa Rica lost more than half of its forest between 1940 and 1989. Loss of primary forest continues in Costa Rica today, but re-growth of previously cleared areas is now reducing net deforestation rates. In the Amazon, less than half of the region's forest is predicted to remain by the year 2030, and deforestation has recently accelerated there with roughly five-million acres cleared annually for pasture or cropland (e.g., for soy). Remaining forest fragments can be too small or isolated to maintain the original complement of bird species. Some tropical forest birds are reluctant to cross open areas such as road cuts, inhibiting natural dispersal. Other threats include the hunting of tinamous, guans, and other large species, as well as exploitation of parrots and finches for the pet trade, and the persecution of raptors. Threats tend to be most severe along rivers and new roads that provide access to the forest and opportunities for settlement. Other threats include habitat degradation from oil drilling and mining that further fragment habitat and contaminate the environment. Despite the attention Amazonia receives, the region has relatively few threatened birds because much habitat still remains intact, and a higher proportion of the species have large ranges compared to those in high-elevation forests.

Conservation: Because the scale of most of these systems is so large, only government-sanctioned parks are typically sufficient to protect significant percentages of their extent. Unfortunately, these parks are often difficult to manage due to their size, remoteness, and the lack of resources available to

Orange-cheeked Parrots, Mealy Amazons, and Chestnut-fronted Macaws at a clay lick at Tambopata in the Peruvian lowlands.

protect them. Carbon cap and trade offers a potentially significant new tool to finance the protection of large areas of habitat, but is complicated by international politics. There are some places where smaller protected areas can be effective in conserving key species and habitats, however, Brazil's Atlantic forests and the forests of the Caribbean islands being primary examples. Lowland forests in these regions are prone to far greater endemism than is Amazonia, and small reserves can protect the entire global ranges of some species (as they often can at higher altitudes).

Actions:

- Protect all lowland tropical forest AZE and Stronghold sites.
- Develop new financial mechanisms, such as carbon markets, to offset greenhouse gas emissions by protecting forest.
- Expand protected areas in intact Amazonian wilderness before pressure to clear these areas increases.

Birds that live high in the canopy can be seen from towers and elevated walkways, such as this one at Rio Napo, Ecuador.

IMPORTANT BIRD AREA
STRESEMANN'S
BRISTLEFRONT RESERVE, BRAZIL

BIRDSCAPE: **TROPICAL LOWLAND FORESTS**
AREA: **1,500 ACRES**

THREATENED SPECIES: **17**
AZE SITE

Zero Extinction
ALLIANCE FOR

Brazil

Remaining patches of Brazil's Atlantic forest contain some of the highest concentrations of endangered vertebrates in the world due to the high levels of endemism and extensive habitat destruction in this region. One site that could perhaps claim more threatened birds per acre than any other in the Americas is the newly established Stresemann's Bristlefront Reserve on the border of the Brazilian states of Minas Gerais and Bahia. This reserve protects a key fragment of Atlantic forest. It has a bird list of 245 species, 37 of which are endemic to Brazil. In addition to being the only known site for the Stresemann's Bristlefront, this is a critically important site for the Endangered Banded Cotinga and the Critically Endangered yellow-breasted capuchin monkey.

Birds: Threatened birds occurring here include the White-necked Hawk, Blue-throated Parakeet, Brown-backed Parrotlet, Red-browed Amazon, Three-toed Jacamar, Hook-billed Hermit, Black-headed Berryeater, Banded Cotinga, Bare-throated Bellbird, Bahia Tyrannulet, Fork-tailed Pygmy-Tyrant, Plumbeous Antvireo, Band-tailed Antwren, Bahia Spinetail, Striated Softtail, and Pink-legged Graveteiro. The Stresemann's Bristlefront is a secretive bird known only from a few specimens until it was rediscovered in 1995 near Una in Bahia, Brazil. It was not seen again there

despite searches, but was rediscovered in 2003 in the Fazenda Balbina forest where the Stresemann's Bristlefront Reserve is now located.

Threats: Like other sites in the Brazilian Atlantic forest, the reserve is primarily threatened by small-scale habitat clearance for pasture, and logging for local construction and fence posts. While these threats may seem minor in comparison to the large -scale land clearance for agriculture and logging that are taking place in parts of the Amazon, it should be noted that the total known extent of the Stresemann's Bristlefront's global range is approximately 5,000 acres whereas the Amazon occupies some 1.7 billion acres.

PHOTO: FUNDAÇÃO BIODIVERSITAS

Though protected here, the Bahia Spinetail is threatened by forest clearance across much of its range.

The Stresemann's Bristlefront Reserve may have more threatened species per acre than any other site in the Americas.

Conservation: With ABC's support, Fundação Biodiversitas acquired 1,500 acres of forested habitat for this reserve. The partners are evaluating the possibility of developing a birding route in Brazil that includes this site, which could bring in sufficient income from birders to cover the management costs of the small reserve. The Lear's Macaw reserve in Bahia could potentially also be a part of this route.

Actions:
- Expand the reserve to 5,000 acres.
- Evaluate the potential for developing birding tourism.
- Signpost the reserve boundary.
- Monitor the bristlefront's population.

The Stresemann's Bristlefront has only been photographed four times to our knowledge. This bird is probably a young male.

GRASSLANDS

AREA: **1,858,634** SQ. MILES
PRIMARY COUNTRIES:
ARGENTINA, BRAZIL, GUYANA,
MEXICO, PARAGUAY, URUGUAY

AZE SITES: **3**
THREATENED SPECIES: **24**

Tropical grasslands occur widely throughout the Neotropics, with the greatest extent in the campos and pampas of central Brazil, Paraguay, Uruguay, and Argentina; large areas also occur in the Gran Sabana of Venezuela and the Guianas, within the Amazon basin, in Mexico, and elsewhere. Several spectacular areas of seasonally flooded grasslands, including the Llanos of Venezuela and Colombia, Argentina's Mesopotamian marshes, Bolivia's Beni savanna, and the Brazilian Pantanal support immense concentrations of wildlife (though parts of the Pantanal in particular are perhaps better considered as wetlands—see p. 330). Cooler grasslands occur in Patagonia (Chile and Argentina). Grasslands vary in structure depending on fire frequency and moisture, and are often interspersed with a variety of arid woodlands to form savannas. Areas where woodlands, palms, and grasslands meet also provide habitat for specialist birds.

Birds: Prominent species groups include tinamous, rheas, quail, ovenbirds, tyrant-flycatchers, pipits, seedeaters, sparrows, finches, and blackbirds. Neotropical migrants, such as the Swainson's Hawk, American Golden-Plover, Buff-breasted

Birds: 1. Bobolink; 2. Rusty-collared Seedeater; 3. Southern Lapwing; 4. Upland Sandpiper; 5. White Monjita; 6. Hyacinth Macaw; 7. White-browed Blackbird; 8. Swainson's Hawk; 9. Greater Rhea; 10. Cattle Tyrant; 11. White-faced Whistling-Duck; 12. White-headed Marsh-Tyrant; 13. Jabiru; 14. Southern Screamer. **Mammals:** 15. maned wolf; 16. capybara. **Threats:** 17. agriculture; 18. deliberate fires; 19. over-grazing; 20. pesticides.

Grasslands Habitats	Condition	Threat Level	Flagship Bird
Seasonally wet grasslands	Fair	High	Bare-faced Ibis
Pampas and campos	Poor	Critical	Pampas Meadowlark
Patagonian steppe	Fair	Medium	Lesser Rhea
Cerrado	Poor	High	White-winged Nightjar
Arid grasslands	Poor	High	Sierra Madre Sparrow

Grasslands Birdscape

Sandpiper, Upland Sandpiper, Dickcissel, and Bobolink winter in South American grasslands.

Threats: Grasslands are among the most endangered and overlooked habitats in the Neotropics, and therefore suffer from a lack of conservation attention, especially compared to Amazonian forests. Much like North America's prairies, neotropical grasslands have been widely cleared for agricultural and grazing lands, or degraded by invasive weeds. For instance, 68% of the pampas has been transformed for human uses. Grasslands in Argentina and Brazil have also been widely converted to pine plantations to supply timber, with tax subsidies in Brazil encouraging this land use. Large areas in Brazil are being converted to soy for food and biofuels. Bird species intolerant of heavily cropped or grazed areas are the most threatened. Only 24% of neotropical grassland birds are found in grazed pastures. Some grassland birds, such as the Red Siskin, are exploited by hunters for the caged bird trade. Others may be taken for food. Grassland birds are intentionally and accidentally poisoned with pesticides, many of which are banned in North

Bunchgrass and steppe vegetation occur across vast areas of windswept Patagonia (Perito Moreno NP, Argentina, pictured).

The Strange-tailed Tyrant is an extravagant, threatened flycatcher of humid grasslands in northern Argentina and adjacent countries.

PHOTO: JAMES LOWEN

America, yet still produced by U.S. companies for use abroad. Fires set by people also damage neotropical grasslands, especially in regions such as the Llanos and pine savannas of Honduras and Nicaragua. Climate change also threatens to alter fire frequencies and precipitation patterns with uncertain consequences for grassland habitats and their dependent birds.

Conservation: Neotropical grasslands are poorly represented in protected areas because they are highly coveted for farming and livestock production. More than 98% of the pampas region is privately owned. Still, protected areas do preserve important grasslands such as the Gran Sabana in Venezuela's Canaima NP. A new reserve has recently been created in Bolivia by Asociación Armonía with, support from ABC, for the Blue-throated Macaw that includes habitat for grassland species such as the Cock-tailed Tyrant and Greater Rhea. Elsewhere, habitat conservation

strategies include protecting native grasslands and improving native habitat on working cattle ranches. To protect grassland birds from pesticide poisoning, ABC successfully lobbied to ban the use of monocrotophos in Argentina, and is currently working towards a similar ban in Bolivia.

Actions:
- Provide further conservation attention for Brazil's Blue-eyed Ground-Dove, and Mexico's Worthen's and Sierra Madre Sparrows (all AZE trigger species); protect Stronghold sites.
- Reduce the use of pesticides in agricultural landscapes and halt the use of pesticides that are banned in the U.S.
- Create and expand grassland protected areas.
- Work with farmers and ranchers to manage grasslands sustainably.
- Slow conversion of grasslands to crops such as soy and sugarcane for food and biofuels.
- Prevent the overuse of fire.

IMPORTANT BIRD AREA
SALTILLO SAVANNA, MEXICO

BIRDSCAPE: **GRASSLANDS**
AREA: **57,600 ACRES**

THREATENED SPECIES: **2**
AZE SITE

Zero Extinction
ALLIANCE FOR

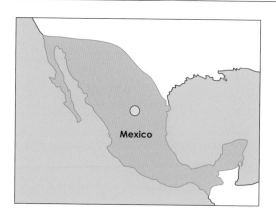

The desert grasslands around the town of Saltillo in northern Mexico are the last refuge for the Endangered Worthen's Sparrow, which qualifies the site for AZE status. The Saltillo area is also important for the Endangered Mexican prairie dog, which creates the region's characteristic short-grass prairie through its grazing activities. The area lies at around 5,000 feet in altitude, and is considered to be one of the most important remnants of remaining grassland habitat in North America.

Birds: Other notable bird species include the Burrowing Owl, Ferruginous Hawk, and several wintering U.S. WatchList species, including the Sprague's Pipit, Mountain Plover, and 15% of the world's Long-billed Curlew population. Golden Eagles and Prairie Falcons are also found here during the winter. Worthen's Sparrows prefer scrubby areas on slopes along the edges of prairie basins. The sparrow was once wider-ranging but appears to have contracted over recent decades to this last stronghold.

Threats: These grasslands are disappearing at a rate of 20,000 acres annually, largely due to the expansion of potato farms, which destroy habitat and draw down water supplies. Water shortages for both prairie and agriculture have been exacerbated by recent ongoing drought conditions. Over-grazing and trampling of vegetation by goats and cattle also severely reduce breeding success for nesting Worthen's Sparrows.

Conservation: Since 2006, ABC, TNC, and other partners have worked with Pronatura Noreste to protect the Saltillo grasslands. To date this partnership has obtained conservation easements on 57,000 acres, acquired 600 acres to establish the Worthen's Sparrow Reserve at El Cercado, and enhanced the reserve's infrastructure with fences to keep out grazing livestock, a visitor center, and a small cabin for reserve guards. Small-scale ecotourism is also planned. The area is close to the main nesting cliffs for the rare Maroon-fronted Parrot, which can also be seen during a short visit to the area.

Actions:
• Expand the El Cercado reserve to 5,000 acres and easements to 60,000 acres.

PHOTO: ANTONIO HIDALGO

The Saltillo Savanna is the last refuge for the Worthen's Sparrow (above) and for the Endangered Mexican prairie dog.

Both resident birds such as the Burrowing Owl (above) and migratory birds such as the Long-billed Curlew occur at this site.

ARID LANDS

AREA: **1,869,703** SQ. MILES
PRIMARY COUNTRIES: **BOLIVIA, BRAZIL, CARIBBEAN, CHILE, ECUADOR, MEXICO, PARAGUAY, PERU**

AZE SITES: **17**
THREATENED SPECIES: **57**

Arid Lands Birdscape

Arid lands occur widely throughout the Neotropics, including along much of the Pacific slope from Mexico to Chile, in the lowlands of the Caribbean islands, in northeastern Mexico and the northern Yucatan Peninsula, in the Galapagos Islands, in inter-Andean Valleys, and across a broad area of the South American interior from northeastern Brazil south through Argentina. Arid habitats are extremely variable, ranging from Chile's Atacama Desert, which is the driest in the world, to the distinctive Tumbesian forests that are dominated by massive deciduous ceiba trees. Arid lands are often vegetated with thorny scrub; the cerrado is an arid savanna-woodland interspersed with gallery forests and grasslands. Other dry woodlands include those of Central America, the caatinga in Brazil, and the chaco which extends from Bolivia to Argentina.

Birds: Arid lands support lower bird diversity than more lush habitats, but often exhibit higher levels of endemism among birds compared to humid forests. For instance, 58% of bird species using arid deciduous forests are restricted to this habitat type, and 90% of those are endemics. The dry forests of the Tumbesian and cerrado regions have particularly high concentrations of endemic species. Prominent species groups include seriemas, parrots, hummingbirds, ovenbirds, crescentchests, tapaculos, tyrant-flycatchers, wrens, mockingbirds, seedeaters, sparrows, buntings, and finches.

Threats: Although some of the great deserts in South America are sparsely inhabited, most neotropical arid lands are sensitive to disturbance and degradation by fires, grazing, wood cutting for charcoal, and land conversion for agriculture or development. For example, the cerrado suffers from rapid habitat clearance and 40% of its endemic birds are threatened. Arid Caribbean lowlands also suffer high clearance rates for development. On Grenada, a new hotel development is clearing habitat within Mount Hartman Estate, which, combined with Mt. Hartman NP, is designated as an AZE site for the Grenada Dove. Climate change also threatens to alter fire frequencies and precipitation patterns with unpredictable consequences for grassland habitats and their dependent

PHOTO: HEINZ PLENGE

Habitat protection and the release of captive-bred birds is helping the White-winged Guan in arid northwest Peru.

The Marañon river carves out a spectacular canyon that harbors endemics such as the Yellow-faced Parrotlet and Marañon Thrush.

PHOTO: MIKE PARR

birds. The dry forests of Central America and the Tumbesian region (coastal southern Ecuador and northern Peru) are also highly threatened by human exploitation.

Conservation: Captive-breeding and future reintroduction remains the last hope for the Spix's Macaw and Socorro Dove, which are both extinct in the wild but persist in captivity. Fundación Cracidae's captive-breeding program has successfully reintroduced the White-winged Guan to protected areas such as the Chaparri Private Conservation Area within its historic range in Peru. Habitat protection remains the best strategy for preventing the need for expensive and complicated captive-breeding efforts though. Fundação Biodiversitas (with help from ABC) has protected Lear's Macaw nesting cliffs and dry forests at the Canudos Biological Station, which contributed to boosting the macaw's population from 455 birds in 2003 to 962 in 2008, and allowing it to be down-listed from Critically

Endangered to Endangered. The University of San Simón in Cochabamba, Bolivia has helped to established municipal protected areas for the last Red-fronted Macaws in Bolivia's Mizque River Valley (again with ABC support). There are many other examples of conservation projects that make a difference for key arid land sites and species, though many more are still needed.

Actions:

- Protect unprotected AZE and Stronghold sites such as Talara for the Peruvian Plantcutter, the Casma and Huarma Valleys for the Russet-bellied Spinetail (both in Peru), and Chile's Azapa and Vitor Valleys for the Chilean Woodstar.
- Although the islands of Guadalupe, Socorro, and Cozumel in Mexico; and the island of Floreana in Ecuador's Galapagos Islands are legally protected, conserving AZE bird species on these islands requires expanded programs to control or eradicate invasive species.

Arid Land Habitats	Condition	Threat Level	Flagship Bird
Deserts	Fair	Medium	Grayish Miner
Tamaulipan thorn scrub	Fair	High	Plain Chachalaca
Caribbean xeric scrub	Fair	High	Yucatan Wren
Dry forests and woodlands	Poor	High	White-winged Guan
Inter-Andean valleys	Poor	High	Bearded Mountaineer

IMPORTANT BIRD AREA
YUNGUILLA AND JORUPE RESERVES, ECUADOR

BIRDSCAPE: **ARID LANDS**
AREA: **4,500 ACRES**

THREATENED SPECIES: **13**
AZE SITE

ALLIANCE FOR Zero Extinction

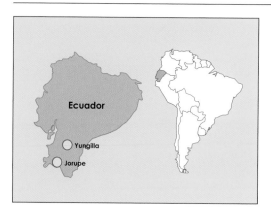

Ecuador

○ Yungilla
○ Jorupe

Fundación Jocotoco has established a network of private protected areas in Ecuador to benefit threatened birds. These include the small Yunguilla Reserve established specifically for the Pale-headed Brush-Finch, and the nearby Jorupe Reserve that protects an important tract of classic arid Tumbesian forest.

The Henna-hooded Foliage-Gleaner is one of many regionally endemic species; it nests in a three-foot-deep earth burrow.

PHOTO: OLÁH JÁNOS

Birds: Threatened species include the Rufous-headed Chachalaca, Gray-backed Hawk, Ochre-bellied Dove, Gray-cheeked Parakeet, Little Woodstar, Henna-hooded Foliage-Gleaner, Blackish-headed Spinetail, Gray-breasted Flycatcher, Gray-headed Antbird, Rufous-necked Foliage-Gleaner, Slaty Becard, Pale-headed Brush-Finch, and Saffron Siskin.

Threats: Most vegetation around the reserve at Yunguilla is degraded by grazing and fires. On-site researchers also discovered that Shiny Cowbirds were parasitizing the brush-finch nests. Forests throughout the Tumbesian region are threatened by grazing, and cutting for charcoal production.

Conservation: The Yunguilla Reserve was established in 1999 (with ABC support) to protect the only known population of the Pale-headed Brush-Finch. This reserve includes a remnant patch of scrub in a valley surrounded by pasture, that is the only place on Earth where the brush-finch is known to occur. A cowbird control program has been initiated, and this has boosted both brush-finch breeding success and population numbers ten-fold to over 225 birds. Fencing has also been erected to exclude grazing animals. Fundación Jocotoco established the Jorupe Reserve to protect a key tract of Tumbesian dry forest in 2004. Here spectacular green-trunked ceiba trees emerge from a dry forest that is home to many endangered and endemic species. A major reforestation effort is underway at the site (40,000 saplings planted to date) and an ecolodge was opened in August 2009.

Actions:
• Expand Jorupe to its target size of 10,000 acres and Yunguilla to its target size of 1,000 acres.
• Continue cowbird control at Yunguilla.
• Expand reforestation and bird tourism programs at Jorupe.

Deciduous ceiba trees with bulbous green trunks are a distinctive feature of Tumbesian dry forests, including those at Jorupe.

WETLANDS

AREA: **UNKNOWN**
COUNTRIES: **ALL**

AZE SITES: **2**
THREATENED SPECIES: **28**

Wetlands Birdscape

Wetlands occur widely throughout the Neotropics. Habitats include a variety of rivers, streams, oxbow lakes, river deltas, desert oases, flooded savannas, saline lakes, highland bogs, and others. Major rivers, include the Amazon, Orinoco, and Paraná among many others. Extensive marshes are found in flooded savannas such as the Llanos and Pantanal (though these systems are really both wetlands *and* grasslands—see grasslands section p. 370), and historically in the highlands of Mexico.

Birds: Neotropical wetlands often support large concentrations of water-birds including flamingoes, grebes, ducks, geese, screamers, herons, storks, ibis, coots, rails, snipe, plovers, gulls, terns, tyrant-flycatchers, wrens, seedeaters, yellowthroats, and blackbirds. Water-birds tend to be highly mobile; 35% of breeding neotropical water-birds are latitudinal migrants, and aquatic habitats are used by 40% of North American migrants.

Threats: Permanent wetlands are threatened by pollution, draining, filling, and development. Most marshes in Mexico's central highlands have been destroyed, causing the extinction of the Slender-billed Grackle. Sugar cane and rice have replaced many coastal marshes in the Caribbean

and Brazil. Increased demand for sugar cane as a biofuel threatens more destruction of native wetlands. Poor water management exacerbates competition for water resources, resulting in pollution and reduced water flow. For example, pollution from urban areas threatens Lake Titicaca (Peru and Bolivia), and Colombia's Bogota marshes, as well as coastal marshes near Lima, Peru. Other contaminants, such as oil spills and toxic waste from mines and industrial facilities, also harm birds. For instance, mine tailings threaten some Andean lakes. Seasonal wetlands, such as those in the Llanos and Pantanal, can be severely degraded by over-grazing during the dry season, when livestock can access them. Unregulated burning is a threat to the Zapata Wren and Zapata Rail in the Zapata Swamp, Cuba. Marsh birds also suffer from exploitation, as both adults and eggs are harvested for food. Bycatch in artisanal gillnet fisheries causes mortality of grebes and other diving birds. Invasive fish can alter wetland ecosystems and reduce food supplies for native birds, and introduced predators (rats, mongooses, cats) affect neotropical wetland birds, especially on Caribbean Islands.

Conservation: Due to the high concentrations of birds and other wildlife attracted to wetlands,

PHOTO: ©ISTOCKPHOTO.

The Scarlet Ibis and Great Egret are two of the abundant water-birds found in the Llanos of Colombia and Venezuela.

PHOTO: MIKE PARR

Lake Junin lies at 13,000 feet in altitude and supports two endemic, endangered birds: the Junin Grebe and Junin Rail.

many sites are protected. The adequacy of this protection is often influenced by the status of lands surrounding the wetlands, however, and as a result, often falls short. The 1971 Ramsar Convention on Wetlands is an international treaty that promotes the conservation of wetlands and near-shore marine habitats, presently involving 159 countries worldwide. In the neotropical realm, over 230 sites in 25 countries are designated under the convention, of which 113 occur in Mexico. Cuba's Zapata Swamp and Peru's Lake Junin are designated as both AZE sites and Ramsar Wetlands of International Importance. Both are also protected by their national governments, but far more work is required to save their threatened species. The fall hunting of migrant shorebirds in Barbados is a much underreported and egregious threat to species such as the American Golden-Plover. Each year, tens of thousands of birds are shot, and it is likely that this hunting is contributing to population declines.

Actions:

- Establish, expand, or strengthen wetland Strongholds for endangered birds, and those where large numbers of birds congregate to breed or stage during migration; extend protections to terrestrial areas bordering wetlands.
- Protect wetlands against pollution from industry, mines, oil spills, and municipal waste from urban communities; improve water management, including water conservation and treatment.
- The Zapata Rail and Zapata Wren need improved population monitoring, combined with a halt to reed burning and the control of introduced predators in Zapata Swamp.
- End shorebird killing in Barbados and elsewhere.

Wetland Habitats	Condition	Threat Level	Flagship Bird
Rivers	Fair	High	Brazilian Merganser
Freshwater marshes	Fair	High	Zapata Rail
Lakes and ponds	Fair	High	Junin Grebe
Salt and playa lakes	Fair	Medium	Andean Flamingo

IMPORTANT BIRD AREA
LAKE JUNIN AND
LAKE TITICACA

BIRDSCAPE: **WETLANDS**
AREA: **2,100,000 ACRES**

THREATENED SPECIES: **3**
AZE SITE (JUNIN)

Alliance for Zero Extinction

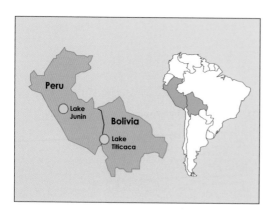

These two high Andean lakes share similar habitats and threats, though Titicaca (at two-million acres) is around twenty times the size of Junin. Both lie at around 13,000 feet in altitude, and have similar bird communities, yet each has a unique and distinctive endemic grebe.

Birds: While high-altitude Andean lakes are often best known for their grebes, these two lakes also harbor populations of other notable bird species such as the Andean Avocet, White-backed Stilt, Giant Coot, Crested Duck, and Chilean Flamingo. The Junin Rail is another highly threatened species that depends on Junin's marshes. Large numbers of Wilson's Phalaropes also stage at these lakes during migration, and Baird's Sandpipers winter at both sites. The flightless Junin and Titicaca Grebes occur at each of the lakes respectively.

Threats: The Junin Grebe is declining due to water fluctuations and contaminants from nearby mines, despite Lake Junin being a Peruvian national reserve. On Lake Titicaca, grebes are regularly drowned as bycatch in gillnets set by fishermen. Introduced fish also affect lake ecosystems, and municipal waste dumped into Lake Titicaca causes eutrophication. Grebes and their eggs are also exploited (especially in Titicaca), and the harvesting and burning of reeds likely impacts grebe reproduction. Lake Titicaca is currently shrinking due to a long-term drought.

Conservation: ABC and ECOAN are working with communities to assess land use in the Junin Lake basin and to identify and reduce sources of pollution to benefit the Junin Grebe and rail. The Peruvian government has established a reserve at the key site for the highest densities of Titicaca Grebes in Peru, and Bolivia is developing a program to treat water from the city of El Alto which flows into the lake.

Actions:
- Reduce mining pollution and regulate water levels in Lake Junin.
- Reduce reed burning at both lakes.
- Improve water treatment from sewage outflows into lake Titicaca from Puno, Peru.

The flightless Titicaca Grebe does not use its vestigial wings even when fleeing an oncoming boat at top speed.

PHOTO: MIKE PARR

The Junin Grebe (right) can be difficult to distinguish from the smaller sympatric Silvery Grebe (left).

PHOTO: MIKE PARR

Reed beds provide habitat for Many-colored Rush-Tyrants at Lake Junín; open waters host thousands of migrating Wilson's Phalaropes.

The Huros live on floating islands on Lake Titicaca; due to hunting and pollution, the Titicaca Grebe is absent from nearby areas.

MARINE

AREA: **MILLIONS OF SQUARE MILES**
COUNTRIES: **ALL EXCEPT BOLIVIA AND PARAGUAY**

AZE SITES: **5**
THREATENED SPECIES: **54**

Marine waters of the Neotropics are generally warmer and less productive than those of temperate regions, with a notable exception created by the cold Humboldt Current that enriches coastal seas off Chile, Peru, and southern Ecuador and creates some of the most productive marine waters on Earth. Mangroves are patchily distributed along coasts in warmer waters. Kelp forests grow in cold waters off southern South America. Other habitats include near-shore and pelagic waters, shorelines, rocky coasts, lagoons, and tidal mudflats.

Birds: Seabirds include penguins, albatrosses, petrels, shearwaters, prions, diving-petrels, boobies, cormorants, tropicbirds, frigatebirds, gulls, and terns. Mangroves support some specialized landbirds and wetland birds, as well as species that are more typically associated with marine habitats, these include some rails, hummingbirds, the Yellow-billed Cotinga, Mangrove Finch, and Yellow-shouldered Blackbird. Finally, coastal mudflats in the Neotropics are important for migrating and wintering shorebirds that breed in North America, such as the Hudsonian Godwit, Red Knot, and Western and Semipalmated Sandpipers.

ILLUSTRATION: C. VEST

Birds: 1. Belcher's Gull; 2. Seaside Cinclodes; 3. Chilean Skua; 4. Peruvian Booby; 5. Whimbrel; 6. Waved Albatross; 7. Peruvian Diving-Petrel; 8. Peruvian Pelican; 9. Inca Tern; 10. Humboldt Penguin; 11. Magnificent Frigatebird. **Mammals:** 12. southern sea lion; 13. bottlenose dolphin. **Threats:** 14. gillnets; 15. oil spills; 16. over-fishing; 17. invasive ungulates.

Marine Habitats	Condition	Threat Level	Flagship Bird
Mangroves	Poor	Critical	Mangrove Finch
Mudflats	Fair	Medium	Hudsonian Godwit
Sandy beaches and dunes	Fair	High	Peruvian Tern
Beach scrub	Poor	Critical	Restinga Antwren
Rocky intertidal	Good	Low	Surf Cinclodes
Sea cliffs	Good	Low	Inca Tern
Marine islands	Poor	Critical	Waved Albatross
Sea ice	Fair	Medium	Adelie Penguin
Pelagic waters	Fair	High	Pink-footed Shearwater

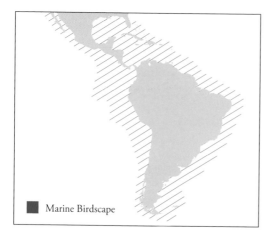

Marine Birdscape

Threats: Threats facing Latin American and Caribbean seabirds include marine pollution, bycatch in fisheries, and invasive predators and grazing animals on breeding islands. Unlike the U.S., most fishing boats are from local or "artisanal" fisheries that involve numerous small vessels. These fisheries are known to cause seabird bycatch, but monitoring is scant, as are legal frameworks and enforcement to address the problem. Seabirds have also suffered as a result of over-fishing. Rats, cats, mongooses, goats, sheep, and pigs are problematic invasive species on neotropical islands. For instance, the Guadalupe Storm-Petrel probably became extinct due to predation by feral cats. Disturbance at nesting colonies includes industrial guano harvesting for fertilizer, and off-road vehicles. Some seabird species, such as the Pink-footed Shearwater are exploited for meat and eggs. Waved Albatrosses are caught at sea for food by fishermen. During El Niño, which occurs every three to ten years, warm water currents expand in the Pacific and disrupt the Humboldt Current. This reduces fish populations contributing to reproductive failure among seabirds, especially the penguins, boobies, cormorants, and pelicans of the Galapagos Islands in Ecuador and the guano islands of Peru. Climate change could increase the severity and frequency of El Niño years. Mangroves have suffered widespread clearance for firewood and charcoal, coastal development, and shrimp farms.

Conservation: Compared to terrestrial parks and reserves, there are relatively few marine protected areas in the Neotropics, particularly in pelagic waters; Galapagos NP and Mexico's Revillagigedo Biosphere Reserve being notable exceptions. Near-shore marine protected areas include Paracas National Reserve in Peru, and the Cienaga Grande de Salamanca in Colombia. The Peruvian government has also recently extended protection to guano islands along its entire coast. Some terrestrial protected areas also support breeding seabirds, such as the Black-capped Petrel which nests

PHOTO: JEFF MANGEL, PRO DELPHINUS

Waved Albatrosses fight over offal discarded from fishing boats; small numbers are caught by fishermen and eaten at sea.

The Peruvian Booby is dependent on anchovies and sardines, which fluctuate with El Niño years and have been over-fished.

PHOTO: MIKE PARR

inland in Haiti's La Visite NP. Seabird conservation requires that strong policies be enforced at sea to reduce bycatch, oil spills, and the dumping of plastics and other garbage. The Convention on the Conservation of Antarctic Marine Living Resources (CCAMLR) requires seabird bycatch avoidance measures in Antarctic fisheries. The Agreement on the Conservation of Albatrosses and Petrels (ACAP) also aims to conserve seabirds, by reducing threats such as bycatch and invasive species. ABC works with Equilibrio Azul in Ecuador, and Pro-Delphinus and APECO in Peru, to monitor and reduce seabird bycatch and the intentional take of Waved Albatrosses. Grupo de Ecología y Conservación de Islas, Island Conservation, and the Juan Fernández Islands Conservancy work to eradicate invasive species on Latin American islands. Galapagos NP has also conducted extensive goat, pig, and cat removal.

Actions:

- Establish protected areas at seabird breeding sites, control or eradicate invasive predators, and restrict the harvest of eggs and chicks. Legal protection is particularly needed for the San Benitos and other unprotected islands of Mexico's Baja Peninsula.

- Increase bycatch monitoring in artisanal and commercial fishing fleets, reduce bycatch through the adoption of bird-scaring streamers and other seabird avoidance methods; reduce intentional catch of seabirds.

- Require oil transport at sea only in double-hulled tankers to reduce risk of spills; develop spill response plans, and "wide-berth zones".

- Protect mangroves and coastal lagoons from development and aquaculture.

IMPORTANT BIRD AREA
PARACAS NATIONAL RESERVE, PERU

BIRDSCAPE: **MARINE**
AREA: **700,000 ACRES**

THREATENED SPECIES: **2**
AZE SITE

Zero Extinction
ALLIANCE FOR

Peru's Paracas National Reserve protects a desert peninsula with several offshore islands that jut out into the Humboldt Current of the Pacific Ocean. This rich environment supports huge numbers of seabirds, sea lions, and other sea life. Unfortunately, past over-fishing and guano harvesting have left seabird populations somewhat reduced over historical numbers, although it is hard to imagine that there once could

have been even more birds there than can be seen today. The main seabird nesting area is on the offshore Ballestas Islands that can be reached by daily tourist boats from the harbor in Paracas.

Birds: Breeding seabirds include the Humboldt Penguin, Peruvian Diving-Petrel, Markham's Storm-Petrel, Peruvian Pelican, Guanay Cormorant, Red-legged Cormorant, Peruvian Booby, Peruvian Tern, and Inca Tern. Migrant seabirds occurring offshore include the Swallow-tailed and other gulls, Chilean Skua, albatrosses, shearwaters, and White-vented and Hornby's Storm-Petrels.

Threats: Guano harvest and artisanal fishing occurs within the reserve. There is a natural gas terminal a few miles north of the main seabird breeding colonies, but this does not appear to have caused any problems for seabirds thus far. El Niño years cause a reduction in seabird breeding success.

PHOTO: MIKE PARR

Inca Terns are abundant at Paracas National Reserve, where they fish in near-shore waters along the coast.

Colonies of breeding Peruvian Boobies and pelicans deposit guano that is still used as fertilizer in some areas.

Conservation: The reserve is protected by the Peruvian government. ABC has been working with local partners to locate and protect Peruvian Tern nesting colonies within the park, and design a trail that will direct vehicle traffic away from nesting areas. ABC has also supported a census and threat assessment of the Peruvian-Diving Petrel, which would benefit from better population monitoring and perhaps from artificial nest burrows.

Actions:
- Determine the status of the Peruvian Diving-Petrel and Markham's Storm-Petrel within the reserve; develop and implement conservation plans for both species.
- Restore Andean Condors to the reserve.
- Restrict vehicle traffic in Peruvian Tern nesting areas.
- End guano mining/harvest in the reserve.

Guano harvests likely reduce habitat for nesting Peruvian Diving-Petrels.

INTERNATIONAL THREATS

Logging, agriculture, grazing, mining, urban and road development, and the alteration of natural flood and fire regimes are major factors that cause reduction of bird habitat in the Neotropics, much as they do in the U.S. Continually decreasing available habitat is affecting 85% of the globally Threatened birds in the Neotropics. Here we briefly discuss how the drivers of habitat loss in Latin American and the Caribbean differ from those in the U.S.

LOGGING AND DEFORESTATION

Latin America and the Caribbean accounts for 22% of the world's forest cover in just 14% of its land area. The region lost seven percent of its forest between 1990 and 2005, and accounted for more than one third of annual global forest loss from 2000 to 2005.

The logging of frontier forests occurred on a large scale when colonists first settled in North America, but has since reduced dramatically. Because a much greater area of old-growth forest exists in

Logging roads open up areas for settlement, providing access to the forest for bush meat extraction and small-scale agriculture.

PHOTO: ©ISTOCKPHOTO.COM

Latin America, frontier logging is still widely pursued in the Neotropics, and often facilitates other human activities such as agriculture. Although logging occurs legally in designated logging concessions, many legal harvests are not managed sustainably and illegal logging is widespread, especially along newly constructed or improved roads. In fact, in cloud forest regions, commercial logging is relatively minimal in comparison to small-scale clearance for pasture and agriculture.

In contrast to many clear-cutting practices in North America, logging operations in tropical forests often take only the high value timber trees, leaving behind damage along logging roads and tree-fall gaps, but much of the forest is left intact. The secondary impacts of opening up forests for high-grade timber extraction are perhaps more damaging because this also provides access for hunters to kill birds, monkeys, peccaries, and deer for food. It also offers access for settlers to clear remaining forests for agriculture. Where primary forests are cleared, forestry operations often plant trees as crops for timber harvesting. Plantations typically use non-native species such as pines and eucalyptus. In the Andes, however, native alders have been widely planted. Some tree plantations replace non-forest habitats. For example, pine plantations have been established on former grasslands in Brazil and other southern cone countries. Pines and eucalyptus are often planted in open habitats such as páramo and puna at high elevations. As in the U.S. forestry industry, certification of neotropical forestry operations can help producers manage healthier forests and help consumers purchase lumber and other forest products that have been produced sustainably.

In the U.S. and Canada, harvested wood is mostly used to produce paper products and lumber for construction and furniture. Timber harvested in Latin America is also used for construction and furniture, although a much higher percentage of wood is burned for energy compared to the U.S.,

Unlike current systems, the Incas concentrated crops on terraces, and planted trees to protect watersheds (Pisaq, Peru, pictured).

PHOTO: MIKE PARR

where other energy sources (electricity, gas, oil) are more widely available and affordable. Firewood is consumed or made into charcoal for daily heating and cooking, or to fire brick kilns. The Royal Cinclodes and Ash-breasted Tit-tyrant lose critical habitat when local communities harvest *Polylepis* in high-altitude Andean forests in southern Peru and northern Bolivia. Fortunately, providing improved wood-burning stoves to communities can significantly reduce their impact on these forests.

AGRICULTURE AND PASTURE

The vast expansion of the agricultural frontier witnessed over one-hundred years ago in the U.S. is now occurring in much of the Neotropics. Many forested areas are selectively harvested for timber, and the rest then felled, burned, and cleared for crops or pasture. New road projects open up previously inaccessible areas to farmers and ranchers who can then ship their produce and meat via these roads to distant markets. Except on volcanic slopes and floodplains, many tropical soils are poor for agriculture. Therefore, farmers might be able to grow crops on newly cleared land for only a few years. After the soils are depleted, farmers convert their lands to pasture or sell them to cattle ranchers, before moving on to the next forest patch to clear and repeat the cycle of "slash and burn" agriculture.

High global demand for palm oil, soy, and corn to feed people and cattle combined with growing numbers of people searching for farmland is driving the expansion of croplands and pastures in the Neotropics. In many countries, government policies encourage people to migrate from crowded areas to settle frontier regions, by providing land rights as incentives to people who clear and occupy lands. Unfortunately, these policies cause the destruction of habitat, sometimes in areas that are unsuitable for agriculture. Some wealthy countries (e.g., China, Saudi Arabia, the United Arab Emirates, and Kuwait) with shortages of arable land or fresh water are now buying farmland or the rights to farmland in poorer countries, to secure supplies of rice, wheat, soy, vegetable oil, and biofuels. Most of these deals involve lands in Asian and African countries, but over 240,000 acres of farmland in Brazil have been acquired by foreign governments, and more such deals could follow. Most beef and soy products exported from Brazil do not go to U.S. markets, but to Europe, Russia, the Middle East, and Asia. U.S.-funded "War on Drugs" programs that spray coca crops with herbicides also damage surrounding forest.

Forest clearance for small farms and pastures has been the main recent threat to Andean forests (Leymebamba, Peru, pictured).

PHOTO: MIKE PARR

Not only is agriculture expanding in Latin America, but it is also intensifying in many regions. Intensification promises high yields, but often requires chemical fertilizers and pesticides which pollute waterways, harm human health, and poison birds. Some crops, such as cacao (for chocolate and cocoa) and coffee, were traditionally grown in forests or under shade trees. Leguminous trees of the genus *Inga* are often planted for shade and are great for birds, providing insects and nectar for many species. Although some understory birds disappear from such agroforestry systems, many canopy species, including neotropical migrants such as the Cerulean Warbler, thrive on these farms. Therefore, such shade-coffee or shade-cacao farms provide important habitat for some birds and other wildlife. Newer varieties of coffee and cacao were developed to produce higher yields of lower quality beans, but require full sun and heavy use of pesticides (some of which are no longer used in the U.S. due to their wildlife impacts). Therefore, the conversion of shade coffee and cacao to sun plantations ruins vast areas of habitat for birds in Latin America. As a solution to this problem, conservation groups (including the Rainforest Alliance, Smithsonian Migratory Bird Center, and ABC) are working to help certify and market shade-grown coffee to help farmers maintain trees on their lands. In the future, planting shade trees might also earn these farmers carbon credits for the additional carbon fixed by their plantations.

Clearing forests for pasture is one of the main drivers of habitat loss in Latin America, but pastures occupying previously forested areas offer prime opportunities for partial reforestation. ABC has studied and promoted silvipasture practices, including maintaining tracts of native trees on cattle farms, planting native trees and grasses in pastures, fencing riparian corridors to reduce damage by cattle, using trees in hedgerows as live cattle fences and windbreaks, and reducing the use of fire and pesticides. These practices provide cattle with additional forage and shade, provide ranchers with timber they can eventually harvest, restore components of the native ecosystem for wildlife, sequester carbon,

PHOTO: MIKE PARR

Pollution from mining often flows straight into rivers, harming birds and people downstream (mine near Ticlio, Peru, pictured).

and improve local water quality. Pastures incorporating these practices in Nicaragua, Costa Rica, and Colombia support a greater abundance and diversity of migrant and resident birds compared to typical pastures of exotic grasses with few trees. Wider adoption of silvipasture practices could help preserve important ecosystem components in degraded landscapes, including creating connectivity between protected areas, and improving buffer zone habitat.

MINING

Mining is big business in Latin America, just as it is in the U.S. In Latin America, mine operators range from large corporations to small artisanal miners, and extract minerals, precious metals, oil, and natural gas. For example, gold is Peru's top export, and is mined at giant sites in the Andes as well as by artisanal miners along lowland rivers. Mining along with the associated roads, pipelines, and other infrastructure harms birds by destroying habitat and by emitting toxic effluent. In much of Latin America, environmental laws governing mining are either weak, or weakly enforced (not unlike the Appalachian region of the U.S.). While the direct footprint of mining may not be that great, some neotropical birds have tiny ranges, and these species are much more vulnerable to individual mining projects or their toxic effluent. For example, mine tailings contaminate Lake Junin and Lake Titicaca, both home to endemic grebe species (see p. 382). Finally, bauxite mining to produce aluminum destroys many Caribbean forests, and is a major threat to the Black-billed Parrot, Jamaican Blackbird, and other endemic species on the island of Jamaica.

DEVELOPMENT

Land development for residential, commercial, and industrial uses displaces natural habitats worldwide. Latin America is home to some of the largest and most vibrant cities in the world. These include the metropolitan areas of São Paulo (22.1 million people) and Rio de Janeiro (14.4 million), Brazil; Mexico City, Mexico (19 million people); Buenos Aires, Argentina (13.3 million people); Lima, Peru (8.4 million people); and Bogotá, Colombia (7.8 million people). Many cities are located along coasts or major rivers, just as they are in the U.S., but unlike the U.S., some of Latin America's largest cities are also located in the highlands or in mountain valleys (e.g., in Mexico, Costa Rica, Colombia, Ecuador, and Bolivia) due to more moderate climates and richer agricultural soils at those elevations. Cities have enormous ecological footprints on local and regional environments through resources imported for consumption and

waste products that pollute water and air. Despite the severe local environmental effects of development, relatively few neotropical species are primarily threatened by habitat loss due to development, but rather are affected by the agricultural expansion and resource exploitation (logging and mining) required to support these urban areas.

Birds most threatened by development occur in wetlands, along coasts, or are restricted to islands. The use of water to supply thirsty cities has severely reduced populations of wetland species in Mexico and Colombia, as well as birds that specialize in coastal scrub, such as the restinga habitat in Brazil. Development in the Caribbean Islands threatens to destroy the last remaining patches of habitat for several species, such as the Grenada Dove, White-breasted Thrasher, and Cuban Sparrow. Developed areas are also sources of invasive predators, such as rats and feral cats that harm island birds. Development is also occurring in Mexico, Central

These small houses in La Paz Bolivia are typical of those in cities around the region, many of which are growing rapidly.

PHOTO: MIKE PARR

The use of fire to clear pasture is degrading ecosystems that rarely burn naturally, such as those here in the Marañon Valley, Peru.

PHOTO: MIKE PARR

America, and the Caribbean to satisfy demands for homes or resorts catering to wealthy nationals as well as retirees and tourists from other countries.

Roads are essential for urban and rural development and are frequently built to connect rural areas to markets. The secondary effects of roads often have a larger impact on forests and bird populations because roads enable colonization by settlers. For example, Brazil's trans-Amazonian highway stretches over 2,000 miles though the Amazon and has opened up large areas to settlement and land clearance.

FIRE AND WATER MANAGEMENT

In the U.S., fire suppression has caused widespread habitat degradation in many grassland and forest ecosystems. In much of the Neotropics, however, increased fire frequency is more of a problem than fire suppression. Most of the arid or semi-arid lands of the Neotropics are affected by intentional burning, which can lead to desertification. People set fires purposefully to clear forest, create pastures, promote fresh grass growth for cattle grazing, and to burn trash. Unlike temperate ecosystems that are fire-dependent, many tropical grasslands and broad-leaf forests naturally burn rarely, if ever. In these areas, fires can severely damage vegetation,

which may recover very slowly. For instance, fires inhibit *Polylepis* regeneration and damage páramo. Increased fire frequency also degrades cerrado habitats and the understory of pine-oak woodlands in Mexico. Fire combined with over-grazing is particularly damaging, affecting dry forests in inter-Andean valleys and many other regions.

Dams, levees, and canals alter rivers, drain wetlands, block migratory fish, and reduce natural floods in the U.S. and in the Neotropics. Fortunately, they are less common in most of the Neotropics compared to the U.S. Major water projects do exist in the Neotropics, however, and more are on their way. For example, major new hydroelectric dams are planned for the Madeira River in Brazil, which is a tributary of the Amazon. Where such dams are constructed, reduced flow degrades flood-dependent habitats downstream, and reservoirs created behind dams inundate terrestrial habitats upstream. For instance, dam building has destroyed much of the fast-flowing river and stream-side forest habitat required by the Brazilian Merganser. The sites where Brazil's Marsh Tapaculo and Venezuela's Carrizal Seedeater were first discovered were flooded behind new dams. Fortunately, the Carrizal Seedeater has since been rediscovered elsewhere, but plans for additional dam construction to supply more water to the city of Curitiba threatens remaining habitat for the Marsh Tapaculo.

SOCIAL AND ECONOMIC ISSUES

The root cause of virtually all environmental issues is growing consumption rates per person combined with growing human populations. Consumption both requires extraction of natural resources as well as disposal of waste products that can pollute the environment. A growing population means that, even if consumption levels are stable, with more people, society as a whole demands more from the environment. Developed nations like the U.S. have relatively low birthrates and mainly grow through immigration, but their citizens have the highest rates of consumption in the world. To satisfy this demand, people in the U.S. import fruits, vegetables, timber, minerals, fossil fuels, and other natural resources. In contrast, many developing countries have much lower *per capita* consumption rates, but high birthrates and rapidly growing populations. Current consumption rates are not stagnant, however, as people around the world strive to attain higher living standards. Globalization and the integration of national economies through free trade agreements are accelerating this process throughout the Americas. For instance, global demand for energy resources is expanding. To meet this demand, more oil drilling, pipelines, gas wells, wind turbines, dams, and crops for biofuels are being produced. When the U.S. increased biofuel production in 2007, many croplands previously growing corn or soy for food were used for ethanol production to meet the new demand. Besides reducing conservation programs on agricultural lands, additional demand for corn and land also raised global prices for both corn and soy used for food and animal feed. Even in Brazil, where there was an established market for domestic biofuels, increased soy prices in the global market stimulated accelerated clearing of Amazonian rainforests, which destroyed habitat and released huge quantities of greenhouse gases. Although growing biofuels in the U.S. may be a positive step towards energy independence, corn-based biofuel production probably causes a net increase in greenhouse gas emissions, particularly if distant rainforests are cut as an indirect result. Globalized markets not only link trade, but also link the environmental consequences of economic policies in complex ways that are not always easy to predict. When development goals are directly linked to reducing the exploitation or extraction of natural resources, however, such projects can benefit conservation. For instance, ECOAN has helped reduce the cutting of *Polylepis* trees for firewood in Peru by providing communities with fuel-efficient stoves that required less fuel, and is providing fuelwood plantations, in exchange for the communities protecting and restoring their *Polylepis* woodlands.

INVASIVE AND OVERABUNDANT SPECIES

Exotic species and pathogens are a threat to native wildlife all over the world. In Latin America, as elsewhere, island ecosystems are most susceptible and provide examples of the most devastating impacts. Invasive predators prey on birds and their offspring and grazing animals destroy habitat. Rats, mongooses, feral hogs, cats, donkeys, sheep, and goats threaten many neotropical species, especially on oceanic archipelagos such as the Galapagos Islands, Juan Fernández Islands, Caribbean Islands, and islands off the west coast of Mexico such as Socorro and those around Baja California. In addition to these, American mink (invasive in southern Argentina and Chile) and feral dogs predate nests and kill native birds. Introduced species can become invasive on the mainland as well. For instance, North American beavers were introduced to southern South America where they damage native temperate forests in Argentina and Chile.

Archipelagos such as Galapagos and Juan Fernández are also vulnerable to invasive diseases, parasites, insects, and plants. Parasitic blow flies (*Philornis downsi*) have severely affected the nesting success of many species of "Darwin's finches" and other passerine land-birds in Galapagos, including the Medium Tree Finch and Mangrove Finch;

PHOTO: MIKE PARR

Sheep have devastated the natural vegetation on Socorro Island off the Mexican coast, reducing habitat for endemic birds.

avian pox has reduced survivorship in some of the same species. After spreading through the "Lower 48" states in the U.S., West Nile virus kept on spreading south into the Neotropics where its impacts on native birds are much harder to measure. Elm-leafed Blackberry, maqui, and other plants are overwhelming native forests on the Juan Fernández Islands, habitats which are essential to the Juan Fernández Firecrown and Tit-Tyrant. In the Galapagos, several species of blackberries, lantanas, and trees such as guava and quinine have conquered extensive areas of natural vegetation, destroying vital habitat for land-birds, and also threatening nesting colonies of seabirds, such as the Galapagos Petrel.

People have unintentionally improved conditions for cowbirds in parts of the U.S., harming bird populations through brood parasitism. No fewer than five species of parasitic cowbirds are native to the Neotropics. Thanks to land-clearance by people, Shiny Cowbirds have invaded most Caribbean Islands and increased in numbers, parasitizing endangered species such as the Jamaican and Yellow-shouldered Blackbirds, and causing population declines among the Puerto Rican Vireo and vulnerable Martinique Oriole. Parasitism by overabundant Shiny Cowbirds in southern cone grasslands may also exacerbate threats facing species such as the vulnerable Saffron-cowled Blackbird. House Sparrows and Rock Pigeons have spread into the Neotropics, and many kinds of pet birds have escaped from captivity to establish feral populations in cities and other degraded areas. The effects of these exotic birds on native birds are as yet poorly known.

COLLISIONS

Few studies have been conducted on bird mortality rates from collisions with glass windows, towers, and other structures in the Neotropics. Collisions certainly occur, however, affecting both resident and migratory species. As cities grow and new wind farms, communications towers, and other forms of modern infrastructure are built, the number of obstacles facing migrant and resident birds will in-

Many songbirds and parrots are threatened by the pet trade, including the Threatened Gray-cheeked Parakeet (above).

PHOTO: MIKE PARR

crease the risk of collisions. It is therefore important that these new structures are designed, sited, and built in ways to minimize collision risk to birds. In particular, large numbers of migrant raptors and songbirds pass through southern Oaxaca in Mexico, where a new wind farm could potentially kill many birds. In addition, rural electrification programs in Latin America should incorporate designs for utility poles that minimize electrocution risk to large raptors and other birds with large wingspans.

EXPLOITATION

Roughly 30% of globally threatened birds worldwide are threatened by overexploitation, victims of hunting for food and trapping for sale as caged birds, particularly in countries such as China and Indonesia, but also in the Neotropics. Direct exploitation and persecution of birds is presently much more problematic in the Neotropics than in the U.S. Following the precipitous declines and extinction of several native birds, U.S. lawmakers enacted the Lacey Act in 1900. As a result, relatively few species in the U.S. are currently hunted. Furthermore, bird hunting in the U.S. is regulated with seasons and bag limits, and habitat is widely managed to benefit game species. Where hunting laws exist in the Neotropics, enforcement is often lacking. A wide variety of birds and their eggs are harvested for food. These include rheas, tinamous, coots, ducks, quail, curassows, guans, chachalacas, turkeys, seabirds at colonies, as well as some less-expected species such as macaws. Still more species are captured for the caged bird trade. These include parrots, parakeets, and macaws; songsters such as solitaires, mockingbirds, thrushes, saltators, and various finches; and colorful birds such as toucans, troupials, orioles, and buntings. Most trapped birds are resident species, but thousands of wintering Painted Buntings and Baltimore Orioles are captured annually in Latin American and Caribbean countries, such as Mexico and Cuba.

Since the 1960s, 175 nations have joined the Convention on International Trade in Endangered Species of Wild Fauna and Flora (CITES). This international agreement regulates and restricts international (but not internal) trade in birds and other wildlife to protect threatened species. In 1992, the U.S. banned the import of wild birds under the Wild Bird Conservation Act, though illegal smuggling continues. In 2005, the European Union temporarily banned the import of wild birds out of concern over avian influenza, but made this ban permanent in 2007 to protect the birds themselves. In 2008, Mexico banned the capture and export of wild parrots, following a report issued by Defenders of Wildlife and Teyeliz A. C. which exposed that at least 78,500 parrots and macaws were captured in Mexico annually, and that over 50,000 of these died before being sold. Mexico's ban is significant in that it protects wild parrots from both the international and domestic caged bird markets. Most other neotropical countries have legal bans or restrictions on the capture of birds for trade, but enforcement is often poor or non-existent. Asociación Armonía has been monitoring bird trade in Bolivia and Peru, showing that there is an active trade route between Brazil, through Bolivia and on to Peru, where many birds leave the continent, presumably for markets in Asia (currently the fastest growing market for caged birds).

Persecution of birds by farmers is also problematic. For example, most of the world's Dickcissels winter in the Llanos of Venezuela, where they consume rice and sorghum crops and congregate in huge roosts at night. Dickcissels are sometimes illegally poisoned as agricultural pests by farmers who aerially spray roost sites at night with organophosphate pesticides. A few of the larger roost sites can contain more than 10% of the global population. Therefore, Dickcissel numbers are vulnerable to declines if major roost sites are targeted by spraying. The Harpy Eagle, Crowned Eagle, and other raptors are frequently shot by farmers in frontier regions afraid for their poultry or livestock. The Red-fronted Macaw, an Endangered species from central Bolivia, is often confused with other parrot species and persecuted by farmers for consuming crops.

Another threat to some rare birds is the often redundant collection of specimens for museums. In most cases, the collection of a few specimens for scientific purposes has no effect on populations, but the collection of some especially rare species may. The collection of any endangered species should be avoided, and there needs to be a strong conservation-related scientific justification for the collection of any specimen.

Bird impacts from exploitation can also be secondary rather than direct. Overexploitation of palms is a threat to some bird species in South America. In the Andes, the Yellow-eared Parrot and Golden-plumed Parakeet depend on palms for nest holes. Local people harvest palm fronds by chopping down trees for Easter celebrations (Palm Sunday), thereby reducing habitat for the parrots. Fundación ProAves, Fundación Jocotoco, and

PHOTO: ROBERT ROYSE

Dickcissels congregate at huge winter roosts in Venezuela, and are targeted with pesticides in rice fields.

PHOTO: MIKE PARR

The central Andes of Peru are heavily mined for metals; La Oroya, is one of the country's most polluted towns.

ABC have worked with churches and communities to end this practice by replacing palm fronds with more sustainable alternatives. In the Amazon, moriche palms form vast swamps. Female trees are often cut down to harvest the fruit, leaving only the male trees standing. Cutting down palms to harvest their fruit is not sustainable and threatens an entire habitat type that is important for several specialist birds (Point-tailed Palmcreeper, Sulphury Flycatcher, Moriche Oriole) and some parrots (Blue-and-yellow Macaw, Red-bellied Macaw) that nest in palm swamps.

Bycatch in fisheries kills birds in the Neotropics as well. Artisanal fishing with gillnets is widespread, causing the accidental death of many grebes, cormorants, and other diving birds. Artisanal longline fleets operate off the coasts of Ecuador and Peru, and large commercial longliners operate off the coasts of Brazil, Uruguay, Argentina, and Chile. These fisheries affect the populations of many seabirds, especially albatrosses and petrels. In Latin America, there is generally less monitoring to track bycatch problems and the lack of a regulatory framework to resolve such problems once detected.

POLLUTION

Use of pesticides harms both resident and migrant birds in neotropical landscapes, particularly on and near farms. Most Latin American countries trail the U.S. in regulating the use of pesticides and testing imported produce for pesticide residues. Many pesticides, such as DDT, carbofuran, and monocrotophos, are banned or severely restricted in Canada and the U.S., but they are still used in much of Latin America. For instance, monocrotophos was removed from U.S. markets in the 1980s, then banned in Argentina in 1996 following ABC intervention after it had killed tens of thousands of Swainson's Hawks there. ABC is currently working with partners to end the use of monocrotophos in Bolivia, where it threatens wintering Bobolinks and other birds on rice farms. Larger quantities of pesticides are also often applied to farms in the Neotropics than to those in the U.S. For instance, less than five pounds of pesticide active ingredient is applied per acre of fruits and vegetables in the U.S., but in Costa Rica, this amount rises to about 18 for most fruits and vegetables, and 40 pounds per acre for bananas.

Heavy metal contamination also harms neotropical birds. Raptors and other scavengers become contaminated by eating carcasses contaminated with lead ammunition fragments. In Argentina, about 1,600 tons of lead shot are deposited by dove hunters in northwestern Córdoba alone, and lead contamination has been documented in Andean Condors, Crowned Eagles, and waterfowl. Mercury, from mine tailings or the burning of fossil fuels is also toxic to birds and has been found in the blood of forest-falcons and other species deep in the Amazon.

Pollution from oil extraction and oil spills at sea also harm neotropical birds. In 2000, over four-million gallons of crude oil burst from a pipeline into Brazil's Iguaçu River, threatening the Brazilian Merganser and other wildlife. Chronic oil pollution from boat ballast killed over 40,000 Magellanic Penguins annually along the coast of Argentina's Chubut province from 1982 to 1991. Tanker lanes were moved c. 100 miles further offshore in 1994; still, the total number of oiled penguins found along the southwestern Atlantic coast of South America has increased since the mid-1990s and is correlated with a rise in oil exports from Argentina. Greater enforcement of national and international regulations prohibiting oil discharge at sea is necessary to protect penguins and other seabirds in Latin America. Oil pits are also problematic, polluting watersheds where oil extraction has occurred. By 1998, Texaco had cleaned up 161 oil pits constructed during the 1960s near Lago Agrio in the Ecuadorian Amazon, but legal battles persist today to determine whether Texaco or the Ecuadorian state-owned oil company Petroecuador should be responsible for damages and clean up associated with hundreds of remaining pits.

Excess nutrients from agricultural runoff, plastics and other garbage, and untreated sewage and municipal waste pollute many wetlands and waterways in Latin America. Although waste production per person is much lower in Latin America than in the U.S., a lack of government-run water treatment and waste disposal programs, combined with littering and dumping of waste in rivers causes widespread problems.

SEVERE WEATHER AND CLIMATE CHANGE

Severe weather events such as hurricanes, droughts, or late freezes can damage habitat or harm bird populations in the U.S. as well as in the Neotropics. Although freezes do not occur in most of the Neotropics (except at high elevations and high latitudes), droughts and severe storms do. Hurricanes form in tropical waters, and therefore are a frequent occurrence in much of the Caribbean and western Mexico. Because of the topography of the land and the scale of deforestation in the Neotropics, mountainous areas where forests have largely been removed are now vulnerable to extreme erosion, especially during severe storms. News stories of hurricanes causing entire mountain sides to collapse into mud-slides that destroy towns and human lives down-slope are tragically predictable. Many neotropical forest soils are fragile, particularly in the lowlands, and the erosion caused by deforestation often renders them rapidly infertile and unsuitable for agriculture.

Birds restricted to small habitat fragments in the Caribbean are more vulnerable to hurricanes or volcanic eruptions. In 1989, Hurricane Hugo damaged forests and nesting trees in Puerto Rico, cutting the population of the Puerto Rican Amazon from 47 to roughly 23 birds. Volcanic eruptions that began in 1995 on Montserrat have produced ash-falls, pyroclastic flows, and acid rain that destroyed 60% of the Montserrat Oriole's habitat.

The El Niño Southern Oscillation (ENSO) phenomenon is a cyclical period of warming surface waters in the eastern Pacific that often drastically reduces food supplies for seabirds. El Niño also causes climatic changes on land in the Neotropics, including warmer and wetter conditions in many areas near the Pacific Coast, but also drier and hotter conditions in parts of Amazonia. La Niña events may occur where colder than normal sur-

PHOTO: ©ISTOCKPHOTO.COM

Climate change is challenging Antarctic birds (King Penguins above), while El Niño events reduce food supplies for tropical seabirds.

face water temperatures spread across the eastern tropical Pacific, causing droughts and unusually cool temperatures on land nearby. These climactic swings can produce booms and busts for land- and seabirds. Although these are natural events, endangered species, especially seabirds, are at greater risk of extinction during ENSO cycles if their populations are already stressed by anthropogenic factors such as over-fishing, bycatch, or introduced predators on breeding islands. Global warming could increase the severity and frequency of El Niño and La Niña years, decreasing the long-term probability of survival for some species.

Climate change is already occurring in much of the Neotropics. The deforestation of lowland forests and their replacement with pastures reduces local humidity and regional rainfall. Cloud forests near deforested lowlands may not receive enough cloud cover and deforested areas may be slower to recover. Global warming caused by greenhouse gas emissions is melting glaciers throughout the world, including those in the Andes, and is also melting Antarctic ice sheets and reducing sea ice.

Such climate changes will require penguins and other seabirds to adjust their foraging and breeding cycles and locations. The resultant sea level rise could inundate low-lying coastal marshes, mangroves, and offshore islands, negatively affecting water-bird and seabird nesting and foraging areas. Lower-elevation cloud forests in Costa Rica are becoming less cloudy as global warming causes clouds to form higher upslope. Disease-spreading vectors such as mosquitoes may also move to higher altitudes. This change of climate also helps lowland bird species move upslope, sometimes displacing montane species. In the case of some high mountain species, there may be no place left to go. Ultimately, climate change could have a huge effect on the distribution of current habitats and bird ranges. With so many more restricted-range species in the Neotropics than in North America, climate change will increasingly challenge conservation efforts as it becomes important to protect not only habitats that are needed now, but also new areas inshore from inundated shorelines and upslope from the present day ranges of some montane species.

INTERNATIONAL CONSERVATION ACTION

While conservation tools are often similar around the world—the need for protected areas, emergency efforts for endangered species, and strong laws to protect the environment—one often-overlooked need is that for capacity and resources among conservation organizations. Without these organizations, and the concern of private citizens they represent, conservation would not have an effective voice in the world. Building this capacity in Latin America and the Caribbean is therefore a high priority for bird conservation. While there are many excellent scientists and individuals working for birds throughout the region, with some notable exceptions (some of which are highlighted earlier in this chapter), there is a paucity of strong national organizations dedicated to birds. Supporting the development of such groups, while perhaps not as immediately rewarding as supporting the purchase of a reserve, will ultimately be critical to the long-term success of conservation.

Land rights are often determined very differently in Latin American countries than they are in the U.S. Land occupation is often demonstrated by environmentally destructive land "improvements," such as clearing forest, farming crops, grazing cattle, or constructing buildings. These actions are often encouraged by national policies aimed at stimulating migration from crowded regions to settle frontier areas and expand territorial control. Such policies can result in controversial land tenure conflicts. For instance, only 14% of privately held lands in the Brazilian Amazon are backed by secure deeds of ownership. In addition to issues surrounding the land rights of individuals, conflicts also involve the land rights of communities and indigenous groups. Indigenous reserves face enforcement problems similar to those of nature reserves, as well as conflicts with government contracts and programs to expand agriculture, oil drilling, and mining. For example, about 70% of Peru's 173 million acres of

rainforest have been offered or granted as concessions for oil and gas exploration, despite some of these areas overlapping tribal lands either claimed by or already titled to indigenous people.

The legal infrastructure defining land rights is just one difference. Legal regulations to maintain clean air and water, and to protect endangered species by requiring review of government development projects are not yet on the books in many Latin American nations. Where laws are written, enforcement is often insufficient to implement these regulations. Law enforcement is a problem for many NPs and reserves with insufficient funding or staff to protect boundaries and prevent hunting or other resource extraction. These are sometimes known as "paper parks" because the protection outlined on maps is not matched by reality on the ground. Despite legal protection, poaching birds and other animals, illegal timber harvests, and other activities, such as mining and clearing land for agriculture, occur within the boundaries of poorly guarded paper parks. Los Haitises NP in the Dominican Republic is just one of many protected areas in Latin America and the Caribbean that has suffered from incursion by people, inadequate land management, and uncertainties surrounding land

PHOTO: MIKE PARR

Peru's Junin National Reserve is protected, but still threatened by water management and mining effluent.

PHOTO: MIKE PARR

Two AZE birds, the Long-whiskered Owlet and Ochre-fronted Antpitta, occur at Abra Patricia, a private reserve owned by ECOAN.

ownership. Despite enforcement problems, parks still do a much better job at protecting nature than areas outside of protected areas.

HABITAT PROTECTION

Protecting terrestrial and marine habitats by restricting human activities within protected areas is critical for preserving biodiversity, preventing species extinctions, and safeguarding ecosystem services such as clean water, carbon stores, and source populations for commercially harvested timber, fish, medicinal plants, or other species.

Just as the U.S. has many national and state parks, neotropical countries have a variety of protected areas including NPs, and regional and municipal reserves. In addition to these government-owned protected areas, there are many privately- or communally-owned protected areas that significantly contribute to bird conservation. These include reserves set aside for indigenous peoples, community

reserves, private reserves owned by individuals or conservation groups, and conservation concessions controlled by leases or easements. Concessions are typically government lands leased to private groups for timber or other resource extraction, but recently conservation groups and ecotourism companies have begun taking on management agreements for such properties.

In some countries where habitat destruction has been extreme (El Salvador, Haiti) parks are about the only places where natural habitat remains. Many other countries in Latin America and the Caribbean experiencing rapid rates of habitat loss are quickly approaching similarly dire circumstances. Despite enforcement problems, habitat loss is still far more likely to occur outside officially designated protected areas, even paper parks. Parks in the U.S. are highly valued by its citizenry and foreign tourists, but these parks often faced resistance at the time of establishment and positive public attitudes required decades to develop. Younger and more recently established protected areas in Latin

Andean communities near Abra Malaga, Peru, are planting thousands of Polylepis *trees to restore habitat for birds.*

PHOTO: MIKE PARR

FINANCING PROTECTED AREAS

At the largest scale, monetary tools can help countries finance the establishment and management of parks and other protected areas. For instance, organizations such as Conservation International, TNC, and the World Wildlife Fund pioneered the use of debt-for-nature swaps to finance large-scale conservation projects. During the 1980s when many Latin American countries faced debt crises, these transactions allowed private organizations to purchase debt, or donor countries to forgive debt in exchange for commitments to establish parks or implement environmental policies. Forgiven debts are instead paid into trust funds that finance conservation activities. In 1998, the U.S. passed the Tropical Forest Conservation Act (TFCA) to provide funds for tropical forest conservation in eligible countries through debt forgiveness. As of 2006, this act had spent $83 million to leverage over $135 million through 12 agreements in 11 countries, including Belize, Colombia, El Salvador, Guatemala, Jamaica, Panama, Paraguay, and Peru. Debt-for-nature swap funds in Peru helped finance ABC and ECOAN's protection of *Polylepis* forests. More recently, a 2007 debt swap forgave $26 million of Costa Rica's debt to finance forest protection programs worth the same amount for 16 years. This deal was facilitated by TNC and Conservation International, which each contributed $1.26 million, with the U.S. government contributing $12.6 million under the TFCA.

Today, the emergence of carbon markets presents enormous opportunities to help advance forest conservation in tropical countries. To reduce greenhouse gas emissions and address global warming, many policy makers advocate cap and trade permits to emitters of carbon dioxide. In such a market, landowners or governments that can store carbon in forest reserves or through forest restoration could be paid by nations, companies, or individuals looking to offset their emissions. Planting new trees and protecting old trees from being cut down is critical to reducing greenhouse

America are still developing public support as tourism and recreational usage of them increases.

Approximately 5% of the world's terrestrial area is formally protected. This percentage is not sufficient to protect many species. Furthermore, much of the land in protected areas was chosen to safeguard stunning scenery of rock and ice of relatively low value for birds and biodiversity. As a result, richer lands such as grasslands and alluvial plains may be under-represented in protected area systems as these lands are highly valued for agriculture. IBAs, AZE sites, and Strongholds are all useful to help prioritize sites for bird conservation in such areas. Habitat management is often required to restore previously degraded areas after land has been acquired for a protected area, or to establish corridors between reserves to improve habitat connectivity. For example, ABC and its partners have established tree nurseries near many of the reserves they have established. These nurseries supply saplings to reforest degraded land within buffer areas surrounding reserves to reduce fragmentation effects. Saplings are also planted to provide shade trees for coffee or cacao plantations and for silvipasture.

gasses. A voluntary carbon market devoted to keeping forests intact already exists and is expanding. This market was estimated to be worth $355 million in 2007, growing to $705 million in 2008. To protect existing forests, the United Nations is leading the Reduced Emissions from Deforestation and Degradation (REDD) initiative, which would direct carbon credit payments to countries that maintain forests. REDD is still being developed, and its successful implementation will depend on accountable governance to ensure that payments are indeed resulting in protected forests.

COMMUNITY CONSERVATION

Local support for protected areas can be enhanced if the loss of economic benefits from restricted activities can be replaced with alternative opportunities. Local people often require job training to take advantage of such alternatives, however. In Colombia, ABC supports Fundación ProAves' Women in Conservation Program, which trains women living near three of ProAves' reserves to make jewelry, belts, shoes, puppets and other products for sale to

visitors. The sale of these products at "eco-shops" at ProAves' reserves and at free-trade stores in the United Kingdom and U.S. has increased the annual household income of participating families by 70-100%, and is raising awareness and support for biodiversity conservation. In exchange for participation, families sign agreements with ProAves promising not to hunt or cut wood within the reserves. ProAves' park guards have reported a drop in these activities as a result. Interest in being part of these groups has grown significantly in each community, as has the demand for their products.

NATURE TOURISM

Making conservation financially sustainable is critical to its success. One tool for financing conservation is ecotourism. Tourism-generated income can support local economies and create jobs related to food, lodging, and nature-guiding; jobs that depend on the natural attractions being protected. Meanwhile, tourists pay park or reserve entrance fees that support protected area management. Tourism based on birdwatching is growing rapidly in the Neotropics. Just as U.S. states such

Tree nursery at Pomacochas Peru growing a food plant (Erythrina) *for the Marvelous Spatuletail.*

PHOTO: MIKE PARR

as Texas and Florida have developed birding trails to boost tourism, similar birding routes are being developed in Latin American countries, including Colombia, Ecuador, and Peru. For private protected areas that cannot be financed by ecotourism or other sustainable activities, endowment funds and carbon finance agreements may be able to support management costs.

Tourism is a huge international industry that is critically important to the economies of many Caribbean Island nations, as well as portions of Mexico, Costa Rica, Peru, and other countries in the Neotropics. Attractions include vibrant cities, unique cultures, ancient ruins (e.g., Peru's Machu Picchu, Guatemala's Tikal, Mexico's Chichen Itza), sandy beaches, and natural wonders like mountain peaks, waterfalls (e.g., Angel Falls in Venezuela, Iguaçu Falls along Argentina and Brazil's border), and wildlife watching.

Tourism can be detrimental to bird conservation where it drives urban expansion in sensitive or rare habitats on islands or along coasts. In contrast, ecotourism is defined by the International Ecotourism Society as "responsible travel to natural areas that conserves the environment and the well-being of local people." Ecotourism is an important tool to help sustainably finance conservation when linked to the protection of birds or natural areas.

Ecotourism based on natural attractions such as wildlife-watching is not an appropriate conservation strategy everywhere. Destinations are more likely to attract ecotourists if they are safe, easily accessible along major transportation routes, offer comfortable lodging and dining services, and provide unique wildlife-watching experiences (such as congregations of wildlife or unique and charismatic species). For instance, even non-birders are drawn to visit congregations of species such as macaws and other parrots at clay licks, penguin and other seabird colonies, and manakins and Cock-of-the-Rocks performing their courtship displays. Birdwatchers are more adventurous than average ecotourists and are often willing to travel further and stay in more rustic accommodation to see birds that are difficult to find elsewhere.

Even ecotourism must be carefully managed to ensure that visitors minimize damage to vegetation and do not disturb wildlife. Trail systems can help restrict visitor activity to minimize erosion. Popular parks can also restrict the number of visitors allowed per day. Blinds are also useful to minimize disturbance to displaying birds. Antpittas can be habituated to visit feeding stations as an alternative to the use of tape playback. Birdwatchers should be extra careful to avoid the use of recorded vocalizations to attract birds in frequently visited areas.

Ecotourism at Paracas, Peru; this is a growing industry that can help to finance park management and support local businesses.

PHOTO: MIKE PARR

ABC is working with partner organizations in Latin America to develop bird tourism to sustainably finance management activities at 18 private re reserves established to protect endangered birds. As part of this effort, ABC launched the Conservation Birding website (www.conservationbirding.org) which lists lodges in Latin America that cater to birders, and whose activities support conservation efforts. Visit this website and consider one of the many fabulous locations to make your next vacation one for the birds!

PHOTO: MIKE PARR

Approximately 40% tree cover is needed to make shade coffee suitable for neotropical migrants.

CONSERVATION COFFEE

One third of the world's coffee is consumed by Americans, therefore our choices as consumers can affect how coffee is produced to benefit birds. Coffee was traditionally grown under a canopy of shade trees, which provide habitat for migrant birds. Beginning in the late 1960s, farms began to clear these trees and to plant new varieties of coffee that grew in open sun. Sun varieties produce higher yields, but require more pesticides, and reduce habitat for migrant birds and other wildlife. In response, conservationists encourage coffee farmers to maintain their coffee in shade-grown systems by helping them market shade-grown coffee for a premium, based on its higher quality and the benefits of preserving habitat for migratory birds. A variety of such programs exist; for example, Rainforest Alliance certifies coffee as shade-grown and fair-trade (where workers are paid decent wages and provided other benefits) through its Sustainable Agriculture Network in Mexico, Central America, Brazil, Colombia, and Peru. Since the late 1990s, the Smithsonian Migratory Bird Center has also marketed certified organic and shade-grown coffee from roughly 2,000 producers on more than

15,000 acres of farmland in eight neotropical countries under the "Bird Friendly" label, much of which is also fair-trade. Such marketing efforts often directly support bird conservation projects. For instance, ABC and Fundación ProAves established the 518-acre Cerulean Warbler Reserve in Colombia, which includes a 25-acre coffee farm used by Cerulean Warblers. This coffee farm was certified by the Rainforest Alliance and produces coffee fetching higher prices from wholesalers. ProAves is also marketing coffee grown at a similar farm close to the reserve to maintain shade trees in the surrounding landscape. You can purchase this coffee to benefit Cerulean Warblers through the Thanksgiving Coffee Company. The emergence of carbon markets may help make shade-coffee more competitive against sun-coffee. ABC is also helping to develop climate-friendly coffee, where shade-coffee farms can market the carbon-sequestration benefits of maintaining shade trees compared to clearing trees for sun coffee.

INTERNATIONAL VEGETATION TERMS

Caatinga arid woodland in northeastern Brazil.
Campos and Pampas open grasslands.
Cerrado savanna woodland in central Brazil.
Chaco arid woodland of interior Argentina, Paraguay, and Bolivia.
Gallery forest riverine forest in arid non-forested regions.
Elfin forest stunted mountain forests.
Llanos and Pantanal seasonally flooded grasslands.
Páramo humid high-elevation shrub-land.
Patagonian steppe cool arid shrub-land intermixed with grassland.
***Polylepis* woodland** high-elevation woodland of *Polylepis* trees.
Puna high-altitude grassland.
Restinga stunted forest of Brazil's sandy coast.
Southern temperate forest cool humid forests in southern South America.
Tumbesian Region deserts, scrubs, and dry forests in N. Peru and S. Ecuador.
White-sand forest forest growing in sandy soils.

CHAPTER 5

STRATEGIES AND ACTIONS

Congratulations on arriving at the most important part of this book. If you have read this far, we hope you are asking yourself "How can I get more involved and make a difference? How can I help solve the problems facing birds and their habitats?" In this chapter, we provide answers to these questions, along with a guide to bird conservation strategies and actions that are already underway—to help you become more engaged in bird conservation efforts. In each preceding chapter we have listed the priority actions that need to be carried out to advance bird conservation (by individuals, groups, governments, and corporations). Here we focus on how we can implement those important tasks.

Conservation happens at global, national, and local scales, and you can influence each one. You can help to address the top priorities for your county or state by managing your own land for conservation if you are a landowner, by joining your state Audubon chapter (or other bird organization) and local land trust and supporting their work, or by joining a "friends of" group that works to protect an Important Bird Area. Laws and ordinances are often important for bird conservation too, such as those relating to the construction of communication towers, wind turbines, and the management of feral cats. Finding out about the introduction of bills and related hearings is often not as easy as it should be, and tracking legislation can be a time-consuming task. Networking with other people concerned about these issues can help get the word out locally, and by becoming a member of ABC (visit www.abcbirds.org) you can learn about national and international campaigns that also need

your support, and the support of your local bird group.

The famous phrase "think globally, act locally" can send a misleading message when it comes to bird conservation. Action at a local scale is important, but we can't hope to conserve migratory birds by restricting our efforts to local communities or even to North America alone; we need to support conservation in Latin America and the Caribbean too. Generally that means by paying for it (since conservation resources to our south tend to be more limited), either through donations, or by visiting reserves that finance the protection of habitat through bird tourism. The good news is that the conservation dollar goes a lot further in South America than it does here in the U.S. (for example, land can often be purchased for a little more than $100 per acre).

ABC receives more calls about local threats to birds than we can ever act on directly, as undoubtedly do many other conservation groups. We can very often provide technical advice, or a letter of support, but we cannot fight every single local battle. For that, we need your help. Usually the best advocate for the conservation of a particular area is someone who lives locally and is willing to put in the hours to make the difference.

Though good things to do, forwarding an e-mail about a bird conservation issue to a friend, filling your feeder, and participating in citizen-science programs won't on their own create habitat for birds, halt threats, or save species from extinction. To really become a "citizen conservationist" you

PHOTO: MIKE PARR

Birders gather at High Island, Texas, one of the best-known hotspots for neotropical migratory birds on the Gulf Coast.

need to get engaged—and there are many good ways to do so. In particular, if you are interested in helping ABC at a state or regional level, please let us know at abc@abcbirds.org.

THE WAY FORWARD FOR BIRD CONSERVATION

"Bird conservation" is often used as a catch-all term for a suite of activities that are potentially beneficial to birds. The core of what bird conservation is about though is preventing the extinction of bird species, protecting habitat for priority birds (those on the WatchList), and eliminating threats to birds. Using this as our definition, we can tackle bird conservation in a strategic way. To underpin this strategic approach, we need to generate sufficient resources and build the necessary partnerships to implement the required actions. Science is also required to help identify priorities. Today, there are millions of active birdwatchers across the U.S., and numerous local, state, and national initiatives that aim to conserve birds

and their habitats. What is lacking, though, is a singular and prominent "smoking gun" of bird plumes, market hunting, DDT, or waterfowl declines that gave rise to the major conservation initiatives of previous decades. However, the toxic cocktail of threats now facing birds is as deadly as it is complex. Habitat loss and fragmentation, invasive species, pesticides, disease, climate change, the impact of fisheries, lighted structures, glass windows, and pollution all combine to make the environment more hazardous to birds than ever. Yet in our complex, problematic, and human-oriented modern society, without a single dramatic bird conservation disaster, these insidious threats have thus far failed to draw a commensurate public response.

The following (next page) Strategic Bird Conservation Framework, which can also be adapted for state or local use, lays out a scheme for bird conservation prioritization based on the four-pronged approach of preventing extinctions, conserving habitats, eliminating threats, and building capacity.

Strategic Bird Conservation Framework

Goals and Targets

Proposed Actions

Halt Extinctions

- AZE and Stronghold sites
- IBAs and habitats for Red WatchList species in the U.S. and migrants internationally
- Endangered marine birds at sea
- Species that are, or should be, ESA listed

Most Endangered

- Create sustainable reserves
- Restore islands
- Stop seabird bycatch
- List species under the ESA; press for full funding

Conserve Habitats

- IBAs and habitats for Yellow WatchList species in the U.S. and migrants internationally

WatchList Species

- Best management practices; direct protection of habitats and IBAs
- Set population targets; promote conservation of WatchList birds among land managers and JVs
- Protect key sites and habitats for migrants internationally

Eliminate threats

- Mortality affecting one-million plus birds per year, or significantly affecting WatchList species
- Precedent-setting causes of mortality
- Unethical practices with broad implications
- Significant threats to habitats
- Federal funding programs that affect birds

All Species

- Direct interventions, public pressure, regulatory change, legal action, funding for solutions, education

Build the capacity and alliances needed to generate the necessary science and conservation action

For bird conservation to succeed, we must acknowledge the need for a new paradigm. For decades, hunters have funded habitat conservation through the purchase of hunting licenses, but hunting is on the decline. Recreational birdwatching, however, is not. One of our challenges, then, is to engage birders in conservation at a much more significant scale. ABC facilitates the Bird Conservation Alliance (BCA), a coalition of more than 200 local, national, and international organizations, and building this alliance is one way we can jointly work to increase the impact of the bird community. Individuals can encourage their local birding group or Audubon chapter to join, and—more importantly—to become active in the Alliance. Membership applications along with a list of current members and priorities are available at www.birdconservationalliance.org.

COLLABORATIVE CONSERVATION

In addition to the BCA, partnerships between NGOs, landowners, states, and the federal government (which owns and manages more than 25% of the nation's land), that use science-based planning tools are working to maintain the diversity and abundance of bird species across the landscape. Habitat conservation opportunities have been increasing as agricultural land becomes available for conservation through landowner incentive programs. By setting land aside from farming; and carefully managing burning regimes, logging rotations, and other periodic habitat adaptations; farmers and other land managers can become important stewards of bird habitats.

Private, state, and federal habitat management programs are also increasingly broadening their work to cover all birds, including conservation priorities advocated by Partners in Flight, the North American Bird Conservation Initiative (NABCI), and the State Wildlife Plans (i.e., managing land for both game and non-game species). One important means of delivering this conservation across the landscape is through bird conservation Joint Ven-

PHOTO: ROBERT ROYSE

Management for waterfowl such as the Green-winged Teal can easily dovetail with efforts to conserve songbirds.

tures (JVs). Originally formed to help implement the North American Waterfowl Management Plan, most JVs now address the conservation of all birds, and some have even been set up in areas with relatively few wetlands (such as the Appalachians and Central Hardwoods) to focus primarily on landbirds. Land Conservation Cooperatives (LCCs) are a new Department of the Interior concept, based on the JV model that aim to generate the science needed to effectively conserve all biological diversity. Incorporating bird priorities into LCC research is an important goal for bird conservation.

Bird conservationists have identified 67 Bird Conservation Regions (BCRs) across North America and in Hawaii where landscape-level bird conservation can be delivered. These BCRs encompass landscapes with similar bird communities, habitats, and resource issues. They are the fundamental biological units through which landscape-scale bird conservation—including planning, implementation, and evaluation—can be delivered (the Birdscapes in this volume are amalgams of related BCRs). Conservation of birds through the BCR model is being coordinated by the NABCI, a powerful coalition for conservation that aims to change the way people think about land management towards an "all birds and all habitats" approach. This large coalition of government agencies, non-governmental and private organizations, and academic institutions, spans the U.S., Canada, and Mexico, and may ultimately extend further southward. NABCI draws together major bird initiatives already operating in North America, including Partners in Flight, which focuses primarily on land-birds, as well as national planning initiatives for waterfowl, shorebirds, upland gamebirds, and water-birds. The emphasis on blending efforts for both game and non-game species makes NABCI a unique model for wildlife conservation that is having far-reaching consequences in the U.S. and elsewhere.

ABC has identified 500 Globally Important Bird Areas, which are the most critical of our nation's bird conservation sites. Fortunately most of these sites are already protected in some way, and 35% are NWRs. That doesn't mean that they don't face conservation issues though. For example, new legislation is badly needed to fund the control and removal of invasive species from refuges. State Audubon chapters are also working on a state-by-state IBA analysis that will identify even more important sites for conservation. Identifying IBA's on private lands remains a challenge, as data concerning bird distributions on private lands are typically less available.

Although much of the necessary work to conserve bird populations can be carried out by protecting IBAs, changing landscape-level habitat management within BCRs, and ensuring that bird protection laws are widely upheld, some threats to birds can only be addressed by changing specific human activities that have direct impacts on birds, including causes of mass mortality. Although most anthropogenic bird mortality is not deliberate, humans are nevertheless at the root of the deaths of hundreds of millions, possibly billions of birds in the U.S. each year. Only by countering and finding solutions to these problems can we be sure that all bird populations will remain healthy in the long-term.

DRIVERS OF CONSERVATION CHANGE

Thanks to environmental campaigns of the past, the U.S. has some of the best conservation laws anywhere. The Endangered Species Act has recovered some of the most endangered birds in the world from the brink of extinction, and clean air and water laws have helped to address some of our most serious pollution issues. We also have the world's premier protected area system with 392 National Parks covering 84 million acres, and 551 National Wildlife Refuges covering 150 million acres. In addition to this, National Forests, BLM lands, state protected areas, lands protected by TNC and many land trusts, and federal landowner incentive programs, ensure that large amounts of natural habitat are conserved (c. 3,000 total sites covering 300 million acres). Despite these successes, unprotected portions of the U.S. landscape (and even some gazetted areas) are still highly threatened, as you have read in the preceding pages.

If you are going to tackle a specific threat or conservation opportunity for birds in your local area, it

National Wildlife Refuges such as Bombay Hook provide habitat for both water-birds and migrant songbirds.

PHOTO: MIKE PARR

pays to be strategic. Conservation is about people changing the way they do things to benefit wildlife and habitats. Broadly speaking, there are relatively few approaches that will succeed in getting people, corporations, or communities to change the way they act. Education is a good starting point, but more is usually needed to transform the behavior of special interest groups with strongly-held beliefs or finances at stake. The primary drivers for effecting bird conservation change (or preventing change) are:

1. Paying for conservation (e.g., land purchase to protect a reserve) or persuading organizations or agencies to protect species and habitats.
2. Taking legal action (e.g., to prevent a project that would harm a species or habitat).
3. Building partnerships around scientific priorities and working jointly towards outcomes (e.g., by encouraging game species managers to also focus on non-game species).
4. Creating a public outcry over a particular issue (e.g., DDT) backed up with policy action.
5. Using scientific discoveries to reveal a major problem combined with any of the above.
6. Developing new technologies or programs that make those that were previously a threat to birds and their habitats obsolete.

Consumer pressure and political action can also help to create the environment for change, though they also usually require one or more of the above actions to effect specific changes.

ABC's experience is that most groups whose actions are negatively affecting birds usually find out when it is too late to easily turn back. Although they are often taken by surprise (e.g., the communication tower industry learning that it was causing bird deaths), and would rather avoid bird impacts, the costs of doing so may be significant, and this frequently leads to major resistance. The first line of defense is to question the science and call for additional research, so it is better to have hard facts based on field studies right from the start. Once the "insufficient science" barrier has been crossed, the next defense tends to be "well, habitat loss is worse, so why pick on _____" to which the answer is:

The Bird Conservation Alliance provides access to lawmakers for the bird conservation community.

PHOTO: MIKE PARR

birds are impacted by a range of threats; we need to address all of them and this one is significant. The last line of defense tends to be: "we are not breaking the law", but for the most part, anything that kills a bird *is* breaking the law. The Migratory Bird Treaty Act prohibits both the intentional and unintentional (but foreseeable) take of any migratory bird without a permit, and can levy fines of up to $500 per incident or jail sentences of up to six months. While the Act has thus-far been underutilized to protect birds affected by major national communications or energy infrastructure, there are signs that this might be changing with recent prosecutions of corporations for drowning birds in oil pits and electrocuting raptors on power lines. The Act does not protect birds' habitat however.

Being prepared for these responses is critical to tackling any threat whether it is national, regional, or local. Being willing to think of and negotiate creative solutions is also something you should be ready for if you are planning on addressing a specific development threat. Some of the best conservation has been accomplished this way (e.g., ESA Safe Harbor Agreements).

BUILDING A NETWORK

Even when you are not tackling a specific issue, you can be an ambassador and advocate for birds

PHOTO: ROBERT ROYSE

The Scarlet Tanager is one of the most striking species found in eastern forests, and is a great species to show to prospective birders.

and their conservation. The best possible way is to introduce your children, friends, or colleagues to birding. Invite people on a spring bird walk in your community. Birds are their own best advocates if people get a chance to see them. Networking with other conservation-minded people such as members of local Audubon chapters and other bird groups can also be a great way to lay the groundwork for future campaigning, and you can get a lot of good birding information that way too.

Building relationships with your elected representatives and other major decision-makers and stakeholders is a very effective way to get things done. Face-to-face meetings are best (going with a group is often best), but phone calls, personal letters, e-mails, and faxes are also effective. The key is to get to know them, and for them to get to know you as someone with conviction, who always provides good information. If you are not sure about the answer to a question, tell them that you will find out and get back to them with that information. Following up gives them a chance to learn that you are a reliable source, and for you to reiterate your

concerns. In your discussions, provide a clear indication of what you would like to see happen, and specifically how you would like the representative to help. If you know that an important decision or vote will take place in the near future, that is a good time to request a meeting, make a phone call or send a message. Everyone can be an effective bird conservation advocate. Please see the Bird Conservation Alliance lobby tips for more advice at www. birdconservationalliance.org.

The organizations and networks listed at the top of the next page can provide information on critical decisions affecting birds at a national level. You can request action alerts that automatically notify you and enable you to write to elected officials in support of important legislation at critical times. You can add value to these automated messages by editing and personalizing the text.

- American Bird Conservancy action page (www. abcbirds.org/action).
- Audubon action center (www.audubonaction. org/site/PageServer?pagename=aa_homepage).

- Bird Conservation Alliance campaign center (www.abcbirds.org/birdconservationalliance/campaigns/index.htm).
- Defenders of Wildlife's Wildlife action center (www.defenders.org/take_action).

For updates on the latest bird conservation news including policy-related information and action alerts, you can also subscribe to ABC's free eNewsletter, BirdWire (www.abcbirds.org/newsandreports/subscribe.html), or the RSS news feed of ABC's Bird News Network (www.abcbirds.org/newsandreports/bnnsubscribe.html).

CONSUMER POWER

Although it is unlikely that a single buying decision will directly benefit bird conservation (unless you are buying an entire company or a controlling interest in one), if enough of us alter the way we shop, then change will inevitably result. Most of the environment-friendly consumer choices you can make are widely-known. These include driving

PHOTO: MIKE PARR

Some fisheries that target "Chilean sea bass" also catch large numbers of seabirds as bycatch.

a fuel-efficient car, carpooling, recycling, printing double-sided; and purchasing organic vegetables, cotton, green power (although note that wind power has yet to address bird collisions effectively), and environmentally sensitive cleaning products.

While the consumer choices mentioned above are not directly aimed at conserving birds, they will help the environment as a whole, which will indirectly benefit bird conservation. More bird-specific consumer choices include buying shade-grown coffee that provides habitat for migrant birds, and basing seafood choices on consumer guides that help you avoid fisheries with a seabird bycatch problem (e.g., guides produced by the Blue Ocean Institute, Environmental Defense Fund, and Monterey Bay Aquarium). Buying paper and construction materials that are either recycled or certified by the Forest Stewardship Council (FSC) or Sustainable Forestry Initiative (SFI) is also likely to be more beneficial to birds. The authors are not aware of any bird-friendly labeled beef as yet, although there are certainly ranches that use more bird-friendly grazing systems, such as fencing cattle out of riparian areas, and rotational grazing. Neither is there an albatross- or seabird-friendly seafood label at the present time, although U.S. Alaskan and Hawaiian longline fisheries could qualify for such a label as they have already implemented successful bird bycatch mitigation measures. There is also no bird-friendly glass that reduces bird collisions widely available, although work is underway on this, and there are some decals and window treatments (e.g., CollidEscape) that can make a big difference already. Ultimately, the option to buy bird-friendly, sustainably-harvested native-prairie biofuels instead of corn ethanol could be a good consumer choice once/if it becomes available.

ON YOUR OWN TURF

Over two-million acres of the U.S. are occupied by residential development. Uniform mowed lawns are appealing to neighborhood associations, but provide little habitat for birds. There are many ways you can enhance your yard for birds, however.

- Keep your cat indoors, or use a leash. Do not feed stray or feral cats, and alert animal control agencies if there is a feral cat colony in your community.
- Reduce or avoid the use of pesticides and fertilizers in your yard to minimize toxins or excess nutrients that can harm birds and their insect prey, and contaminate water supplies. If you can't avoid pesticide use altogether, spray just weeds or trouble spots rather than the full lawn.
- Landscape with trees, shrubs, and plants native to your region. These plants will provide more natural habitat for birds and other wildlife. Native plants often provide birds with flowers, fruit, nesting sites, and other resources, so don't cut down vegetation while birds are breeding. Creating a brush pile, and allowing at least some areas of grass to grow naturally are also beneficial. In arid or water-stressed areas, landscaping with native plants (xeriscaping) is particularly important to reduce water drawdown from local aquifers and wetlands. You can also get your yard or native plant garden certified as wildlife habitat by the National Wildlife Federation (see www.nwf.org/gardenforwildlife/create.cfm) or other environmental groups with similar programs.
- Providing bird feeders, bird baths, and nest boxes are probably the most common ways people enhance their yards for birds. They bring birds into our daily lives, and are great for helping people form connections with them. Bird feeders can also offer critical resources to birds during migration or in winter. The real key to providing food for birds, however, is in maintaining insect populations. Even seed eaters rely on the protein in insects and other invertebrates to feed growing young. To really encourage birds, you must allow the plants you grow to harbor insects (by selecting native species and avoiding pesticides). Nest boxes attract cavity-nesters, and can help bird populations in areas where natural cavities are scarce. Bird feeders, baths, and nest boxes each require proper placement and cleaning to minimize the risk of collisions with nearby windows, to limit disease transmission between birds that visit feeders, and to prevent mosquitoes from breeding in standing water.
- Window collisions are a problem that can mostly be avoided. Densely grouped paper silhouettes may help to alert birds to the presence of the window, mylar strips or CollideEscape (www.collidescape.citymax.com) can also be used, and for windows that seem to have a high rate of bird strikes, a taut net or mesh mounted a few inches from the glass will reduce mortality. Feeders should either be very close to the window (so that birds can't generate enough speed for serious injury), or at least 30 feet away.

More resources available online include: ABC's Top Ten Tips for Bird-Friendly Living at www.abcbirds.org/newsandreports, the NRCS website (www.nrcs.usda.gov/feature/backyard), and Backyard Birdcare: www.backyardbirdcare.org/steps.html.

Note that disease and poisoning problems can be a concern for backyard bird feeders. The eye disease *Mycoplasma gallisepticum*, which primarily affects House Finches (though not always with fatal consequences), has spread from the Mid-Atlantic States, where it was first observed in 1994, to Canada, Texas, and the Midwest. The presence of aflotoxins, which are produced by soil fungi, can be fatal to birds even in minute quantities (legal maximum is 20 parts-per-billion for commercial

Bird feeders might be lucky to catch a glimpse of a Pine Siskin, a species that can irrupt into southern areas in some years.

PHOTO: BILL HUBICK

seed). These are most likely to be found in old or improperly stored seed, especially seed kept under excessively damp conditions. The vast majority of bird feeders are free from problems, but be sure to properly store seed, clean feeders in a bleach solution at least once a week, and remove spilled seeds from the ground. If an outbreak of disease is noticed, the feeder should be temporarily removed so the disease won't spread to more birds.

A few "backyards" that are hundreds or thousands of acres in size could become Important Bird Areas if managed appropriately. If you have a chance to do this, contact ABC, or the Partners in Flight or water-bird expert in your state wildlife agency.

Even modestly-sized properties can provide habitat for some priority birds. Owners of private forest-, farm-, and ranch-lands can manage their properties to maximize habitat for birds by protecting buffers around streams and wetlands, grazing cattle on a rotational basis, and maintaining snags along with a variety of tree species, shrubs, or other natural vegetation. Utilizing long harvest rotations, harvesting outside bird breeding seasons, and minimizing the use of pesticides can also help birds.

Resources are available online to help land managers improve the quality of habitat for birds and other wildlife on their property. Information on a variety of conservation programs for private landowners can be found at the USDA Natural Resources Conservation Service website at www.nrcs.usda.gov/programs/farmbill/2008/index.html. Also see the Partners in Flight web page on best management practices at www.partnersinflight.org/pubs/BMPs.htm. Additionally:

- The American Forest Foundation offers a series of conservation handbooks focused on southeastern pine ecosystems, available online at www.forestfoundation.org/ccs_publications.
- ABC provides examples of how landowners can manage their lands for cavity-nesting birds in ponderosa pine forests in a booklet available at www.abcbirds.org/newsandreports.
- The Wildlife Habitat Council provides a man-

Lighted buildings can attract migrants at night, and kill tens of thousands of birds in total each year due to collisions.

PHOTO: MIKE PARR

agement guide for landowners for riparian areas of the Great Basin at www.wildlifehc.org/ewebeditpro/items/O57F9025.pdf.
- Golf course managers (and others) can visit http://acsp.auduboninternational.org to find out how to provide good bird habitat on site.

OUT AND ABOUT

Americans love the great outdoors. When recreating outdoors, we can do a few simple things to benefit birds and the environment:

- Hunt and fish responsibly without introducing toxic lead into the environment or discarding monofilament line or other trash that can entangle and harm birds and other wildlife.
- Keep pet dogs on the leash at beaches where birds congregate or nest.
- Avoid off-road vehicles, snow-mobiles, power-boats, and other high-impact vehicles that can disturb native wildlife and degrade natural habitats. Similarly, wakes from speeding powerboats can flood the nests of water-birds and degrade shoreline habitats.
- At the office, turn off the lights at night to both save on energy costs and reduce the risk of nighttime collisions for migratory birds.

BIRDING WITH A PURPOSE

We often go birding in hopes of seeing a target species, witnessing a wave of migration, or just for an enjoyable time outdoors. Today, many birders can also participate in a variety of citizen science projects that collect their bird observations. In this manner, birdwatchers can assist scientists in monitoring bird populations by gathering data across geographic areas and seasons. These data can help build an understanding of which bird populations are increasing, stable, or declining, and where such changes are occurring. Some of the best-known surveys include the Breeding Bird Survey, the Christmas Bird Count, and regional atlas projects (see ABC's website www.abcbirds.org). One problem, however, is a lack of consistency between the many data-gathering projects in both methods and analysis. ABC has produced a guide to monitoring for the Northeast region called the Northeast Monitoring Handbook www.nemonitor.org. This handbook is also applicable to other regions, and can help with study design for bird research projects.

Birding website (www.conservationbirding.org) for a list of fantastic birding locations where your visits will help support habitat protection. Fundación ProAves in Colombia (www.eco-volunteer.com and www.proaves.org) also has a program through which volunteers can help with monitoring, conservation, and management activities for birds at private reserves.

When you go birding, let people know "a birder was here." Don't forget to sign guest books at wildlife refuges, and other birding locations. This helps parks and refuges estimate the number of people visiting for birds, and helps them justify budgets and park usage based on visitation rates. You can also leave behind calling cards at businesses you patronize during your birding trips, to tell the owners that birders support them and will return if the local birds and habitats are maintained. A "Birders Support Businesses and Jobs card" is available for download from ABC's website at www.abcbirds.org.

There are many listserves that primarily carry rare bird reports which can also be a useful resource for networking. You can view posts to lists organized by region at http://birdingonthe.net/birdmail.html. The American Birding Association also offers a range of resources on its website (www.aba.org), including directories of bird clubs, birding festivals, birding trails, and additional listserves. A plethora of both volunteer and paid positions can also be found at the Ornithological Societies of North America's ornithological jobs webpage (www.osnabirds.org/on/ornjobs.htm).

If you are traveling abroad, stay at a lodge that supports bird conservation. See the Conservation

When out birding, please adhere to the American Birding Association's code of ethics (available at www.aba.org/about/ethics.html). This code focuses on not stressing birds or their habitats through birding activities, as well as respecting other birders, property owners, and laws while birding. Birdwatchers who are unfamiliar with this code of ethics should take a moment to read it over.

NATIONAL BIRD CONSERVATION PRIORITIES—STATE BY STATE

O
n this and the following pages we provide a summary of the top national bird conservation priorities for each of the 50 U.S. states, focused on protecting populations of the rarest species. Each state clearly has many, many, bird priorities (including for example, protecting Important Bird Areas, and conserving large areas of core habitat) and we can't hope to list them all here. What we do hope to do though is to provide a list of national-level bird priorities that states should absolutely ensure are incorporated in their bird conservation plans, so that species and habitats that have national or global significance are given adequate priority alongside the many projects and actions that are competing for limited attention and funds. We also aim to expand on these online and look forward to your feedback.

ALABAMA (AL) Red WatchList Species: Mottled Duck, Yellow and Black Rails, Piping Plover, Least Tern, Red-cockaded Woodpecker; Bachman's, Henslow's, and Seaside Sparrows. **Most Endangered Habitats**: Longleaf pine forests, freshwater marshes, sandy beaches, saltmarshes. **Threats**: Conversion of older pine forests to younger loblolly plantations, fire suppression, sea level rise. **Key Actions**: Protect key longleaf pine, freshwater, and saltmarsh sites (especially saltmarshes that could move inland); restore natural fire regimes to longleaf areas; reduce disturbance on beaches used by Piping Plovers.

ALASKA (AK) Red WatchList Species: Steller's Eider, Spectacled Eider, Sooty Grouse, Laysan Albatross, Black-footed Albatross, Short-tailed Albatross, Pink-footed Shearwater, Rock Sandpiper, Buff-breasted Sandpiper, Ivory Gull, Kittlitz's Murrelet. **Most Endangered Habitats**: Coastal tundra, sea ice, glaciers and scree, offshore islands, marine waters. **Threats**: Climate change and sea level rise, invasive predators on islands with seabird colonies, pollution and increased nest predation associated with oil development and transport, road construction through Izembek NWR, old-growth logging. **Key Actions**: Ensure that development avoids critical areas for birds (e.g., Teshekpuk Lake, Izembek, and Arctic NWR); halt logging of old-growth in the south; remove invasives from islands; determine causes of eider declines.

ARIZONA (AZ) Red WatchList Species: California Condor, Mountain Plover, Thick-billed Parrot, Spotted Owl, Lewis's Woodpecker, Gilded Flicker, Bell's Vireo, Bendire's Thrasher, Black-chinned and Baird's Sparrows. **Most Endangered Habitats**: Grasslands and riparian forests. **Threats**: Habitat loss and degradation to expanding development, water withdrawal, over-grazing, construction of new solar power facilities, mining expansion, border wall construction, lead poisoning of condors, climate change. **Key Actions**: Ban use of lead ammunition; prevent extinction of "Masked" Bobwhite in U.S.; fence key riparian areas and rotate grazing.

ARKANSAS (AR) Red WatchList Species: Least Tern, Bell's Vireo, Bachman's and Henslow's Sparrows. **Most Endangered Habitats**: Pine forests, grasslands, rivers, freshwater marshes. **Threats**: Habitat loss and degradation to agriculture; flood control. **Key Actions**: Restore large tracts of bottomland hardwoods; restore natural fire to pine forests.

CALIFORNIA (CA) Red WatchList Species: Sooty Grouse, Black-footed and Laysan Albatrosses, Pink-footed and Black-vented Shearwaters; Ashy, Black, and Least Storm-Petrels; California Condor, Black Rail, Mountain Plover, Rock Sandpiper, Least Tern, Xantus's and Craveri's Murrelets, Spotted Owl, Lewis's Woodpecker, Gilded Flicker, Bell's Vireo, Bendire's Thrasher, Black-chinned Sparrow, Tricolored  Blackbird. **Most Endangered Habitats**: Oak woodlands, grasslands, coastal scrub and chaparral, sagebrush, alpine tundra, riparian woodland, saline lakes, beaches, mudflats, saltmarshes, freshwater marshes, marine waters. **Threats**: Fire suppression, invasive diseases (e.g., oak rust), bark beetle outbreaks, expanding urban development, agricultural intensification, over-grazing (alpine meadows), invasive predators (e.g., rats on islands), climate change, eutrophication (e.g., Salton Sea). **Key Actions**: Protect large colonies of Tricolored Blackbirds; restore San Francisco Bay and riparian woodlands in the south; restore natural fire regimes; limit grazing in alpine areas; protect central valley wetlands and water flow; restore the Salton Sea; protect the "Island" Loggerhead Shrike.

COLORADO (CO) Red WatchList Species: Gunnison Sage-Grouse, Greater and Lesser Prairie-Chickens, Black Rail, Piping and Mountain Plovers, Least Tern, Spotted Owl, Lewis's Woodpecker, Bell's Vireo, Bendire's Thrasher. **Most Endangered Habitats**: Sagebrush, grasslands, alpine tundra, freshwater wetlands. **Threats**: Climate change, fire suppression, invasive plants, bark beetle outbreaks, expanding urban development, over-grazing. **Key Actions**: Protect and restore sagebrush for sage-grouse in Gunnison Basin; restore habitat and limit grazing and development in prairie systems.

CONNECTICUT (CT) Red WatchList Species: Black Rail, Piping Plover, Least Tern, Golden-winged Warbler, Saltmarsh and Seaside Sparrows. **Most Endangered Habitats**: Coastal saltmarshes, beaches, freshwater wetlands, mudflats, marine waters, early successional forest. **Threats**: Sea level rise is predicted to inundate many saltmarshes and other coastal habitats for shorebirds and sparrows. **Key Actions**: Protect coastal habitats and manage for sea level rise where possible, create early- to mid-successional habitats for Golden-winged Warbler, reduce beach disturbance.

DELAWARE (DE) Red WatchList Species: Piping Plover, Least Tern, Black Rail, Saltmarsh and Seaside Sparrows. **Most Endangered Habitats**: Grasslands, marshes, saltmarshes, mudflats, beaches, marine waters. **Threats**: Sea level rise is predicted to inundate many coastal saltmarshes; beach nesting birds disturbed by pedestrians, pet dogs, and feral cats. **Key Actions**: Restrict horseshoe crab harvests until crab numbers and shorebird numbers recover, minimize risk of oil spills, protect and improve management of saltmarshes for rails and sparrows.

DISTRICT OF COLUMBIA (DC) Red WatchList Species: Cerulean Warbler (rare). **Threats**: Collisions with glass windows and lit buildings kill nocturnal migrants, overabundant white-tailed deer in Rock Creek Park degrade forest understory. **Key Actions**: Reduce collisions by initiating Lights Out efforts and mandating new standards for glass windows in federal buildings; cull deer herds in Rock Creek Park.

FLORIDA (FL) Red WatchList Species: Mottled Duck, Magnificent Frigatebird, Reddish Egret, Yellow and Black Rails, Whooping Crane, Piping Plover, Least Tern, White-crowned Pigeon, Red-cockaded Woodpecker, Florida Scrub-Jay; Bachman's, Henslow's, Saltmarsh, and Seaside Sparrows. **Most Endangered Habitats**: Longleaf pine forests, Florida oak scrub, sub-tropical forests, tropical hammocks, grasslands, sawgrass marshes, lakes, beaches, mudflats, offshore islands for breeding seabirds, and marine waters. **Threats**: Suburban development and agriculture (e.g., citrus); invasive diseases, plants, and animals (including feral cats); sea level rise is predicted to eventually inundate much of southern Florida (e.g., keys, Everglades), pesticides, fire suppression. **Key Actions**: Restore the Everglades from Lake Okeechobee to the Gulf of Florida; protect remaining Florida scrub with viable scrub-jay populations; reduce beach disturbance.

GEORGIA (GA) Red WatchList Species: Black-capped Petrel, Yellow and Black Rails, Piping Plover, Least Tern, Red-cockaded Woodpecker; Bachman's, Henslow's, Saltmarsh, and Seaside Sparrows. **Most Endangered Habitats**: Longleaf pine forests, freshwater marshes, saltmarshes, mudflats, marine waters. **Threats**: Expanding urban development (e.g., Atlanta). **Key Actions**: Restore natural fire regimes to longleaf systems; protect key saltmarsh areas and manage them for rails and sparrows; reduce beach disturbance.

HAWAII (HI) Red WatchList Species: Hawaiian Goose, Hawaiian and Laysan Ducks; Laysan, Black-footed, and Short-tailed Albatrosses, Hawaiian Petrel, Newell's Shearwater, Band-rumped and Tristram's Storm-Petrels, Hawaiian Hawk, Hawaiian Coot, Hawaiian Crow, Elepaio, Millerbird, Kamao, Olomao, Omao, Puaiohi, Laysan and Nihoa Finches, Ou, Palila, Maui Parrotbill, Oahu and Kauai Amakihis, Anianiau, Nukupuu, Akiapolaau, Akikiki, Hawaii Creeper, Oahu and Maui Alauahios, Akekee, Akepa, Iiwi, Akohekohe, Poo-uli. **Most Endangered Habitats**: Rainforests, dry forests, shrub-lands, grasslands, beaches, ponds, offshore islands for breeding seabirds, marine waters. **Threats**: Invasive diseases, plants, and mammals; warming climate, extensive habitat loss, seabird collisions with structures, albatross mortality caused by plastics and lead poisoning. **Key Actions**: Greatly increase efforts to control invasive species and diseases; restore forest habitat; boost bird populations through captive-breeding, translocations, and reintroductions; remove lead from Midway.

IDAHO (ID) Red WatchList Species: Lewis's Woodpecker. **Most Endangered Habitats**: Sagebrush, pine forests, pinyon woodlands, wet conifer forests, spruce-fir forests, aspen woodlands, pinyon-juniper woodlands, alpine tundra, rivers. **Threats**: Habitat loss and degradation to altered fire regimes, over-grazing, invasive plants, bark beetle outbreaks, energy development, climate change. **Key Actions**: Ensure energy development avoids areas occupied by lekking grouse, manage ponderosa pine forests to help cavity nesters (e.g., Flammulated Owl).

ILLINOIS (IL) Red WatchList Species: Greater Prairie-Chicken, Least Tern, Bell's Vireo, Golden-winged Warbler (historically bred here but now almost extirpated), Henslow's Sparrow. **Most Endangered Habitats**: Grasslands, deciduous forests, mixed forests, bottomland hardwoods, rivers. **Threats**: Most grassland already converted to agriculture, collisions in Chicago kill migrants. **Key Actions**: Restore and protect grasslands and early-successional habitats; expand efforts to reduce collisions.

INDIANA (IN) Red WatchList Species: Bell's Vireo, Golden-winged Warbler, Henslow's Sparrow. **Most Endangered Habitats**: Grasslands, deciduous forest, rivers. **Threats**: Most grassland already converted to agriculture. **Key Actions**: Restore and protect grasslands.

IOWA (IA) **Red WatchList Species:** Piping Plover, Buff-breasted Sandpiper, Least Tern, Bell's Vireo, Henslow's Sparrow. **Most Endangered Habitats:** Grasslands, shrub-lands, marshes, lakes, rivers. **Threats:** Warming climate predicted to dry up many wetlands. **Key Actions:** Maintain river flows to restore terns and plovers; restore grasslands.

KANSAS (KS) **Red WatchList Species:** Greater and Lesser Prairie-Chickens, Black Rail, Whooping Crane, Piping Plover, Buff-breasted Sandpiper, Least Tern, Bell's Vireo, Henslow's Sparrow. **Most Endangered Habitats:** Grasslands, saline lakes, freshwater marshes, rivers. **Threats:** Most grassland already converted to agriculture; expansion of wind farms and power lines threaten prairie-chickens. **Key Actions:** Ensure wind energy development avoids prairie-chicken areas and the Whooping Crane migration corridor; reduce prairie-chicken collisions by marking wire fences; enhance state protection for Lesser Prairie-Chicken; ensure adequate water flows to wetland reserves (e.g. Cheyenne Bottoms).

KENTUCKY (KY) **Red WatchList Species:** Least Tern, Golden-winged Warbler, Henslow's Sparrow. **Most Endangered Habitats:** Mixed forests, grasslands, rivers. **Threats:** Mountaintop removal mining; poor water management and fire regimes. **Key Actions:** Halt new mining operations and restore abandoned mine-lands; create early successional habitats for Henslow's Sparrow and Golden-winged Warbler; restore natural fire regimes.

LOUISIANA (LA) **Red WatchList Species:** Mottled Duck, Reddish Egret, Yellow and Black Rails, Piping Plover, Least Tern, Red-cockaded Woodpecker, Bell's Vireo; Bachman's, Henslow's, and Seaside Sparrows. **Most Endangered Habitats:** Longleaf pine forests, freshwater marshes, saltmarshes, marine waters. **Threats:** Flood control reduces silt deposited in the Mississippi Delta and contributes to loss of marshes; excess nutrients from agricultural runoff create a huge "dead zone" off the coast; collisions kill nocturnal migrants. **Key Actions:** Protect and restore coastal marshes; restore natural fire to longleaf pine systems; reduce beach disturbance.

MAINE (ME) **Red WatchList Species:** Yellow Rail, Piping Plover, Least Tern, Bicknell's Thrush, Saltmarsh and Seaside Sparrows. **Most Endangered Habitats:** Spruce-fir, saltmarshes, offshore islands for breeding seabirds, marine waters. **Threats:** Overabundant gulls, beach disturbance, sea level rise. **Key Actions:** Continue efforts to restore seabird colonies on offshore islands; protect key Bicknell's Thrush sites; reduce beach disturbance near plover nest sites or potential nest sites; protect and restore saltmarshes.

MARYLAND (MD) **Red WatchList Species:** Black Rail, Piping Plover, Least Tern, Golden-winged Warbler; Henslow's, Saltmarsh, and Seaside Sparrows. **Most Endangered Habitats:** Grasslands and early- to mid-successional forests, saltmarshes, mudflats, beaches, barrier islands, marine waters. **Threats:** Excessive nutrients from agricultural runoff degrade the Chesapeake Bay; invasive pests kill trees and invasive plants degrade wetlands; development along the coast and Piedmont; sea level rise. **Key Actions:** Restore grasslands and early successional habitat in Appalachians; restore the Chesapeake Bay; protect key saltmarsh areas.

MASSACHUSETTS (MA) Red WatchList Species: Piping Plover, Least Tern, Golden-winged Warbler, Saltmarsh and Seaside Sparrows. **Most Endangered Habitats**: Beaches, saltmarshes, early- to mid-successional forests, mudflats, barrier islands, marine waters. **Threats**: Overabundant gulls, pedestrians, pet dogs, and feral cats disturb or predate beach-nesting birds. **Key Actions**: Restore early successional habitats for Golden-winged Warbler; reduce beach disturbance; protect and manage key saltmarsh areas for sparrows (especially saltmarsh with room to move).

MICHIGAN (MI) Red WatchList Species: Yellow Rail, Piping Plover, Golden-winged and Kirtland's Warblers. **Most Endangered Habitats**: Jack pine, grasslands, freshwater marshes, lakes, beaches. **Threats**: Habitat loss and degradation to agriculture, fire suppression; overabundant cowbirds parasitize Kirtland's Warbler nests. **Key Actions**: Continue habitat management and cowbird control efforts to maintain or increase Kirtland's Warbler numbers; increase early- to mid-successional habitat for the Golden-winged Warbler.

MINNESOTA (MN) Red WatchList Species: Greater Prairie-Chicken, Yellow Rail, Piping Plover, Bell's Vireo, Golden-winged Warbler, Henslow's Sparrow. **Most Endangered Habitats**: Grasslands, freshwater marshes, early- to mid-successional forests, lakes. **Threats**: Wetland loss to cranberry farming, peat harvesting, and drainage. **Key Actions**: Restore early successional habitats and grasslands for Henslow's Sparrow and Golden-winged Warbler; protect plover breeding sites.

MISSISSIPPI (MS) Red WatchList Species: Mottled Duck, Yellow and Black Rails, Piping Plover, Least Tern, Red-cockaded Woodpecker, Bell's Vireo; Bachman's, Henslow's, and Seaside Sparrows. **Most Endangered Habitats**: Longleaf pine forests, rivers, freshwater marshes, saltmarshes, mudflats, beaches, marine waters. **Threats**: Water drawdown for irrigation, sea level rise, fire suppression. **Key Actions**: Restore natural fire regimes to longleaf systems; protect coastal marshes (especially those with a chance to move inland with sea level rise).

MISSOURI (MO) Red WatchList Species: Greater Prairie-Chicken, Least Tern, Red-cockaded Woodpecker (small numbers), Bell's Vireo, Bachman's and Henslow's Sparrows. **Most Endangered Habitats**: Grasslands, pine forests, rivers. **Threats**: Forests and grasslands fragmented or converted to agriculture. **Key Actions**: Improve forestry management; restore grasslands, longleaf systems, and bottomland hardwoods.

MONTANA (MT) Red WatchList Species: Yellow Rail, Piping and Mountain Plovers, Lewis's Woodpecker, Baird's Sparrow. **Most Endangered Habitats**: Sagebrush, pine forests, rivers, prairie. **Threats**: Invasive plants, altered fire regimes. **Key Actions**: Protect and restore areas of sagebrush for Greater Sage-Grouse and other sage-dependent species; enhance ponderosa pine forests for cavity-nesting species (e.g., Flammulated Owl).

NEBRASKA (NE) Red WatchList Species: Greater Prairie-Chicken, Whooping Crane, Piping and Mountain Plovers, Buff-breasted Sandpiper, Least Tern, Bell's Vireo, Henslow's Sparrow. **Most Endangered Habitats**: Grasslands, rivers, playa lakes. **Threats**: Water management; grasslands converted to agriculture. **Key Actions**: Strengthen farm bill programs to protect habitat on private agricultural lands.

NEVADA (NV) Red WatchList Species: Lewis's Woodpecker, Bell's Vireo, Bendire's Thrasher, Black-chinned Sparrow. **Most Endangered Habitats**: Sagebrush, alpine tundra. **Threats**: Invasive plants, altered fire regimes, over-grazing, climate change. **Key Actions**: Protect and restore sagebrush ecosystems.

NEW HAMPSHIRE (NH) Red WatchList Species: Piping Plover, Least Tern, Bicknell's Thrush, Golden-winged Warbler, Saltmarsh and Seaside Sparrows. **Most Endangered Habitats**: Spruce-fir forests, beach, saltmarsh, early- to mid-successional forest. **Threats**: Acid rain, climate change. **Key Actions**: Strengthen Clean Air Act rules to further reduce acid rain; strengthen partnerships with groups in Hispaniola to conserve wintering areas for Bicknell's Thrush; protect beaches and saltmarsh.

NEW JERSEY (NJ) Red WatchList Species: Black Rail, Piping Plover, Least Tern, Saltmarsh and Seaside Sparrows. **Most Endangered Habitats**: Beaches, saltmarshes, mudflats. **Threats**: Sea level rise; beach nesting birds disturbed by walkers and their dogs, or killed by feral cats. **Key Actions**: Maintain restrictions on horseshoe crab harvests until crab and shorebird numbers recover; control feral cats near beaches and nesting birds.

NEW MEXICO (NM) Red WatchList Species: Lesser Prairie-Chicken, Mountain Plover, Least Tern, Spotted Owl, Lewis's Woodpecker, Bell's Vireo, Bendire's Thrasher, Black-chinned and Baird's Sparrows. **Most Endangered Habitats**: Riparian forests, deserts, pinyon-juniper woodlands, grasslands. **Threats**: Over-grazing degrades deserts and grasslands; water withdrawals threaten riparian forests. **Key Actions**: Restore riparian forests; reduce grazing pressure on public arid lands.

NEW YORK (NY) Red WatchList Species: Black Rail, Piping Plover, Least Tern, Bicknell's Thrush, Golden-winged Warbler; Henslow's, Saltmarsh, and Seaside Sparrows. **Most Endangered Habitats**: Spruce-fir forests, saltmarshes, beaches, mudflats. **Threats**: Acid rain harms forests and wetlands in the Adirondacks; sea level rise is predicted to flood many saltmarshes; warming climate may threaten species at southern extent of their range (e.g., Gray Jay, Boreal Chickadee). **Key Actions**: Strengthen Clean Air Act rules to further reduce acid rain; strengthen partnerships with groups in Hispaniola to conserve wintering areas for the Bicknell's Thrush.

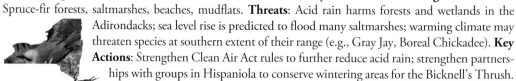

NORTH CAROLINA (NC) Red WatchList Species: Bermuda Petrel, Black-capped Petrel, Black Rail, Piping Plover, Least Tern, Red-cockaded Woodpecker, Golden-winged Warbler; Bachman's, Henslow's, Saltmarsh, and Seaside Sparrows. **Most Endangered Habitats**: Pine forests (longleaf), saltmarshes, beaches, mudflats, barrier islands, marine waters. **Threats**: Conversion of longleaf pine to commercial loblolly plantations or agriculture, expanding development, sea level rise, disturbance to beach-nesting birds. **Key Actions**: Protect and restore healthy longleaf pine ecosystems; identify and protect key saltmarsh sites that can move inland.

NORTH DAKOTA (ND) Red WatchList Species: Greater Prairie-Chicken, Yellow Rail, Piping Plover, Least Tern, Bell's Vireo, Baird's Sparrow. **Most Endangered Habitats**: Grasslands, sagebrush, rivers, lakes. **Threats**: Conversion of native prairie to crops; global warming is predicted to dry up some pothole wetlands. **Key Actions**: Strengthen farm bill programs to enhance habitats on private lands; repeal mandates for corn ethanol that drive the conversion of fallow grasslands to crops.

OHIO (OH) Red WatchList Species: Golden-winged Warbler, Henslow's Sparrow. **Most Endangered Habitats**: Prairies, early- to mid-succesional forests. **Threats**: Most grasslands and forests have been fragmented or converted to agriculture; mining. **Key Actions**: Restore abandoned mine-lands; maintain and expand wetlands as stopover sites for migratory waterfowl and shorebirds.

OKLAHOMA (OK) Red WatchList Species: Greater and Lesser Prairie-Chickens, Piping and Mountain Plovers (small numbers of the latter), Least Tern, Bell's and Black-capped Vireos, Henslow's Sparrow. **Most Endangered Habitats**: Grasslands, pine forests, playa lakes. **Threats**: Habitat loss or degradation to agriculture, over-grazing, altered fire regimes, water withdrawals. **Key Actions**: Enhance protections for prairie-chickens, including marking wire fences and restricting wind farms or transmission line construction near leks.

OREGON (OR) Red WatchList Species: Sooty Grouse, Black-footed Albatross, Pink-footed Shearwater, Yellow Rail, Rock Sandpiper, Xantus's Murrelet, Spotted Owl, Lewis's Woodpecker, Tricolored Blackbird (small numbers). **Most Endangered Habitats**: Wet conifer forests, oak woodlands, grasslands, marine. **Threats**: Logging old-growth forests. **Key Actions**: Halt logging of old-growth forests; enhance protection for "Streaked" Horned Lark.

PENNSYLVANIA (PA) Red WatchList Species: Golden-winged Warbler, Henslow's Sparrow. **Most Endangered Habitats**: Early successional deciduous forest, grasslands. **Threats**: Maturation of early successional forests to more mature forests, mining, wind development along mountain ridges. **Key Actions**: Restore abandoned mine-lands; restore early successional habitats; maintain and expand wetland habitats as stopover sites for migratory waterfowl and shorebirds.

RHODE ISLAND (RI) Red WatchList Species: Piping Plover, Least Tern, Saltmarsh and Seaside Sparrows. **Most Endangered Habitats**: Saltmarshes, beaches, mudflats. **Threats**: Sea level rise. **Key Actions**: Protect saltmarshes and manage for sea level rise where possible; reduce beach disturbance.

SOUTH CAROLINA (SC) Red WatchList Species: Black-capped Petrel, Yellow and Black Rails, Piping Plover, Least Tern, Red-cockaded Woodpecker; Bachman's, Henslow's, Saltmarsh, and Seaside Sparrows. **Most Endangered Habitats**: Longleaf pine forests, saltmarshes. **Threats**: Longleaf pine forest conversion to loblolly plantations, altered fire regimes, sea level rise. **Key Actions**: Restore large tracts of mature longleaf pine; identify and protect key saltmarsh areas.

SOUTH DAKOTA (SD) Red WatchList Species: Greater Prairie-Chicken, Piping Plover, Least Tern, Lewis's Woodpecker, Bell's Vireo, Baird's Sparrow. **Most Endangered Habitats**: Grasslands, sagebrush. **Threats**: Habitat loss and degradation to agriculture, expanding energy development, invasive plants. **Key Actions**: Strengthen farm bill programs to enhance habitats on private lands and repeal mandates for corn ethanol that drive conversion of grasslands to crops; limit wind and other energy development at key sites.

TENNESSEE (TN) Red WatchList Species: Least Tern, Bell's Vireo, Golden-winged Warbler, Bachman's Sparrow. **Most Endangered Habitats**: Early- to mid-successional deciduous forests, bottomland forests, rivers. **Threats**: Habitat loss and degradation to agriculture, mining, dams. **Key Actions**: Restore abandoned mine-lands; stop mountaintop removal coal mining.

TEXAS (TX) Red WatchList Species: Mottled Duck, Greater and Lesser Prairie-Chickens, Reddish Egret, Yellow and Black Rails, Whooping Crane, Piping and Mountain Plovers, Least Tern, Green Parakeet, Red-crowned Parrot, Spotted Owl, Red-cockaded Woodpecker, Bell's and Black-capped Vireos, Golden-cheeked Warbler; Bachman's, Black-chinned, Baird's, Henslow's, and Seaside Sparrows. **Most Endangered Habitats**: Grasslands, ashe-juniper woodlands, oak scrub, sub-tropical forests, riparian forests, saltmarshes, beaches. **Threats**: Urban development, fire suppression; over-grazing, invasive plants and fire ants, sea level rise. **Key Actions**: Expand Balcones Canyonlands NWR; improve growth planning in Lower Rio Grande Valley and Edward's Plateau; prevent extinction of "Attwater's" Prairie-Chicken; ensure adequate water flows from Guadalupe River to Aransas NWR.

UTAH (UT) Red WatchList Species: Gunnison Sage-Grouse, Mountain Plover, Spotted Owl, Lewis's Woodpecker, Bell's Vireo, Bendire's Thrasher, Black-chinned Sparrow. **Most Endangered Habitats**: Sagebrush, alpine tundra, saline and playa lakes. **Threats**: Habitat loss and degradation to over-grazing, agriculture, invasive plants, energy development, water pollution, water withdrawals, and climate change. **Key Actions**: Enhance water management to provide more fresh water to the Great Salt Lake and reduce risk of botulism outbreaks; reduce disturbance of Snowy Plovers from off-road vehicles; restore sagebrush in Gunnison Basin.

VERMONT (VT) Red WatchList Species: Bicknell's Thrush, Golden-winged Warbler. **Most Endangered Habitats**: Spruce-fir forests, early-successional deciduous forests. **Threats**: Acid rain, climate change. **Key Actions**: Strengthen Clean Air Act rules to further reduce acid rain; strengthen partnerships with groups in Hispaniola to conserve wintering areas for Bicknell's Thrush; create more early- to mid-successional habitat.

VIRGINIA (VA) Red WatchList Species: Black Rail, Piping Plover, Least Tern, Red-cockaded Woodpecker (tiny population), Golden-winged Warbler; Bachman's, Saltmarsh, and Seaside Sparrows. **Most Endangered Habitats**: Pine forests, saltmarshes. **Threats**: Forest habitat loss to agriculture and urban development; sea level rise. **Key Actions**: Protect coastal habitats and manage these for rising sea levels where possible.

WASHINGTON (WA) Red WatchList Species: Sooty Grouse, Black-footed Albatross, Pink-footed Shearwater, Rock Sandpiper, Xantus's Murrelet, Spotted Owl, Lewis's Woodpecker. **Most Endangered Habitats**: Old-growth conifer forests, sagebrush, grasslands, rivers. **Threats**: Old-growth logging, persecution of Caspian Terns and other fish-eating birds, dams, invasive plants, over-grazing. **Key Actions**: Halt logging of old-growth forests; restore rivers through dam removal; stop plans to relocate Caspian Tern colony from East Sand Island; enhance protection for "Streaked" Horned Lark; control invasive cordgrass in coastal areas.

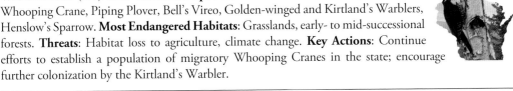

WEST VIRGINIA (WV) **Red WatchList Species**: Golden-winged Warbler, Henslow's Sparrow. **Most Endangered Habitats**: Early successional and mature deciduous forests, grasslands. **Threats**: Habitat loss and degradation to coal mining. **Key Actions**: Restore abandoned mine-lands; stop mountaintop removal coal mining.

WISCONSIN (WI) **Red WatchList Species**: Greater Prairie-Chicken, Yellow Rail, Whooping Crane, Piping Plover, Bell's Vireo, Golden-winged and Kirtland's Warblers, Henslow's Sparrow. **Most Endangered Habitats**: Grasslands, early- to mid-successional forests. **Threats**: Habitat loss to agriculture, climate change. **Key Actions**: Continue efforts to establish a population of migratory Whooping Cranes in the state; encourage further colonization by the Kirtland's Warbler.

WYOMING (WO) **Red WatchList Species**: Mountain Plover, Lewis's Woodpecker. **Most Endangered Habitats**: Grasslands, sagebrush, pine forests. **Threats**: Habitat loss and degradation to invasive plants, over-grazing, altered fire regimes, and expanding energy development. **Key Actions**: Restore large tracts of prairie and sagebrush; minimize impacts of energy development; mark wire fences to reduce sage-grouse collisions; enhance management of ponderosa pine forests for cavity-nesting birds.

Key Conservation Plans

The table on the following page enables you to find out which Bird Conservation Regions (BCRs) are found in your state. For bird conservation plans and contacts for each BCR visit: www.nabci-us. org/map.html. Other key plans include:

- The Canadian Shorebird Conservation Plan is available at www.cws-scf.ec.gc.ca/publications/ AbstractTemplate.cfm?lang=e&id=318.

- Joint Ventures have implementation plans and other planning documents on their websites, the URLs for which can be found on the FWS website under the JV directory at www.nabci-us.org.

- The North American Colonial Waterbird Plan is available at www.pwrc.usgs.gov/nacwcp/nawcp. html.

- North American Waterfowl Management Plan (NAWMP) documents can be found at www.fws. gov/birdhabitat/NAWMP/index.shtm. This plan is currently under revision to be completed in 2011-2012.

- PIF's bird conservation plans for states and physiographic areas, and the North American Landbird Conservation Plan are available from PIF's website (www.partnersinflight.org).

- State Wildlife Action Plans are available for most states in the U.S. at www.wildlifeactionplans. org/about/action_plans_text.html.

- The U.S. Shorebird Conservation Plan and regional shorebird plans are available at www.fws.gov/ shorebirdplan.

Use this table to see which Bird Conservation Regions (BCRs) occur in your state. Red cells indicate overlap between states and BCRs. BCR numbers 6-8 and 38-66 occur exclusively in the provinces and states of Canada and Mexico, and are not included in this tabe. See p. 435 for a map of BCRs.

BCR legend:
1 Aleutian/Bering Sea Islands · 2 Western Alaska · 3 Arctic Plains And Mountains · 4 Northwestern Interior Forest · 5 Northern Pacific Rainforest · 9 Great Basin · 10 Northern Rockies · 11 Prairie Potholes · 12 Boreal Hardwood Transition · 13 Lower Great Lakes/St. Lawrence Plain · 14 Atlantic Northern Forest · 15 Sierra Nevada · 16 Southern Rockies/Colorado Plateau · 17 Badlands and Prairies · 18 Short-grass Prairie · 19 Central Mixed-grass Prairie · 20 Edwards Plateau · 21 Oaks and Prairies · 22 Eastern Tallgrass Prairie · 23 Prairie Hardwood Transition · 24 Central Hardwoods · 25 West Gulf Coastal Plain/Ouachitas · 26 Mississippi Alluvial Valley · 27 Southeastern Coastal Plain · 28 Appalachian Mountains · 29 Piedmont · 30 New England/Mid-Atlantic Coast · 31 Peninsular Florida · 32 Coastal California · 33 Sonoran and Mohave Deserts · 34 Sierra Madre Occidental · 35 Chihuahuan Desert · 36 Tamaulipan Brushlands · 37 Gulf Coastal Prairie · 67 Hawaii

BCR#	1	2	3	4	5	9	10	11	12	13	14	15	16	17	18	19	20	21	22	23	24	25	26	27	28	29	30	31	32	33	34	35	36	37	67
Alabama (AL)																					●			●	●	●									
Alaska (AK)	●	●	●	●	●																														
Arizona (AZ)													●																	●	●	●			
Arkansas (AR)																					●	●													
California (CA)					●	●						●																	●	●					
Colorado (CO)													●	●	●																				
Connecticut (CT)										●																	●								
Delaware (DE)																										●	●								
District of Columbia (DC)																										●									
Florida (FL)																								●				●							
Georgia (GA)																								●	●	●									
Hawaii (HI)																																			●
Idaho (ID)						●	●						●																						
Illinois (IL)																			●	●	●														
Indiana (IN)																			●	●	●														
Iowa (IA)								●											●	●															
Kansas (KS)														●	●	●		●	●																
Kentucky (KY)																					●			●	●										
Louisiana (LA)																						●	●	●										●	
Maine (ME)											●																								
Maryland (MD)																									●	●	●								
Massachusetts (MA)											●																●								
Michigan (MI)									●										●	●															
Minnesota (MN)								●	●										●	●															
Mississippi (MI)																							●	●											
Missouri (MO)																			●		●														
Montana (MT)							●	●						●																					
Nebraska (NE)								●						●	●	●		●																	
Nevada (NV)						●						●	●																						
New Hampshire (NH)											●																								
New Jersey (NJ)																									●	●	●								
New Mexico (NM)													●		●																	●			
New York (NY)										●	●														●		●								
North Carolina (NC)																								●	●	●									
North Dakota (ND)								●																											
Ohio (OH)										●									●		●														
Oklahoma (OK)															●	●		●	●		●	●													
Oregon (OR)					●	●	●																												
Pennsylvania (PA)										●															●	●									
Rhode Island (RI)																											●								
South Carolina (SC)																								●	●	●									
South Dakota (SD)								●						●	●	●		●																	
Tennessee (TN)																					●			●	●										
Texas (TX)															●	●	●	●		●												●	●	●	
Utah (UT)						●	●						●																	●					
Vermont (VT)										●	●																								
Virginia (VA)																								●	●	●	●								
Washington (WA)					●	●	●																												
West Virginia (WV)																									●										
Wisconsin (WI)									●										●	●															
Wyoming (WO)							●						●	●	●																				

BIRD CONSERVATION DIRECTORY

KEY AGENCIES AND PROGRAMS RELATED
TO BIRDS AND THE ENVIRONMENT

The **U.S. Department of the Interior (DOI)** oversees the management of millions of acres of public land in the U.S. through the Bureau of Indian Affairs, Bureau of Land Management, Bureau of Reclamation, Fish and Wildlife Service, National Parks Service, Geological Survey, and Minerals Management Service.

The **U.S. Fish and Wildlife Service (FWS)** is a federal agency within the DOI, dedicated to working to conserve, protect, and enhance fish, wildlife, and plants and their habitats for the benefit of the American people. FWS Divisions particularly involved with bird conservation include the Office of Endangered Species, Division of Bird Habitat Conservation, Office of Migratory Bird Management, and Office of International Affairs.

The **U.S. National Park Service (NPS)** is a bureau within the DOI, established in 1916, and which now cares for a system of 392 national parks, seashores, battlefields, and monuments covering more than 84 million acres in the U.S. and its territories (of which 55 million acres are in Alaska). With the establishment of Yellowstone NP in 1872, this system of national parks became the first of its kind, and served as a model for establishing protected areas throughout the world. These parks provide visitors with recreational opportunities and protect unique geologic features, cultural sites, and habitats for birds and other wildlife.

The **Bureau of Land Management (BLM)** is part of the DOI and manages 256 million acres and the subsurface mineral rights of 700 million acres in the U.S., totaling 13% of the U.S. land surface and 40% of federally managed lands. Most of these lands are in Alaska and western states. BLM manages the Public Lands System for multiple purposes, including recreation, wildlife habitat, energy production, mineral extraction, timber, and grazing. More than 150 million acres are managed for grazing livestock, and 57 million acres are subject to timber harvest, including over two-million acres of old-growth forests in Oregon. The management of the Public Lands System is especially important for western birds that inhabit forests, sagebrush, grasslands, and deserts.

Partners for Fish and Wildlife Program: Over two-thirds of the land in the U.S. is privately owned, including an estimated three-quarters of the remaining wetlands. In recognition of the importance of private lands to conservation, the FWS sponsors the Partners for Fish and Wildlife Program to promote the restoration, improvement, and protection of fish and wildlife habitat on these lands while leaving the land in private ownership. Through this program, the Service can provide landowners with assistance concerning wetland restoration and protection, food and shelter for fish and wildlife, pesticide use reduction, native plant restoration, and soil management

The **Safe Harbor Program** of FWS recruits private landowners to sign agreements to restore habitat for endangered species with a guarantee that no new restrictions will be incurred by the landowner should endangered species take up residence in this restored habitat. The program does not release the landowner from responsibilities for endangered species already on their property, however. This program began in 1995 to benefit the Red-cockaded Woodpecker in North Carolina. Landowners interested in this program should contact FWS to determine the baseline responsibilities and actions they intend to perform on the property.

The **U.S. Geological Survey – Biological Resources Division (USGS – BRD)** was created as a result of the transfer of FWS research functions to USGS, and includes offices such as the Patuxent Wildlife Research Center that concentrate on biological data storage and analysis.

The **U.S. Department of Agriculture (USDA)** oversees the U.S. Forest Service as well as many conservation programs on privately-owned agricultural lands through conservation programs under the Farm Bill.

The **U.S. Forest Service** was established in 1905 and is an agency within the USDA. The Forest Service manages 193 million acres of land comprising the National Forests and National Grasslands. Therefore, it has an important stewardship role for birds and other wildlife inhabiting areas under its management. The Forest Service is active in various bird conservation coalitions (including PIF and NABCI, which are further described below).

The **Natural Resources Conservation Service (NRCS)** is part of the USDA, administering many conservation provisions of the Farm Bill. Its sister agency, the **Farm Services Agency (FSA)**, administers the Conservation Reserve Program and all of the commodity components of the bill.

The **U.S. Department of Defense** protects many important habitats for birds on its military bases. Examples include Fort Hood, Fort Bragg, Eglin Air Force Base, Camp Pendleton, Fort Sill, Fort Polk, and Fort Huachuca.

The **U.S. Army Corps of Engineers** manages dikes, levees, and other water management infrastructure that influence wetland bird habitats through floods and sedimentation, including sandbars required by Least Terns and other water-birds for nesting.

The **Environmental Protection Agency (EPA)** is responsible for regulating and registering pesticides for use in the U.S., as well as regulating emissions and pollution to protect our air, water, and soils. These include emissions that cause acid rain, climate change, and other problems that affect birds as well as human health. The EPA is often involved with watershed and wetland restoration efforts, such as in the Great Lakes, Chesapeake Bay, and Long Island Sound. The **Office of Pesticide Programs (OPP)** is the division in the EPA directly responsible for pesticide issues. The **Environmental Fate and Effects Division (EFED)**, under OPP, is directly responsible for the science concerning the ecological impacts of the registration and re-registration of pesticides.

The **Association of Fish and Wildlife Agencies** represents fish and wildlife professionals in the 56 U.S. states and territories, and the U.S. federal agencies, in addition to many provinces of Canada and states of Mexico. Founded over 100 years ago, its goal is to promote sound management and conservation and address important fish and wildlife issues. Among its core functions are inter-agency coordination, legal services, international affairs, conservation and management programs, and legislation.

The **National Fish & Wildlife Foundation (NFWF)** is a non-profit group created by Congress in 1984 to direct public spending towards conservation issues. The Foundation's board is confirmed by the White House. NFWF has awarded over 10,000 grants to more than 3,500 organizations in the U.S. and abroad, spending more than $600 million to leverage over $1.5 billion for conservation projects. NFWF prioritizes grants by keystone initiatives that include Hawaiian birds, seabirds, shorebirds (focused on the American Oystercatcher along the Atlantic Coast), and cleaning up marine debris.

The **U.S. Department of Commerce** includes the **National Oceanic and Atmospheric Administration (NOAA)**, which regulates commercial activities in federal waters, such as industrial fishing. NOAA also administers several Marine Protected Areas and often shares responsibility for bird conservation on islands with FWS. The agency also has responsibility for implementation of the Endangered Species Act for marine species.

Environmental Laws and Conservation Regulations

The **Clean Air Act** was originally passed in 1963, with important amendments in 1970 and 1990. Under this law, the EPA sets limits on certain air pollutants emitted within the U.S. from sources such as chemical plants, utilities, and steel mills.

The **Clean Water Act** was passed in 1972 and gives the EPA the authority to regulate the discharge of pollutants into the lakes, streams, and rivers of the U.S. by setting waste-water standards for industry, and funding the construction of sewage treatment plants. By protecting water quality for people, this act also helps protect wetland and other aquatic habitats for birds and other wildlife.

Conservation Easements are legally binding agreements that prohibit certain types of development (typically residential or commercial) on privately owned property to protect the resources (natural or other) associated with the parcel. Federal income, state property, and estate tax advantages can accrue to property owners in compensation for lost uses of the property. Typically, easements are held by local government agencies, land trusts, or not-for-profit organizations.

The **Endangered Species Act (ESA)** was passed in 1973 and authorizes the FWS to list populations of plants and animals as Endangered or Threatened, to designate and acquire Critical Habitat for listed populations, and to create management plans for the recovery of listed populations. It also prohibits individuals, businesses, and governments from the unauthorized take (killing, injuring, capturing, or otherwise harming), possession, sale, and transport of listed species in the U.S. Section 7 of the Act requires federal agencies to consult with FWS on actions that may harm endangered species. Candidate species can also benefit from Candidate Conservation Agreements. There are 76 species, subspecies, or populations of birds in the U.S. currently listed under the ESA. Successes include the delisting of the Aleutian population of the Cackling Goose, Brown Pelican, Peregrine Falcon, and most Bald Eagle populations (all except Sonoran). Although recovery and delisting of all listed species is the goal of the ESA, conservationists are realizing that many "conservation-reliant species" will require ongoing management to sustain their populations.

Federal Migratory Bird Hunting and Conservation Stamps, known as "**Duck Stamps**", are produced by the U.S. Postal Service for the FWS. Created in 1934 as the federal license required for hunting migratory waterfowl, Duck Stamps have generated close to $700 million in revenue for the purchase or lease of wetlands for the National Wildlife Refuge System, which now totals some 5.2 million acres, including many Important Bird Areas in the U.S. The Migratory Bird Conservation Commission oversees the spending of funds raised through duck stamps. Currently, there is proposed legislation in Congress to raise the price of duck stamps in order to provide more funding for these habitat conservation programs.

The **Federal Insecticide, Fungicide, and Rodenticide Act (FIFRA)** allows the EPA to regulate the use and sale of pesticides to protect the environment and human health. It was first passed in 1947 to allow USDA to register pesticides and establish labeling provisions, but has been amended significantly since 1972 to allow the EPA to enforce compliance against the use of unregistered and banned pesticides. Under this authority, EPA has banned the use of many pesticides that are highly toxic to birds.

The **Lacey Act** of 1900 was the first federal law protecting wildlife. It prohibits the sale and transportation across state lines of animals or animal products that are protected under state or other laws. It helped end illegal commercial hunting that threatened many game bird species in the U.S. It has been amended (as recently as 2008) to also cover trade in plants, which helps prevent the importation or spread of invasive species, as well as timber from illegal logging operations in foreign countries.

The **Land and Water Conservation Fund (LWCF)** was established by Congress in 1965. It requires that a portion of receipts from offshore oil and gas leases be placed into a fund for state and local conservation, as well as for the protection of parks, forests, and wildlife areas. This fund has invested more than $13 billion to help acquire nearly seven-million acres of land. Unfortunately, most of the receipts from offshore leases intended for this fund have been used for other things, including debt reduction. A bill to guarantee that the $900 million in financing authorized for the LWCF goes to conservation is currently going through Congress.

The **Magnuson-Stevens Fishery Conservation and Management Act** is the primary law governing marine fishery management in U.S. federal waters. Amendments to the act mandate rebuilding depleted fisheries, protecting fish habitat, and reducing bycatch of protected living marine resources, which include non-target fish, turtles, marine mammals, and now seabirds. The act also provides funding to develop bycatch reduction gear for all species, including seabirds.

The **Migratory Bird Treaty Act** of 1918 regulates the hunting, trade, and transport of any body part, nest, or egg belonging to a migratory bird. This law helped end the unsustainable hunting of many shorebirds and waders (egrets and herons) that were being decimated by market hunters along their migration routes or at their rookeries. The law currently provides the broadest legal protection for birds in the U.S. As part of this act, treaties protecting migratory birds were signed with Canada, Mexico, Japan, and Russia. See also the Neotropical Migratory Bird Conservation Act below.

The **National Environmental Policy Act (NEPA)** of 1969 required all federal agencies to consider the environmental impacts of their actions in formal environmental assessments prior to conducting any action. By formally reviewing the potential environmental harm of projects prior to implementation, these projects could either be prevented or changed to reduce or mitigate environmental damage. NEPA is one of the most important U.S. environmental laws, and has served as a model for similar legislation in other countries. NEPA helps birds by requiring environmental review of the impacts of dams, wetland drainage projects, and other activities that could affect bird populations—but construction of communication towers is exempt from NEPA.

The **National Forest Management Act** of 1976 established a forest planning process that enables citizens to have input on how each National Forest is managed. The law also requires that the Forest Service maintain "viable" populations of bird species across their range. This viability requirement has been instrumental in protecting habitat for Spotted Owls, Red-cockaded Woodpeckers, and many other species of birds inhabiting the National Forest System.

The **Neotropical Migratory Bird Conservation Act** of 2000 established a matching grants program to support projects that conserve neotropical migratory birds and their habitats in the U.S., Canada, Latin America, and the Caribbean. This law stipulates that at least 75% of the total funding available for grants each fiscal year be used to support projects outside the U.S. Grant recipients must match every federal dollar with three they raise from elsewhere. In this way, more than $21 million in NMBCA grants have leveraged over $95 million in partner contributions. Current congressional legislation being considered seeks to increase funding under this act to $15-25 million by 2015, and to $35 million by 2020.

The **North American Wetlands Conservation Act (NAWCA)** of 1989 funds projects to conserve wetlands, wetland birds, and the implementation of the North American Waterfowl Management Plan (NAWMP). Money for this program is generated from Pittman-Robertson taxes (see below), and Coastal Zone Management Plan funds, and is supplemented with congressional appropriations. A large percentage of funds are earmarked for waterfowl surveys in Canada, a smaller amount for Mexico, and the remainder for the U.S.

The **Pittman-Robertson Wildlife Restoration Act** of 1937 provides federal aid to states for the management and restoration of wildlife. Funds for these activities are raised through an excise tax on sporting arms and ammunition. These funds support a variety of wildlife projects, including acquisition and improvement of wildlife habitat, and wildlife management research. For example, these funds helped establish Cheyenne Bottoms Wildlife Area and other important refuges for birds. A similar Act (Dingle-Johnson) aimed at restoring sport fisheries was passed in 1950.

The **Wilderness Act** of 1964 created the National Wilderness Preservation System, to prevent roads or other permanent structures from being developed on federal lands that were "untrammeled by man." The Wilderness Act initially designated 9.1 million acres of national forest land as wilderness, and this has increased to more than 100 million total acres on a variety of federal lands at present.

USDA Conservation Programs for Private Landowners

The following are voluntary USDA programs that incentivize private landowners to conserve habitat on their lands. More information on these and other programs can be found at www.nrcs.usda.gov/Programs.

Conservation Reserve Program (CRP): Leases over 36.8 million acres of highly erodible cropland for 10-15 years from 400,000 farmers. Landowners maintain stands of perennial vegetation and keep land idle for the length of the lease.

Conservation Security Program (CSP): Provides technical and financial assistance for the conservation of soil, water, air, energy, and habitat on tribal lands, ranches, and farms. This program is implemented across selected watersheds.

Environmental Quality Incentives Program (EQIP): Provides technical and financial assistance to farmers and ranchers to improve soil conservation, water quality, nutrient management, and habitat for fish and wildlife.

Farm and Ranch Lands Protection Program (FRLPP): Provides up to 50% of the fair market easement value to help non-governmental organizations purchase easements on farms and ranches as an economic alternative to development.

Grassland Reserve Program (GRP): Uses 30-year rental agreements or permanent easements to protect grasslands and pastures, thereby helping many grassland birds.

Healthy Forest Reserve Program (HFRP): Provides funds under cost-sharing agreements and easements to restore forests, enhance forest habitat for endangered species, and promote carbon sequestration.

Wetlands Reserve Program (WRP): Provides technical and financial support to landowners who protect, restore, or enhance wetlands on their property.

Wildlife Habitat Incentives Program (WHIP): Enrolls over 2.8 million acres of habitat for fish and wildlife in 18,000 five to ten year contracts.

BIRD CONSERVATION AND RELATED COALITIONS

The **Bird Conservation Alliance (BCA)**, facilitated by ABC, is a network of more than 200 member organizations, including bird conservation organizations and other environmental groups. Originally formed in 1994 (as the American Bird Conservancy Policy Council), the BCA unites bird groups and conservation professionals to benefit bird conservation efforts by coordinating lobbying activities and broadening awareness of bird conservation issues. Members benefit from representation in Washington DC, and the opportunity to engage in national bird conservation campaigns to affect policy. See www.birdconservationalliance.org for more information.

Joint Ventures (JVs) are coalitions of government and non-governmental bird conservationists who coordinate conservation activities. JVs are either regionally focused, such as the Appalachian Mountain JV, or are organized around species or species groups, such as the Sea Duck JV. Since 1987, JVs have been coordinating and implementing bird conservation projects, directing $4.5 billion in spending to enhance more than 13 million acres of habitat for birds in the U.S. A directory of JV's in the U.S. and Canada is available on the FWS webpage at www.fws.gov/birdhabitat/JointVentures/Directory.shtm. Most recently, the DOI has begun establishing Landscape Conservation Cooperatives based on the JV model.

Partners In Flight (PIF) is an international coalition of government and non-governmental organizations formed in 1990 in response to concerns about declining neotropical migrants. PIF has since broadened its focus to include all terrestrial bird species in North America. It produces conservation plans for birds at the continental level, as well as for states and physiographic areas. For more information and PIF publications, see www.partnersinflight.org.

North American Bird Conservation Initiative (NABCI) is a coalition of Canadian, U.S., and Mexican government and private organizations formed in 1998 to coordinate bird monitoring and conservation planning for all groups of birds. NABCI coordinates efforts under four separate conservation plans: the North American Waterfowl Management Plan, the PIF North American Landbird Conservation Plan, the U.S. Shorebird Conservation Plan, and the North American Colonial Waterbird Plan. NABCI has delineated and mapped 67 Bird Conservation Regions (BCRs) in Canada, the U.S., and Mexico. See www.nabsci-us.org for more information.

The **Western Hemisphere Shorebird Reserve Network** conserves shorebirds and their habitats through a network of key sites. See www.whsrn.org.

The **National Pesticide Reform Coalition** is a group of NGOs led by ABC that work on pesticide-related issues. The group has regular meetings with the EPA to air concerns and exchange information. See www.abcbirds.org/abcprograms/alliances/NPRC.html.

The **Alliance for Zero Extinction** includes more than 60 biodiversity conservation organizations around the world working to identify and protect critical sites for the most endangered species. See www.zeroextinction.org.

U.S. Bird Conservation Region Map and Plans

See table page 428 to cross-reference BCRs with states. For an online version of this map with links to bird conservation plans and contacts by BCR, see: www.nabci-us.org/map.html. For state by state wildlife action plans, see: www.wildlifeactionplans.org/index.html. For information about geographically-based Joint Ventures by region, see: www.nabci-us.org/jvmap.html. For additional links to taxonomically-based Joint Ventures, see: www.fws.gov/birdhabitat/JointVentures/Directory.shtm. For recovery plans for ESA-listed species, see: www.fws.gov/endangered/wildlife.html. For conservation plans by bird groups (e.g. shorebirds, waterfowl, etc.), see: www.nabci-us.org/plans.html.

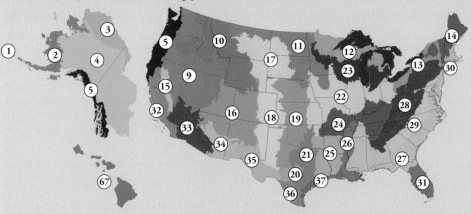

1. Aleutian/Bering Sea Islands
2. Western Alaska
3. Arctic Plains and Mountains +
4. Northwestern Interior Forest +
5. Northern Pacific Rainforest +
6-8. Exclusively in Canada
9. Great Basin +
10. Northern Rockies +
11. Prairie Potholes +
12. Boreal Hardwood Transition +
13. Lower Great Lakes/St. Lawrence Plain +

14. Atlantic Northern Forests
15. Sierra Nevada
16. Southern Rockies/Colorado Plateau
17. Badlands and Prairies
18. Shortgrass Prairie
19. Central Mixed-grass Prairie
20. Edwards Plateau
21. Oaks and Prairies
22. Eastern Tallgrass Prairie
23. Prairie Hardwood Transition
24. Central Hardwoods

25. West Gulf Coast Plain/ Oachitas
26. Mississippi Alluvial Valley
27. Southern Coastal Plain
28. Appalachian Mountains
29. Piedmont
30. New England/Mid-Atlantic Coast
31. Peninsular Florida
32. Coastal California *
33. Sonoran and Mojave Deserts *

34. Sierra Madre Occidental *
35. Chihuahuan Desert *
36. Tamaulipas Brushlands *
37. Gulf Coastal Prairie *
38-66. Exclusively in Mexico and/or Central America
67. Hawaii

Key
+ Extends into Canada
* Extends into Mexico

Selected Bibliography

Allsop, M., A. Walters, D. Santillo, & P. Johnston. 2006. *Plastic debris in the world's oceans.* Greenpeace International, Amersterdam, Netherlands.

American Bird Conservancy. 2009. Hawaiian Birds: Challenges and Second Chances. *BirdConservation.* Spring:7-23.

American Bird Conservancy. 2008. The state of our seabirds. *BirdConservation.* Fall:7-23.

American Bird Conservancy. 2006. An Endangered Species Act Success Story. *BirdConservation.* Spring:1-20.

American Bird Conservancy. 2006. Top 20 most threatened bird habitats in the U.S. Part 1: Counting Down Habitats 20-11. *BirdConservation.* Summer: 6-27.

American Bird Conservancy. 2006. Top 20 most threatened bird habitats in the U.S. Part 2: The Top 10. *BirdConservation.* Fall: 6-27.

American Bird Conservancy. 2002. *Sudden Death on the High Seas. Longline Fishing: A Global Catastrophe for Seabirds.* American Bird Conservancy, The Plains, VA.

Askins, R. A. 2000. *Restoring North America's Birds: Lessons from Landscape Ecology.* Yale University Press, New Haven & London.

BirdLife International. 2000. *Threatened Birds of the World.* Lynx Edicions and BirdLife International, Barcelona and Cambridge, UK.

BirdLife International. 2008. *Important Bird Areas in the Caribbean: key sites for conservation.* BirdLife Conservation Series No. 15. BirdLife International, Cambridge, UK.

Bonney, R., S. Carlson, & M. Fisher (Eds.). 1995. *Citizen's Guide to Migratory Bird Conservation: A project of Partners in Flight.* Cornell Laboratory of Ornithology with the National Audubon Society, Ithaca, NY.

Brinkley, E. S. & P. J. Baicich. 2004. Changing Seasons: Rome burning? *North American Birds.* 58:474-485.

Brower, M. & W. Leon. 1999. *The Consumer's Guide to Effective Environmental Choices: Practical Advice from the Union of Concerned Scientists.* Three Rivers Press, NY.

Brown, S., C. Hickey, B. Harrington, & R. Gill (Eds.). 2001. *U.S. Shorebird Conservation Plan, 2nd ed.* Manomet Cntr for Cons. Sciences, Manomet, MA.

Butcher, G. S., D. K. Niven, A. O. Panjabi, D. N. Pashley, & K. V. Rosenberg. 2007. The 2007 WatchList for United States birds. *American Birds.* 61:18-25.

Brush, T. 2005. *Nesting Birds of a Tropical Frontier: The Lower Rio Grande Valley of Texas.* Texas A & M University Press, College Station, TX.

Carver, E. 2009. *Birding in the United States: A demographic and Economic Analysis.* Addendum to the 2006 National Survey of Fishing, Hunting, and Wildlife-Associated Recreation. U.S. Fish & Wildlife Service Report.

Chipley, R. M., G. H. Fenwick, M. J. Parr, & D. N. Pashley. 2003. *The American Bird Conservancy Guide to the 500 Most Important Bird Areas in the United States: Key Sites for Birds and Birding in all 50 States.* Random House, NY.

Crossett, K. M., T. J. Culliton, P. C. Wiley, & T. R. Goodspeed. 2004. *Population Trends Along the Coastal United States: 1980-2008.* National Oceanic and Atmospheric Administration.

Elphick, C., J. B. Dunning Jr., & D. A. Sibley (Eds.). 2001. *The Sibley Guide to Bird Life & Behavior.* Alfred A. Knopf, NY.

Evans Ogden, L. J. 1996. *Collision Course: The Hazards of Lighted Structures and Windows to Migrating Birds.* World Wildlife Fund Canada, Toronto, Ontario.

Erickson, W. P., G. D. Johnson, & D. P. Young Jr. 2005. *A summary and comparison of bird mortality from anthropogenic causes with an emphasis on collisions.* USDA Forest Service Gen. Tech. Rep. PSW-GTR-191. 1029-1042.

Faaborg, J. 2002. *Saving Migrant Birds.* University of Texas Press, Austin.

Food and Agriculture Organization of the United Nations. 2009. *State of the World's Forests 2009.* FAO, Rome, Italy. Pp. 168.

Food and Agriculture Organization of the United Nations. 2006. *Global Forest Resources Assessment 2005.* FAO Forestry Paper 147. FAO, Rome, Italy.

Gillihan, S. W. 2006. *Sharing the land with pinyon-juniper birds.* Partners in Flight Working Group. Salt Lake City, Utah.

Hames, R. S., K. V. Rosenberg, J. D. Lowe, S. E. Barker, & A. A. Dhondt. 2002. Adverse effects of acid rain on the distribution of the Wood Thrush *Hylocichla mustelina* in North America. *PNAS*. 99:11235-11240.

Howell, S. N. G. & S. Webb. 1995. *A Guide to the Birds of Mexico and Northern Central America*. Oxford University Press, Oxford, UK.

Kushlan, J. A., M. J. Steinkamp, K. C. Parsons, J. Capp, M. Acosta Cruz, M. Coulter, I. Davidson, L. Dickson, N. Edelson, R. Elliot, R. M. Erwin, S. Hatch, S. Kress, R. Milko, S. Miller, K. Mills, R. Paul, R. Phillips, J. E. Saliva, B. Sydeman, J. Trapp, J. Wheeler, & K. Wohl. 2002. *Waterbird Conservation for the Americas: The North American Waterbird Conservation Plan, Version 1.* Waterbird Conservation for the Americas, Washington, DC, USA.

Leonard Jr., D. L. 2008. Recovery expenditures for birds under the US Endangered Species Act: The disparity between mainland and Hawaiian taxa. *Biological Conservation*. 141:2054-2061.

Lubowski, R. N., M. Vesterby, S. Bucholtz, A. Baez, & M. J. Roberts. 2006. *Major Uses of Land in the United States, 2002.* United States Department of Agriculture, Economic Research Service. Economic Information Bulletin No. 14.

Morrison, R. I. G., B. J. McCaffery, R. E. Gill. S. K. Skagen, S. L. Jones, G. W. Page, C. L. Gratto-Trevor, & B. A. Andres. 2006. Population estimates of North American shorebirds, 2006. *Wader Study Group Bulletin.* 111:67-84.

North American Bird Conservation Initiative, U.S. Committee. 2009. *The State of the Birds, United States of America, 2009.* U.S. Department of Interior: Washington, DC.

NSRE. 2007. *National Survey on Recreation and the Environment: Bird watching trends in the United States, 1994-2006.* USDA Forest Service, Athens, Georgia.

Olson, D. M, E. Dinerstein, E. D. Wikramanayake, N. D. Burgess, G. V. N. Powell, E. C. Underwood, J. A. D'amico, I. Itoua, H. E. Strand, J. C. Morrison, C. J. Loucks, T. F. Allnutt, T. H. Ricketts, Y. Kura, J. F. Lamoreux, W. W. Wettengel, P. Hedao, & K. R. Kassem. 2001. Terrestrial Ecoregions of the World: A New Map of Life on Earth. *BioScience.* 51:933-938.

Pashley, D. N., C. J. Beardmore, J. A. Fitzgerald, R. B. Ford, W. C. Hunter, M. S. Morrison, & K. V. Rosenberg.

2000. *Partners in Flight Conservation of the Land Birds of the United States.* American Bird Conservancy, The Plains, VA.

Pimentel, D., R. Zuniga & D. Morrison. 2005. Update on the environmental and economic costs associated with alien-invasive species in the United States. *Ecological Economics.* 52:273–288.

Rich, T. D., C. J. Beardmore, H. Berlanga, P. J. Blancher, M. S. W. Bradstreet, G. S. Butcher, D. W. Demarest, E. H. Dunn, W. C. Hunter, E. E. Iñigo-Elias, J. A. Kennedy, A. M. Martell, A. O. Panjabi, D. N. Pashley, K. V. Rosenberg, C. M. Rustay, J. S. Wendt, T. C. Will. 2004. *Partners in Flight North American Landbird Conservation Plan.* Cornell Lab of Ornithology. Ithaca, NY.

Richkus, K. D., K. A. Wilkins, R. V. Raftovich, S. S. Williams, & H. L. Spriggs. 2008. *Migratory bird hunting activity and harvest during the 2006 and 2007 hunting seasons. Preliminary estimates.* Division of Migratory Bird Management, Branch of Harvest Surveys. Laurel, Maryland.

Ricketts, T. H., E. Dinerstein, T. Boucher, T. Brooks, S. H. M. Butchart, M. Hoffmann, J. F. Lamoreux, J. Morrison, M. Parr, J. D. Pilgrim, A. S. L. Rodrigues, W. Sechrest, G. E. Wallace, K. Berlin, J. Bielby, N. D. Burgess, D. R. Church, N. Cox, D. Knox, C. Loucks, G. W. Luck, L. L. Master, R. Moore, R. Naidoo, R. Ridgely, G. E. Schatz, G. Shire, H. Strand, W. Wettengel, & E. Wikramanayake. 2005. Pinpointing and preventing imminent extinctions. *PNAS.* 102:18497-18501.

Safina, C. 2002. *Eye of the Albatross: Visions of Hope and Survival.* Henry Holt and Company, LLC, New York, NY.

Salvo, M. 2008. *The Shrinking Sagebrush Sea: Spatial Analyses of Threats to Sagebrush-Steppe and Greater Sage-Grouse.* WildEarth Guardians, Santa Fe, NM.

Savacool, B. K. 2009. Contextualizing avian mortality: a preliminary appraisal of bird and bat fatalities from wind, fossil-fuel, and nuclear electricity. *Energy Policy.* 37:2241-2248.

Sibley, D. A. 2000. *The Sibley Guide to Birds.* Alfred A. Knopf, New York, NY.

Simberloff, D. 1996. Impacts of introduced species in the United States. *Consequences.* 2:1-13.

Stattersfield, A. J., M. J. Crosby, A. J. Long, & D. C. Wege. 1998. *Endemic Bird Areas of the World: Priorities* for Biodiversity Conservation. BirdLife Conservation Series No. 7. BirdLife International, Cambridge, UK.

Stotz, D. F., J. W. Fitzpatrick, T, A. Parker III, & D. K. Moskovits. 1996. *Neotropical Birds: Ecology and Conservation.* University of Chicago Press, Chicago.

Stutchbury, B. 2007. *Silence of the Songbirds.* Walker & Company, New York, NY.

Terborgh, J. 1989. *Where have all the birds gone?* Princeton University Press. Princeton, NJ.

Terborgh, J., C. van Schaik, L. Davenport, & M. Rao. (Eds.) 2002. *Making Parks Work: Strategies for Preserving Tropical Nature.* Island Press, Washington.

U.S. Department of the Interior, Fish and Wildlife Service and U.S. Department of Commerce, U.S. Census Bureau. 2006. *National Survey of Fishing, Hunting, and Wildlife-Associated Recreation.* Washington, DC.

U.S. Environmental Protection Agency (EPA). 2008. *EPA's Report on the Environment.* National Center for Environmental Assessment, Washington, DC; EPA/600/R-07/045F.

U.S. Fish & Wildlife Service. 2002. *Migratory bird mortality: many human-caused threats afflict our birds.* U.S. Fish and Wildlife Service, Division of Migratory Bird Management. Arlington, VA.

U.S. Fish & Wildlife Service. 2006. *Revised Recovery Plan for Hawaiian Forest Birds.* Region 1, Portland, OR.

Weidensaul, S. 2007. *Of a Feather: A Brief History of American Birding.* Harcourt, Inc., Orlando, FL.

Weidensaul, S. 2005. *Return to Wild America: A Yearlong Search for the Continent's Natural Soul.* North Point Press, New York, NY.

Weidensaul, S. 1999. *Living on the Wind: Across the Hemisphere with Migratory Birds.* North Point Press, New York, NY.

Wells, J. V. 2007. *Birder's Conservation Handbook: 100 North American Birds at Risk.* Princeton University Press, Princeton and Oxford.

Wuethner, G. & M. Matteson. 2002. *Welfare Ranching: The Subsidized Destruction of the American West.* Island Press, WA.

Wilcove, D. S. 1999. *The Condor's Shadow: The Loss and Recovery of Wildlife in America.* W. H. Freeman and Company, New York, NY.

For more, see www.abcbirds.org.

SELECTED INTERNET RESOURCES

American Bird Conservancy: www.abcbirds.org.

Audubon 2007. List of Top 20 Birds in Decline: http://stateofthebirds.audubon.org/cbid/browseSpecies.php.

Bird Conservation Alliance: www.birdconservationalliance.org.

BirdLife International Data Zone: www.birdlife.org/datazone/index.html.

Boreal Songbird Initiative: www.borealbirds.org.

Igl, L. D. 1996. Bird Checklists of the United States. Jamestown, ND: Northern Prairie Research Center. www.npwrc.usgs.gov/resource/birds/chekbird/index.htm (Version 12May2003).

Integrated Bird Conservation in the United States. North American Bird Conservation Initiative: www.nabci-us.org/main3.html.

Invasive Species Specialist Group (ISSG) of the IUCN Species Survival Commission. Global Invasive Species Database: www.issg.org/database/welcome.

National Oceanic and Atmospheric Administration (NOAA) 2009. Marine Debris: http://marinedebris.noaa.gov.

National Park Service: www.nps.gov/index.htm.

Neotropical Birds Online: http://neotropical.birds.cornell.edu/portal/home.

Partners In Flight: www.partnersinflight.org.

PRBO Conservation Science: www.prbo.org.

Rocky Mtn Bird Observatory: www.rmbo.org/v2/web.

The Birds of North America Online: http://bna.birds.cornell.edu.

The State of the Birds: www.stateofthebirds.org.

The Nature Conservancy: www.nature.org.

U.S. Department of Agriculture, Forest Service: www.fs.fed.us.

U.S. Department of the Interior, Bureau of Land Management: www.blm.gov.

U.S. Fish & Wildlife Service. National Wildlife Refuge System: www.fws.gov/refuges.

Western Hemisphere Shorebird Reserve Network (WHSRN): www.whsrn.org.

INDEX

Bold page numbers indicate major entries,
italics indicate separate illustrations or photographs. See p. IX for glossary of acronyms.

JOIN AMERICAN BIRD CONSERVANCY

American Bird Conservancy's (ABC's) mission is to conserve native wild birds and their habitats throughout the Americas. ABC is the leading, partner-based, U.S. nonprofit organization focused solely on bird conservation. ABC acts to safeguard the rarest bird species, protect and restore habitats, and reduce threats to all wild birds.

ABC delivers the best bird conservation results in the Americas. With our partners, we have created or expanded 37 reserves that harbor many of the rarest species on Earth, led in the development of nationwide partnerships for bird conservation, taken on the greatest threats to birds (such as collisions, free-roaming cats, and pesticides), and become birds' best advocate with landowners and government—all with terrific results.

We can do even more with your help.

For $40 you can join American Bird Conservancy and receive our newsletter *Bird Calls* and magazine *Bird Conservation*:

- Donate online at www.abcbirds.org where you can also learn more about our programs;

- Send your tax-deductible contribution to American Bird Conservancy, P.O. Box 249, The Plains, VA 20198, or call 1-888-247-3624.

ABC members get great value for their conservation dollars. For every $1 you send to ABC, 82 cents goes directly to conserve wild birds and their habitats. ABC is consistently rated a four-star, "exceptional" charity by the independent group Charity Navigator. Only one percent of all U.S. charities have achieved this status, which differentiates ABC from our peers, and demonstrates to the public that we are worthy of their trust.

AMERICAN BIRD CONSERVANCY